Advance Praise for *The School G*

The School Garden Curriculum addresses something near and dear to my heart and probably the most critical issue in society today—the need for garden, food and ecology education for school-age children. This book will help transform society with each new generation. Well done!

— Zach Loeks, author, *The Permaculture Market Garden*

Not just for teachers, this hands-on, down in the dirt, fun, imaginative and eminently practical guide should be in the hands of anyone involved with kids and gardening. The author's depth of teaching experience shows in her well-thought-out lessons and projects suitable for each age group. This is a wonderful resource for introducing children to science as they discover ecological principles for themselves along with the rewards of gardening.

— Linda Gilkeson, author, *Backyard Bounty*

Early and regular interaction with nature is critical to healthy childhood development. With half of all people living in urban environments, school gardens offer one of the best opportunities for children to connect with the earth. Kaci Rae Christopher's *School Garden Curriculum* shows us how to make the most of lessons of ecology and gardening, to provide an intimate experience with nature and skills that will last a lifetime.

— Darrell Frey, author, *Bioshelter Market Garden*,
co-author of *Food Forest Handbook*

Ever been enthusiastic about a school garden project only to come up against typical pitfalls—summer maintenance, how beds are allocated, it's one more chore, etc.? This book to the rescue! Simple, well-thought out solutions to better school yard gardening have arrived. Not only will you build soil, grow edibles and learn how to forage, be prepared for a healthy harvest of community building, rituals, empathy, and gratitude.

— Lindsay Coulter, David Suzuki's Queen of Green

A school garden makes the perfect outdoor classroom. As learners step off the edge of chemistry into the new lands of biology, new experiences evoke paradigm shifts and epiphanies. Life begins to write itself! For all the talk of teaching methods, learning styles, and motivational techniques, the best teachers simply point the student toward an edge. No better place to do that than in the garden. *The School Garden Curriculum* shows you how.

—John Wages, Publisher, Permaculture Design magazine

The school gardening movement has long been a laboratory of educational innovation. By fusing permaculture with science, Kaci Rae Christopher takes it one step further in effectively promoting systems thinking and garden ethics. Her teacher-friendly new book offers hundreds of lessons for fall, winter and spring that can be accomplished within a single class at each grade level from K-8. Highly recommended!

—Tim Grant, editor emeritus, Green Teacher magazine

the SCHOOL GARDEN CURRICULUM

An Integrated **K-8 Guide** for Discovering Science, Ecology, and Whole-Systems Thinking

Kaci Rae Christopher

I acknowledge that I inhabit a colonized land and that powerful Indigenous sciences, Traditional Ecological Knowledge (TEK), and agricultural traditions are appropriated and entrenched in the gardening practices and ethics I have learned. I stand in support of food sovereignty movements that seek to decolonize and decentralize unjust systems of power and privilege, work to empower the marginalized, and promote social justice through self-determination and ecological regeneration. TEK and Indigenous sciences continue to instruct Western gardening practices and permaculture principles, and I acknowledge my privilege in acquiring and applying this knowledge.
I am not an expert in these matters, but a constant learner,
and my teachers are the land and those who care for it.

To all my teachers and guides,
who taught me that
wonder is contagious
and at the heart of learning,
no matter our age.

Cover design by Diane McIntosh. Cover Images ©iStock
Interior images: p. 1 archideaphoto; p. 15 © Carpierable; p. 52: Olivier Le Moal; p. 64 idesign2000; p. 136 Crazy nook; p. 142 izumikobayashi; p. 178 justaa; p. 227 Lorelyn Medina; p. 242 pinkcoala; p. 278 sommaria / Adobe Stock.

Printed in Canada. First printing April 2019.

Any other inquiries can be directed by mail to:

New Society Publishers
P.O. Box 189, Gabriola Island, BC V0R 1X0, Canada
(250) 247-9737

LIBRARY AND ARCHIVES CANADA CATALOGUING IN PUBLICATION

Title: The school garden curriculum : an integrated K-8 guide for discovering science, ecology, and whole-systems thinking / Kaci Rae Christopher.

Names: Christopher, Kaci Rae, 1989– author.

Description: Includes bibliographical references and index.

Identifiers: Canadiana (print) 20189069317 | Canadiana (ebook) 20189069325 | ISBN 9780865719057 (softcover) | ISBN 9781550926989 (PDF) | ISBN 9781771422949 (EPUB)

Subjects: LCSH: Environmental education—Activity programs. | LCSH: Gardening—Study and teaching (Elementary) | LCSH: Science—Study and teaching (Elementary) | LCSH: Ecology—Study and teaching (Elementary) | LCSH: Place-based education.

Classification: LCC GE77 .C57 2019 | DDC 372.3/57044—dc23

Funded by the Government of Canada | Financé par le gouvernement du Canada | Canada

New Society Publishers' mission is to publish books that contribute in fundamental ways to building an ecologically sustainable and just society, and to do so with the least possible impact on the environment, in a manner that models this vision.

Contents

School Garden Curriculum Worksheets
are available for free download at:
https://tinyurl.com/SGC-Worksheets

Acknowledgments

I want to extend tremendous gratitude to the people who have supported and inspired this book and guided my teaching over the years: my mother, Tarri, who first planted my feet into warm soil; my father, Kent, who taught me to spot an elk's breath in the early morning; my husband, Sam, who shares wonder with me every day. I thank my mentors and friends who continually advise and support me: Jeff Gottfried, Katie Boehnlein, Alison Pollack, and Hannah Hostetter.

I am grateful to the staff and community of Springwater Environmental Sciences School. My deepest thanks to Dawn Bolotow in allowing me the creative freedom to learn and experiment in the garden, and Jon Vogel, who guided me as an educator and taught me how to have fun while capturing the attention of children.

Confluence Environmental Center and their innovative AmeriCorps support program gave me the opportunity to do work I love to do. Community supporters—such as Matt Brown of Food|Waves, Cathy McQueeney of Clackamas County Soil and Water Conservation District, and Rick Sherman of Oregon Department of Education—have remained advocates of my outdoor and garden education work, and their own passions for garden education inspire me continuously.

I am grateful for Calypso Farm and Ecology Center and Educational Recreational Adventures for the opportunities to learn and explore my passion for generating opportunities for wonder in children. And thank you to all the local teachers, nonprofits, and community members at work around me—your commitment to revitalizing children's connection with nature inspires me daily.

I am especially thankful for New Society Publishers for offering me this opportunity to share my work. I am grateful to be a part of a community that inspires deeply thoughtful, engaging, and community-building conversations. Thank you to Rob West, Sue Custance, Greg Green, Ingrid Witvoet, Sara Reeves, EJ Hurst, and all the incredible staff members who work hard to generate and support work worth doing. I am deeply grateful for the work of Betsy Nuse in copy editing my manuscript and providing a sounding board to me, as a writer and educator.

Thanks to the teachers, parents, volunteers, and garden enthusiasts who inspire me daily with their diligence and commitment to fostering new generations of leaders and global changemakers. Together, our work makes all the difference.

School Garden Curriculum Worksheets
are available for free download at:
https://tinyurl.com/SGC-Worksheets

INTRODUCTION

WHERE PERMACULTURE AND SCIENCE MEET

How can children produce food, learn science skills, and develop ecological values at the same time? This may seem to ask a lot from a single garden space. Over years of teaching garden education, I have compiled a model of what collaboration between permaculture-inspired practices, ecological mind-sets, and science-based activities can be in a school or youth learning garden. I developed the garden activities in this book when I worked as a School Garden Coordinator and Outdoor Educator for progressive schools in Oregon, as well as for educational farms, nonprofits, and small businesses. These lessons offer a framework for integrating a garden program into a school or community, cultivating place-based garden science studies for five- through fourteen-year-olds, and fostering a unique and valuable garden culture.

There is an endless collaboration between science and permaculture, in practice and philosophy. Both require a systems-based mind-set to guide individual and group work; both utilize observation, data gathering, and analysis to understand how diverse pieces create an interconnected whole. They encourage prioritizing objectivity over subjectivity and using a *beginner's mind* to experience wonder, generate questions, and explore ecology and natural systems. In the words of Bill Mollison:

> Permaculture is a philosophy of working with, rather than against nature; of protracted and thoughtful observation rather than protracted and thoughtless labor; and of looking at plants and animals in all their functions, rather than treating any area as a single product system.[1]

Youth learning gardens are a powerful place for children to cultivate the skills and mind-set of science and permaculture. It is where children learn to *Observe and Interact* and develop positive cooperation skills, inclusivity, and inquiry. When science and permaculture are focused together in one place for children—a youth garden—then the ecosystem studies and science skills that they develop and delve into are incredible. When "the landscape is the textbook," food, nutrition, and gardening skills extend naturally into a child's life experience.[2]

In an age where many children are growing disconnected from nature, where farmland is being saturated with chemicals, where the global climate is unpredictable and extreme, and where wild habitats are being replaced by urban development, our communities need creative and innovative leaders to change our world. We need children to learn and engage in a new land ethic that inspires Care for Self, Care for Others, and Care for the Land and cultivates new generations of changemakers and environmentally conscious global citizens.

We accomplish this by encouraging children to engage in ecosystems and create spaces for cooperative stewardship and interaction. We provide opportunities to practice building community, creative problem-solving, healthy lifestyles, personal responsibility, craftsmanship, and engagement in nature through science. Gardens are a powerful place to learn these valuable skills and foster strong environmental and social literacy.

The lessons in *The School Garden Curriculum* are permaculture-inspired, place-based, and hands-on extensions that are supplemental to, or fulfill, Next Generation Science Standards (see Appendix A at the back of the book). In a school, these standards provide structure for even more immersive and engaging classroom units. The lessons in *The School Garden Curriculum* were also inspired by the Oregon Environmental Literacy Program which engages students in nutrition, literature, writing, art, and practical life skills.[3] *The School Garden Curriculum* helps children develop an ecological and naturalist perspective through science projects and permaculture gardening techniques.

I have taught all these lessons in onsite school and youth gardens, but the activities are easily adaptable to indoor projects, container planting, or window boxes. Many of the winter projects presented here illustrate gardening can be taught indoors and with limited materials or a budget. *The School Garden Curriculum* can be utilized by outdoor educators and teachers, especially those beginning to incorporate hands-on unit extensions into a school garden, as well as garden coordinators, after-school programmers, volunteers, and parents.

Cultivating a Gardener's Mind

One valuable and practical aspect of this garden program is the ritualization of certain garden chores. Each season, children of every age group participate in garden activities that help care for and develop the school, youth, or home garden by sowing seeds, mulching, planting starts, and composting. And even though some activities are repeated every year, they do not lose their meaning. Rather, they enhance the children's gardening knowledge by utilizing new themes for every grade and more complex engagement from different perspectives.

During visits to other school gardens, I have noticed teachers or coordinators who delegate certain tasks based on age. The job of one lower-grade class will be to plant garlic, but in the next grade the students move on to mulching and never return to nurture and observe their garlic as it grows. While I realize value in having students do age-appropriate activities—there are some things a five-year-old will struggle with—I've seen students lose appreciation for a garden activity when they no longer engage with it in following years. Some schools have their students garden for the first few years and then participate in only a couple of gardening projects when they get older. I would challenge these schools to do more.

Repetition instills more than a "practice makes perfect" model—though the children do improve planting, harvesting, and garden techniques each time they practice them. When children engage in seasonal garden activities every year as they grow up, they develop valuable intrinsic knowledge of how seasons change and participate in the human agricultural tradition which tells them, "Now is the time to save seeds" or "Plant your seeds for summer food." I want children to watch leaves change colors and get an itchy sense nagging at them, from deep in their bodies and minds, that now is the time to prepare the garden for winter. This is the gift I can pass onto them so that years later, even if they forget many of the techniques of gardening, they are connected to the natural seasons and know intrinsically how to grow a garden.

The Garden as an Ecosystem

Practice letting go of any notion of what a garden should look like. What can a children's garden bloom into? With an open mind, evaluate the intentions of garden spaces and the community members participating in it. What will a children's garden become when science, observation, wonder, and thoughtfulness are practiced? It will be an ecosystem, a science lab, and a place of discovery. Adults can be helpful but also a damaging influence on school or youth gardens. Full of idealistic images and lacking a *beginner's mind*, willful adults can salt slugs, build sharp-edged raised beds, spray aphids, remove edible so-called weeds, and expose bare soil to the heat of the summer sun. Their priority is on a product, not a process.

In *The Sense of Wonder*, Rachel Carson observed, "A child's world is fresh and new and beautiful, full of wonder and excitement. It is our misfortune that for most of us that clear-eyed vision, that true instinct for what is beautiful and awe-inspiring, is dimmed and even lost before we reach adulthood."[4] Children are brimming with the capacity for wonder and awe. When student goals are to observe, study patterns and complex relationships, ask questions, conduct experiments, and learn from direct experience, they perceive the garden as an ecosystem and their wonder at the natural world is magnified. Youth gardens should be created from the minds and hands of children, who build up the space and their ownership of it, rather than act as occasional visitors who perform stand-alone activities. For all adults, allowing wonder to thrive requires practice in letting go.

> "The desire to solve problems, to experiment and to design is one of the defining characteristics of the permaculture gardener."[5]

Whether within one year of gardening regularly or nine years, children who work with permaculture, ecology, and science in mind will understand how soil organisms relate to decomposition and plant growth, consumer insects, predators, pollinators, seed sprouting, and so much more. Their knowledge of the garden as an ecosystem will be vast and punctuated by wonder, a willingness to experiment, and a deep appreciation for the value of a whole and complex system. The learning garden will look a little wild, but children will be committed to it, own it, and own their learning.

Children in Charge Everywhere

When children are allowed to be in charge of a garden, pick up slugs without prejudice, watch how they bite into leaves, and determine their own solutions from an ecological perspective, then they will naturally piece together an understanding of a holistic garden ecosystem.

They will also engage in valuable experiences of failure and apply trial and error, inquiry, and direct experience to their learning.

A method to enhance these opportunities is for an entire school or community to share the garden. This means that children work with the whole garden and overlap their plantings with other groups while interacting in the same spaces. Many schools are in favor of assigning a garden bed for each grade level as a way to delegate work in the garden. While this is successful for some schools, I have found that it pushes too many children into too small of a space, limits their engagement in activity for briefer amounts of time, and inhibits a sense of commitment to the work students perform. Students who plant tomatoes in one bed and then graduate to a new class and garden bed are not encouraged to see the relationships between those tomatoes and the plants or soils they are currently tending.

In many ways, this engagement is dependent on a school having access to a few garden beds, a plot of land, or part of a blacktop to grow plants on. Communities that plant in containers, window boxes, or inside classrooms have very different space requirements and opportunities for whole-school interaction. But if it is possible to have students return to their past container gardens and interact with them in the following seasons, I would encourage educators to make the time for those opportunities.

Letting children garden and interact with a whole space provides more opportunities for observing the garden's needs and its changes, while increasing a child's relationship and connections with the land. The space will change over time, and more opportunities will arise in an organic succession of student learning and natural spaces. When children are in charge, they can creatively challenge and innovate preconceptions of right and wrong and develop new, sustainable value systems in the garden.

Inspiring a Garden Culture

A *garden culture* is unique to the needs of every community and grows naturally from a commitment to common goals and visions. Educators and their school support staff or active community members often inspire initial gardening initiatives. Strengthen this initiative right away by taking a collaborative look at the garden program's long-term goals. What goal, or yield, will each teacher and class give to the school at the end of the year? How will they contribute to harvesting, work and labor, planting seeds, or offering scientific conclusions after a season of studying in the garden? Every community will have unique goals and site-specific solutions. Common positive goals can successfully create and respect the "local culture of place."[6] Communities who set realistic goals that fit their needs create a gardening structure that is collaborative, manageable, and sustainable. Remember to both start small and dream big.

A whole school or community commitment to a gardening initiative means that each class can commit to a yearly yield which will benefit the children and the whole community. The gardening activities I offer provide a structure for these yields in the form of developing science skills, application of organic gardening solutions, building structures, planting seeds, mulching, weeding, and more. But before integrating the gardening lessons, I recommend that communities discuss their group needs, site-specific projects, goals for the year, and how these can be integrated into the gardening program. With everyone working cooperatively to accomplish small pieces of a larger vision, the work will become much lighter for everyone.

The sustainability of a gardening program is tied to a school-wide commitment to the garden space or gardening initiative. School garden programs struggle if there is only one teacher integrating garden activities in their class. In order to truly offer the students a science-based ecosystem experience that transforms their environmental literacy, let all of the school be involved. The garden then becomes a place where students build knowledge by applying their lessons in a physical way. But how can staff and school boards be motivated to buy in on the garden and commit to a yield? By experiencing the garden not as an additional chore but as the focal point of hands-on and science-based enhancement of the students' indoor studies, as well as an indispensable part of community culture, seasonal celebrations, and social events.

A school and community garden culture can also be enhanced through continuous use and engagement, such as talking about the garden publicly, reading about gardening themes in class, and studying other subjects in the space. Unstructured sensory exploration is also critical for fostering a culture that celebrates a positive land ethic and a child's relationship to land. As Rachel Carson suggested in *The Sense of Wonder*, visceral connection and engagement inspires a desire for knowledge.[7] Facts without emotional and experiential context hold little meaning. Leaving the garden open for student retreat and recovery, feeding families, making friends, magic and wonder powerfully foster positive garden culture.

In some communities, a youth learning garden is a place where different philosophies meet and clash, especially among adults. Addressing garden "pests" is one example of where different opinions come into contact. Every gardener has a different idea of what a pest is and what plants or insects fall into that category. Many school gardeners and educators share my experience in receiving a plethora of unasked-for advice on how to get rid of the garden's aphids, "too many" ladybugs, or ravenous slugs.

My own gardening philosophies are open to change, but rooted in permaculture and ecological ethics, organic solutions, and systems-based thinking. *Integrated pest management* is a key part of this mind-set and promotes student observation skills and ecosystem studies. It sees that "the problem is the solution" and perceives pests as opportunities or "surpluses of nature, which need to be used rather than destroyed."[8] When we value all the aspects of a system for their relationships to each other, then we begin to interact with "care, creativity, and efficiency."[9] Adhering to common goals and broad yet inclusive expectations can transform the garden itself and the culture into a space for sharing, exploration, open-minded conversations, and collaboration.

Care for Self, Care for Others, Care for the Land

Behavior expectations are important in setting a standard for community use and respect of the garden. The three expectations that inspired me at the first school garden I built are derived from the Ethical Principles of Permaculture. *Care for Self, Care for Others, and Care for the Land* (setting limits to consumption and redistributing the surplus) are my framework for ecological-based garden expectations and student behavior. I've heard it used in other school gardens using different language. The remaining permaculture tenets and ecological ethics fit into these three broad categories, and are stated in words that children can easily aspire to understand and use.

These three ethics support the inclusive ecosystem-perspective of the garden, including the humans in it, and help foster socially and environmentally minded action. They guide children on how to treat each other in and outside of the garden. And they are flexible to children's growth by providing an empathetic framework for day-to-day activities. The language can be incorporated into wellness programs in schools and communities who focus on nutrition and physical activity.

Each beginning fall lesson in *The School Garden Curriculum* starts with a review of the behavior expectations for students and community members while in the garden space. The students discuss what *Care for Self, Care for Others, and Care for the Land* means when exploring the garden or engaging with an activity. This behavior could look like keeping low voices so as not to scare the living creatures in the garden, walking instead of running, and respecting each other's personal space. By creating a social contract within the class and making a list of behaviors that the class can refer to, students play their roles as a community members, gardeners, and scientists responsibly and respectfully.

School Gardens in the Summer

The summertime reality for every school is different. Some schools continue classes throughout the summer, and others take a few months for break. Schools with greenhouses experience scorching temperatures when left unattended, and it can be exhausting to garden after noon on some summer days. While winter and water-loving plants thrive during cooler seasons, they require extra love and nurturing in the early summer to become established. But who will water these plants over the summer, diligently, every week? Who will pick the beans, cucumbers, and tomatoes so that the harvest will continue into the school year? How will the garden survive for fall classes and food production?

The schools I've worked with all face summer concerns and issues, but I've developed some strategies that fit the model and focus of an ecologically minded garden. These solutions will change as a school garden becomes more established, more perennial, and more productive.

Water and Soil

Every community garden has different structures, needs, and resources. But all can benefit from a strong commitment to its garden soil. A thriving and rich soil structure that is filled with organic matter and decomposers at work will retain water longer throughout the summer. I prefer to build gardens where the paths and beds are inspired from children's explorations and wonder. And many watering systems don't quite fit into this creative model without careful planning. In order to cultivate a kid-planted, interactive, edible, biodynamic jungle, I have had to research and experiment with water-conserving methods that can be applied to a whole garden. Inspired by permaculture solutions and dry land farming techniques, I have encouraged students and schools to focus intensely on soil health throughout the school year.

In all the gardens I work with, I build experimental systems in different garden beds and compare the success of different methods over the seasons. For example, I have students build Hügelkultur beds in a section of a garden, and in other beds, we dig in raw compost or simply rotate cold compost systems until the beds are ready to plant. And, rather than import soil, I have all the students actively explore varying methods of building a healthy soil ecosystem. Students learn how to grow and nourish soil as a living natural resource.

These experiments and techniques are more adapted to gardening in raised beds or direct soil, but can be extended to gardening in large containers. While it is not uncommon for container gardeners to purchase new soil every year, the methods of this curriculum encourage educators to strive to build their soil, feed and nourish a living system, and for student exploration and ecosystem study.

"Within every terrestrial ecosystem the living soil—which may only be a few centimeters deep—is an edge or interface between nonliving mineral earth and the atmosphere. For all terrestrial life, including humanity, this is the most important edge of all. Deep, well-drained and aerated soil is like a sponge, a great interface that supports productive and healthy plant life."[10]

In fostering life-rich soils, students focus on the importance of various soil particle sizes, bacteria and mycelium, humus, oxygen, water, and lessening soil disturbance. With every class prioritizing soil health, the garden flourishes in all seasons, especially in the summer. It retains seasonal moisture for longer amounts of time, and the soil slows and holds water in a way that promotes optimal plant health, thus limiting a community's stress about summer watering. Having all student energies focused on fostering soil health gets the plants off to a great start in the spring, which will create even stronger and healthier plants by summertime. In repeating certain seasonal activities, mulching and composting, and engaging in new themes, children who study these lessons keep the garden soil health in mind.

"The creation of soil...is greater than the sum of the parts. Soil must be understood and managed from this holistic perspective. Soil management is ecosystem management. Ecosystem management in turn is relationship management."[11]

Planting Perennially and All-season

Another perspective that helps garden programs thrive—and limits the stress of summer maintenance—is planting perennials and planting throughout every season. For schools that have summer breaks, garden education really begins in the fall, which is also the end of many annual plant life cycles. The first garden lesson begins in a time of transition. While many other gardens are "being put to bed" to wait for spring planting, fall can be an active season in the youth learning garden.

Permaculture and ecological gardens grow many of the same plants as a more conventional one, but prioritize perennials and all-season, cold-loving plants as well. Throughout the fall, winter, and early spring, a garden can provide taste tests of brassicas, edible "weeds," root vegetables, and plants grown under student-made cold frames or cloches. As a perennial garden becomes more established, natural succession and changes among plant systems will occur. Fruit trees with grow to provide shade, food, and leaf mulch. The seasonal foods and fruits will change, and new plant species may need to be introduced to thrive in changing environments or niches. Crop extension tools such as cold frames, cloches, frost-cloth, and mini-greenhouses can contribute to all-season food production even more. As more perennial or multi-season plants become established, the amount of labor is often reduced because of the hardiness of these plants. A focus on perennial plants can slowly reduce the stresses of schools over summer garden maintenance.

Summer Maintenance

My summer message for communities is always the same: harvest, eat, and harvest some more! Gardeners can promote longer fruiting periods by harvesting regularly from many annual plants. And the opportunity to harvest free food is an activity some people love to do and will go out of their way to enjoy.

Because of the intense summer heat and wind that often sweeps across the region I live in, I don't usually ask volunteers to pull weeds. The only cool hours of the day are well before noon, so I say that if the work isn't done by the end of the school year, then "oh well!" Another reason that I don't ask volunteers to pull weeds is because the vast subjectivity about what is a weed. Some so-called weeds are foods that students love to eat, or they provide excellent medicine or increase soil vitality and habitat.

> "Weeds are perhaps landscape 'pests', but they are not landscape predators. In fact substantial research suggests just the opposite, that weeds are caretakers of the soil. From the ecological perspective, weeds are plants like all others."[12]

If I don't want certain "weeds" to go to seed, I will trim back potential seed heads and consider their removal in a less energy-draining season. But after months of intensive gardening and science experiments, the heat of the summer is a school gardener's version of "winter rest." Now is the time to rely on all the careful work the school has done in the garden and see how successful it was when the students return in the fall.

In my experience, when the students do return, it is to a lush, jungle garden. Most plants survive, a few do not—but what a great learning opportunity! And every year, the students can come back into the garden to a ready feast. They will spend months munching on tomatoes, beans, edible flowers, cucumbers, rhubarb, and so much more. And because of the vibrant soil health, these plants last far into the fall and provide months of produce for the community to eat.

Share the Surplus

A school garden reality that I don't address in these lessons is what to do with the surplus produce from the garden. I note opportunities for seasonal taste testing at the end of almost every lesson, and a few harvest celebrations and events that are possible. Taste testing is a great way to explore new foods and flavors. The ritual of sharing a small meal after class allows for regular conservations about flavor, preference, recipes, and changing taste buds. Foraging is my preferred method of student taste test experience, though the ability to forage-and-eat in a school garden may depend on regional and district regulations. However, when children forage for their food, then they take ownership of their choices and are more open to new experiences, seasonality, and experimentation.

> "If we expose very young children to the delight of foraging food in a garden, they are more likely to grow up with a deep and intuitive understanding of our dependence on nature and its abundance."[13]

School Garden Curriculum Outline

	Garden Theme	Fall: Patterns and Change	Winter: Discovery and Observation	Spring: Community and Interdependence
Kindergarten	An Introduction to Gardening	Getting to Know the Garden	Pollination and Seeds	Insects and Garden Species
1st Grade Grade One	Seeds	The Edible Parts of a Plant	Traveling Seeds	Sprouting Seeds Experiment
2nd Grade Grade Two	Pollinators and Cycles	Seasons and Garden Changes	Beneficial Insects and Bird Pollinators	Insect Pollinators
3rd Grade Grade Three	Becoming Soil Scientists	The Garden Ecosystem	Soil Vitality Experiment	Fostering Healthy Soils
4th Grade Grade Four	Vermicomposting	Becoming a Worm Expert	Mini-worm Bin Project	Worms Help Grow a Better Garden
5th Grade Grade Five	Composting	Composting Basics	Compost in a Jar Experiment	Healthy Soils and Garden Plants
6th Grade Grade Six	Rain Gardens	The Way Water Moves	Rain Garden Design	Building a Rain Garden
7th Grade Grade Seven	Renewable Resources	Catching and Storing Energy	Valuing Renewable and Nonrenewable Re- sources	Using Small and Slow Solution
8th Grade Grade Eight	Leadership and Stewardship	Energy Cycles and Conservation	Designing Systems	Supporting Biodiversity

Honoring the permaculture design principle of *Obtaining a Yield* should be a priority in every season. Each community has different goals and needs in their garden, so the methods of processing produce will vary. Where student hunger isn't a concern, I've seen some schools donate their harvests to local food banks. I've used a school garden as a "free grocery store" for rural families and students who live a distance from a store with fresh produce. Other schools work alongside their food services and nutritionists to incorporate the garden produce into student meals. Find the needs in your school, discuss the dreams and goals of the community around the garden, and acknowledge the reality of accomplishing or building this vision.

> "The experience of abundance encourages us to distribute surplus beyond our circle of responsibility (the earth and people) in faith that our needs are provided for."[14]

The First Lesson

For the first organized lesson at any level, I introduce the space to the students and allow them time to become familiar with the changes that have taken place over the season. I may point out areas of key interest, but mostly I observe and enjoy the students' discoveries, like the unexpected vine that liberally grew up the side of the fence, beautiful pie pumpkins dotted along the path, or the rich smell of fennel on the wind when students brush by. Before students embark on their quiet exploration, we first sit together outside and come to an agreement about student behavior and communal land ethic.

I introduce each expectation, *Care for Self, Care for Others, and Care for the Land*, and ask the students to brainstorm examples of what each means. As I receive answers, I write them down on a list that will be posted on garden signboards outside of the main gates, in the classroom, or in another public space for all the students and community to see. Since the students envisioned these specific behaviors together, there is a higher expectation to follow those rules as a community.

As students sit in the garden, ladybugs crawl on their shoulders and spiders occasionally fall on their heads. Learning in the garden is a new or difficult experience for many children. Being outside and studying in the space can offer tough transitions and sensory distractions. I find these conditions to be optimal because they give me a chance to address students concerns and their behaviors toward insects, for example, or messy surprises that may await them in the garden. This is also a great time to judge the interests of the group for garden education during the year. What are they excited, curious, or even skeptical about?

A Final Note

The lessons in *The School Garden Curriculum* provide a valuable framework for whole school and community gardening integration, but also space for innovation and additional activities by educators. As an Outdoor and Garden Educator, I typically have only 45 minutes with each class, including transitions, and it is never enough to engage in all the complexities of any subject. I especially wish to explore the cultural diversity of garden foods with children more, but struggle to balance sit-down learning time with the value of hands-on work.

Some teachers are more open to extending garden integration into their own lessons by altering their reading lists and expanding garden themes into their units. When staff step up to integrate garden lessons into their curriculum, I observe remarkable differences in student engagement and knowledge. With children reading, writing, and doing math, and physical activity all around a gardening subject in their classroom, then my lessons provide the real-world experience and application of those subjects. Gardening is no longer a supplemental subject but part of a student's whole learning.

My advice for in-class teachers, educators, volunteers, and parents is to use the activities I provide as hands-on and place-based opportunities, but to extend from them even more. Explore incorporating multicultural nutrition education, agricultural history, advocacy, and food sovereignty into broader classroom curricula and children's learning. Practice community engagement and support local farmers through volunteerism, field trips, and interviews. Bring experts, leaders, and elders into your learning spaces to cultivate awareness, growth, and a community of voices.

A teacher, a parent, or a volunteer can teach these weekly lessons, though having a garden coordinator is a great force for change in any school. The lessons are most successful to student learning if the entire staff or community is dedicated to supporting garden-based education in and out of the classroom. As with all teaching tools, this document is for inspiration and adjustment to each community's needs.

The lessons follow an organic and seasonal flow that respects the opportunities that arise each season in the garden, with an intention to develop science skills in young learners. Generally, the fall lessons will involve harvesting, seed saving, and introducing the students to their garden theme. In the winter, when the garden is mostly "asleep" and the weather often doesn't permit as much outdoor garden exploration, the lessons focus on inquiry and exploration into scientific methods. And the spring lessons utilize this developing knowledge and bring it back out to the garden for application and work.

I hope you can adapt and gather inspiration from these activities and extend the lessons I have learned into your own gardens and classrooms. May you and your students find fulfillment in your gardening adventures and may your gardens grow abundantly.

KINDERGARTEN

AN INTRODUCTION TO GARDENING

This year of discovery in the garden is a truly kinder garten experience for young students as they begin to interact with the basic values of a permaculture-inspired garden and healthy eating practices. When I am able to implement taste tests each week with this age group and work with them on how to discuss and experience new foods, then I see tremendous behavior differences in later school years. Students who taste spicy, sweet, bitter, and sour garden foods (raw turnips, beets, sprouted seeds, and strange flavor combinations) grow into remarkably adventurous and open-minded eaters.

Beyond taste testing, this year is the class's first introduction to an ecosystem, in the school garden and community. The students identify all the parts of the garden that they will work with in the future: compost bins, garden

FALL
Getting to Know the Garden

1. Garden Exploration
2. Seed Saving
3. Flower Bulbs
4. Planting Garlic Bulbs
5. Mulching
6. The Soil Ecosystem
7. Life in the Garden
8. Creating Compost
9. Compost to Soil
10. Soil Health and Roots

WINTER
Pollination and Seeds

1. Seed Discovery
2. Mini-greenhouses
3. Seed Survival
4. Sprouts
5. The Parts of a Plant
6. What is Pollination?

7|8. Butterflies, Birds, and Flower Shapes

9. Bees at Work
10. Planting Seeds for Spring

SPRING
Insects and Garden Species

1. Slugs and Pests
2. Decomposers
3. Consumers
4. Planting Insect Forage
5. Predators
6. Sowing Seeds
7. Planting Flower Starts
8. Mulching and Weeding
9. Spring Harvest
10. Celebration

beds, secret hiding spots, giant sunflowers, and garden tools. They come to understand how each person, place, and living thing work and relate to each other in the garden, and learn to cherish and honor every element through careful observation and interaction. By getting to know key players like worms, slugs, birds, snakes, the students practice the values of the garden: Care for Self, Care for Others, and Care for the Land.

Older students can be especially influential when it comes to leading by example and passing on these garden values. With older student partners, Kindergartners receive one-on-one guidance for how to interact with insects and living things, how to work like good, careful gardeners, and how to think like scientists with wonder and excitement. Behaviors that may take teachers months to instill in a class of Kindergartners will take a small amount of time when an older student directly engages with a younger one. I have found this collaborative teaching to be an incredible way to introduce the culture of a school garden to a new generation.

© 445017 / Adobe Stock

And by having these young students observe pollinators at work, harvest and taste garden foods, plant seeds and watch them sprout, then they are inspired to participate in a positive culture, act with constructive behaviors, and grow a gardening mind-set.

Next Generation Science Standards

Over the year, the students will develop an understanding of what plants and animals need to survive as well as the complex relationships and balances in the garden ecosystem. The class will also engage in the beginnings of the scientific method through observations and wonderings, asking questions and investigating, creating working models and solutions to larger problems, and recording and interpreting data.

Classroom extensions for science, reading, writing, and storylines of this year's units include: mapping weather patterns, recording seasonal changes, and studying what seeds, pollinators, and garden ecosystems members need to survive.

Permaculture Principles

- Observe and Interact
- Catch and Store Energy
- Obtain a Yield
- Produce No Waste
- Use Small and Slow Solutions
- Use and Value Diversity
- Use Edges and Value the Marginal

Fall: Getting to Know the Garden

1 Garden Exploration

Time Frame: 30 minutes

Overview: The students will learn about behavior expectations in the garden and what are the key features.

Objective: To begin learning good garden behaviors and have the students practice taste testing and talking about food.

Introduction (5–10 minutes)

1. Begin introducing the students to what the garden class structure will be like, what activities they will do during class, what an outdoor classroom means, and the special environment that is the garden. Allow plenty of room for questions. Today is about practicing good behaviors and learning to act in garden class as they would in their other classrooms. The guideline for these discussions should revolve around three basic permaculture-inspired tenets that are the centerpiece for garden values: Care for Self, Care for Others, and Care the Land.

2. What kinds of creatures live in the garden? What should students do if they come across them? How should they treat the plants in the garden? When is it permissible to pick plants in the garden? What behaviors, words, and actions should students use in the garden toward each other, the plants, and the creatures that live there?

3. For today's exploration, the students will focus on how to walk carefully through the garden, stay on the paths, how to respond to insects and exciting things they may see, and when to pick or taste test things in the garden. Solicit their suggestions about how to accomplish these goals to the best of their abilities.

Activities (20 minutes)

1. Students will take a tour of the garden space with a teacher or adult volunteer as guides. Key features can be pointed out and discussed with the students (the greenhouse, garden shed, outdoor classroom spaces, compost bins). These spaces should be ones that the students will regularly interact with in the garden or have special rules around them.

2. If there is time for the students to have a free exploration, then they can return to places of interest with a classmate, or discover new features—like a vibrant pumpkin or large tomato. They can also share their favorite places and what features are most exciting for them. They will practice good garden behavior during their exploration. Students can try to identify plants they see and look for the items

> In the garden, I discuss with the students how to avoid words like "Gross" or "Bleh!" Rather, I want them to think about flavors and textures by using phrases such as "This isn't a flavor I enjoy" or "Perhaps this would be better cooked." I want students to realize that gardeners put a lot of effort into growing these plants—and that taste buds change over time. Another phrase I learned from a teacher was "Don't yuck my yum!" If students didn't want to finish their taste test, I encourage them to calmly walk over to the compost bins and dispose of the food.

they will be taste testing (tomatoes, kale, cucumbers).

3. Before the students taste test, have them discuss how to try something new, use words to describe taste and texture, and what to do and say if they like or dislike the item.

4. Finish the tour at the compost bins and allow students to practice putting their food waste in the right spot.

Assessment (1–2 minutes)

1. Before students leave, conduct a verbal survey of the garden and behavior expectations, as well as the students' favorite discoveries.

2. What new foods did the students eat and enjoy? What was the most exciting or astounding thing they discovered?

Students carefully harvest sunflower seeds and experience the diverse textures of a seed head.

2 Seed Saving

Time Frame: 30 minutes

Overview: Students will save dry seeds, such as beans or sunflowers, from the school garden.

Objective: To practice a valuable fall skill and explore touching and describing different textures.

Vocabulary

1. Seedpod
2. Harvest
3. Dormant
4. Seed Saving

Introduction (5 minutes)

1. Introduce students to the idea of saving seeds, or collecting seeds and storing them away until springtime to plant. Discuss the value of doing this practice in the fall, as seeds become available, and mimicking winter conditions by storing them in a dark, cool place where they can stay dormant.

2. What is a seed? Why is it important to gather them and store them away? How will seasonal changes affect the seeds if they aren't harvested? What other animals collect seeds for winter? What textures do students expect to experience with the seeds and seedpods? What words can they use to describe different textures?

3. Demonstrate which seeds the students will be picking and how to identify them. With beans, use two examples: one seedpod that is ready to be harvested and another that is not ready or is still green. Have students determine what the differences are between the two examples and which one has traits that indicate the seeds are dry and ready to be gathered. Students should use these traits to identify similar seedpods in the garden (dry, brown, crispy, rather than green and moist). Also demonstrate how to open seedpods and gather the seeds. If possible, this is a good activity to invite older students to join in and have them work with the younger students to identify dry seedpods and carefully harvest them.

Activities (20 minutes)

1. Using collection cups, the students will explore the garden in pairs or individually to find, identify, and gather dry bean seeds. They should practice careful picking—with two hands—and display focused work, rather than being rushed and competitive.

2. When the students' cups are full or it is the end of the activity, have them deposit the seeds in a labeled paper bag or envelope, and make sure to keep each seed variety separated. The class can gather the empty seedpods and put them in the compost pile.

3. After cleaning up from the activity, have the students reflect on what they discovered during their exploration and their experiences with the seedpods and seed textures.

Assessment (5 minutes)

1. What textures did the students experience that they did not expect? What discoveries did they make? What did the bean pod feel like compared to the bean seed?

2. Finally, take time to make a weather map with the students in their classroom, where they can record observations about the weather changes over the course of the school year (temperature, extreme weather). They can make these observations on the days that they have gardening class. The goal is for students to develop a sense of seasonality and how changing weather conditions relate to the seasonal garden activities, such as saving seeds. *How Groundhog's Garden Grew* by Lynne Cherry has great illustrations for changing seasons and the importance of saving seeds for spring.

Winter

is coming the squirrels say. Time for

hibernation. Time for plants to leave

and decay. Time for snow to fall.

Time for gardens to go to bed. Time

for thunder. Time for hail. Time for

rain... Now time for Spring again.

— Anonymous Student Poem

Preparation

1. Collection cups
2. Seed/seedpod examples
3. Paper bags for storing seeds
4. A marker
5. Weather map

Resources

- *How Groundhog's Garden Grew* by Lynne Cherry.

NGSS and Activity Extensions

Further in-class studies can include making "observations of local weather conditions to describe patterns over time" (K-ESS2-1: Earth's Systems).

3 Flower Bulbs

Time Frame: 30 minutes

Overview: With the help of older students, the Kindergarteners will plant flower bulbs.

Objective: Younger students will learn how deeply to plant a bulb and build community with another class and for older students to share the culture and values of the school garden.

Vocabulary

1. Bulb
2. Depth

Introduction (5 minutes)

1. The goal of this activity is for older students to learn by teaching and for younger students to learn the garden's cultural values of Care for Self, Others, and the Land by interacting and observing older students at work. The older students will focus on teaching how deeply to plant a seed or a bulb.

> Planting depth varies on seed size, but I simplify this for my students by teaching them to plant seeds twice as deep as the size of the seed. A large tulip and daffodil bulb are great visuals for this message. Gardeners don't usually have measuring tape or rulers on hand, so when students use their feet, hands, and fingers, then gardening becomes an excellent application of mental and intuitive math.

2. Before the activity, have the students brainstorm as a class: What is a bulb? Is it alive? Why do students plant these seeds in the fall? How will they survive the winter? What does is mean when something is dormant? How deep should the bulbs be planted? Which end is the root and which is the stem?

Activities (20 minutes)

1. In pairs—an older student with a younger student—the class will plant flower bulbs in a garden bed or an otherwise color-lacking area of the garden and school grounds.

2. Older students should focus on two goals: teaching their partner how deeply to plant seeds and building community with their Kindergartener by getting to know their name and staying with their partner through the entire activity.

3. Younger students will also focus on staying with their partner and getting their hands dirty by planting the bulbs—the older student shouldn't do all the work!

4. Before cleaning up, students will make sure all the holes they dug are filled in (by hands, not feet) and any extra bulbs returned. Finally, the partners can wash their hands, return their tools, and explore the garden to taste test their favorite plants, with the older students demonstrating plant identification and proper harvesting techniques.

Assessment (5 minutes)

1. Have the older students test their partners. Can younger students thank their partners by name and recall how deeply to plant a seed?

2. Older students should also recall their partner's name and look over their planting site to make sure the bulbs are planted at the correct depth.

Preparation

1. Flower bulbs
2. Shovels

4 Planting Garlic

Time Frame: 30 minutes

Overview: Students will plant garlic bulbs in the garden and practice the lessons they learned from their older student partners during the previous activity.

Objectives: To learn that some plants can grow in cold weather, while others lie dormant under the soil until spring.

Introduction (5 minutes)

1. Introduce the students to a garlic plant by naming the parts of it. What part of a plant is a bulb? What kind of foods and recipes use garlic as an ingredient?

2. Explain how to plant the garlic in the soil and how deeply to plant it. What side of the bulb goes down into the soil? Where are the best places to plant garlic in the garden? Students should strive to plant the bulb twice as deep as the bulb is large and to cover each bulb with enough soil. Have the students recall the lessons they learned about planting bulbs during the previous activity.

Activities (20 minutes)

1. Have students count out five garlic bulbs to start out with, considering the best place to plant them and taking their time to space them apart and cover each one with soil. They can also work in a designated garden bed with pre-dug furrows and should work with their classmates to make sure the bulbs are spaced apart far enough (6 inches). This planting activity can be done in containers too.

2. The students can use their hands or shovels if the soil is too compacted. Adult volunteers can help the students in planting the bulbs upright and with enough soil to cover them. The students can repeat this process for the activity period. They should focus on careful gardening, rather than competitive work.

Assessment (5 minutes)

1. Have the students go back over the places they planted and make sure all the bulbs are covered with soil. Quality control is important. They should make sure the bulbs are planted deeply enough and in the right direction.

2. For the end of class, students can have a taste test opportunity of raw or green garlic (if available), as well as fresh nibbles from the garden foods.

Preparation

1. Garlic bulbs
2. Taste test for students
3. Adult volunteers
4. Shovels

Resources

- A song about planting: *Inch by Inch: The Garden Song* by David Mallett.

NGSS and Activity Extensions

Further in-class studies can include the effect of sunlight of the Earth's surface, identifying seasonal changes, and the Earth's yearly rotation (K-PS3-1 Energy).

5 Mulching

Time Frame: 30 minutes
Overview: Students will "put the garden to sleep" by tucking in the soil and plants with mulch.
Objective: To learn how to protect the garden soil from winter rains, provide habitat for decomposers, and create more nutritious soil.

Vocabulary

1. Mulch
2. Decomposers
3. Hibernate

Introduction (5–10 minutes)

1. Introduce the class to the practice of mulching. One of the last and important fall tasks to do in the garden is lay mulch over the soil. Straw and leaves capture and slow winter moisture, provide a home for important decomposers, and build soil organic matter. Mulch also protects the soil from winter wind, rain, and snow. Just like different mammals and amphibians hibernate during the winter, so do gardeners help the garden hibernate, or "go to sleep," until the springtime.

2. It is important to do careful gardening during this task and be careful of the plants that are still alive during the winter. Messy, competitive, and sloppy work will not help the garden or the creatures living in it. Each student should focus on "Caring for the Land" by being careful of living plants and not wasting any mulching materials as best they can. Demonstrate the best method to cover the soil and tuck mulching materials in around plants.

Activities (20 minutes)

1. In pairs, students will gather small amounts of straw or leaves and tuck them around still-living plants and over mulched dead plants or bare soil. They will work slowly and do careful work to spread the mulch out on the beds. It is most effective for all students to work in one garden bed at a time. When students have finished mulching with straw, they can go on a taste test tour of the garden with their partners.

2. This is an activity that is excellent for older student and younger student partnerships. With older students helping, the Kindergarteners are productive and focused throughout the whole activity while receiving one-on-one gardening lessons. More experienced student gardeners can reinforce the importance of mulching for the soil and building habitat.

Assessment (5 minutes)

1. Have student groups share their experiences with another group. Did the students discover any worms or decomposers when they were mulching? What plants were still alive in the garden? What plants had already died?

2. Students can finish the activity with a taste test in the garden, either pre-picked or of a specific seasonal plant.

Preparation

1. Straw or gathered leaves
2. Older student partners
3. Volunteers

6 The Soil Ecosystem

Time Frame: 30 minutes

Overview: Students will explore soil samples with an older student partner, sketching and identifying what they see.

Objective: Younger students learn how to behave and interact with insects as well as identify new soil creatures.

Vocabulary

1. Ecosystem
2. Decomposer
3. Predator
4. Consumer

Introduction (5 minutes)

1. Brainstorm with the older and younger students, in turn, before introducing the activity: What kinds of creatures have students encountered in the soil before? What are the best ways to treat and interact with insects and bugs? What is a decomposer, a predator, and a consumer?

2. The goal of this activity is for older students to pass on the important skills of how to best treat insects and living things that make their home in the garden and play valuable roles as decomposers, consumers, and predators. The older students should understand that their role is to show the Kindergarteners through their actions and conversations how to respect living things according to permaculture-inspired principles. The older students can also exhibit how to think like a scientist by asking questions and generating "wonderings" throughout the activity and reinforce the concept of an ecosystem.

3. Before the activity, discuss what an ecosystem is with the Kindergarteners. They should understand that a soil ecosystem is vital to the health of a school garden. The greater the diversity of relationships within an ecosystem,

A *Polyphemus* caterpillar, in the giant silk moth family, travels safely across the shelter of mulch and layered garden plants.

the better balanced and more stable it often is. In the garden, students should celebrate all the insects and bugs they encounter. Sometimes unfamiliar creatures make children uncomfortable, but the students know that they shouldn't be fearful of them. Instead, the students can practice being calm and careful with the creatures they encounter. They are encouraged to think like scientists and ask questions about what they see.

Activities (20 minutes)

1. In partners, the students will work on The Soil Ecosystem worksheet to draw and identify different soil creatures. They can use pencils and magnifying glasses to find and sketch what they see.

2. If they can't identify the insect or bug, then they can use available resources to identify it and/or come up with their own name for it as a group. When the groups have filled up their worksheet with sketches, then they can color their sketches while older students lead a discussion with their partner and write down the questions they generated about the creatures they found.

Assessment (5 minutes)

1. Gathering back together as a group, the students can share one creature they found or participate in a quick survey of what they learned on how to treat insects and bugs. What parts of the soil ecosystem did they discover? What creatures were they unfamiliar with? How many different creatures did they find, and did they discover more of one type than another? What predators, consumers, and decomposers did they identify?

2. When the groups have cleaned up their learning stations, then they can enjoy a taste test in the garden with their partners.

Preparation

1. The Soil Ecosystem worksheets
2. Magnifying glasses
3. Pencils, colored pencils, crayons
4. Clipboards
5. Garden soil samples
6. Tray or plate for soil observation
7. Insect identification books

7 Life in the Garden

Time Frame: 30 minutes

Overview: To the best of their ability, students will search for and identify five key insects in the garden.

Objective: To practice how to behave around the insects in the garden and discover what roles they play in the garden ecosystem.

Vocabulary

1. Scientist
2. Observe
3. Insects
4. Habitat
5. Behavior

Introduction (10 minutes)

1. Begin with a brainstorm and quick share of what types of mammals and insects the students have seen in the garden. What creatures do they expect to see on a "safari" or exploration of the garden? What creatures live under the logs and rocks? What will students see flying in the garden?

2. The students can review what insects and creatures they discovered in the last activity with their older student partners, how they responded when they found something that made them uncomfortable or excited, and what behavior they used when interacting with the insects. How did they speak, move, or touch the insects? How did they Care for Self, Others, and the Land?

> The insects and creatures in the permaculture garden are valued as a part of the ecosystem. Students may encounter bugs and insects that concern them, such as slugs and snakes, but they should understand that all these creatures have a role in the garden ecosystem, even if they don't know or understand them yet. I encourage students to speak softly, not yell, and be very gentle with insects if they pick them up.

3. Students should put on their "scientist glasses" for this activity; let them observe their surroundings with wonder, questions, and without judgment (such as "Eww!" or "Gross"). If students feel uncomfortable about an insect, bug, or living thing, they should calmly walk away to a new spot without screaming, shouting, or making loud noises. Students should understand that it is OK to feel uncomfortable, but that they need to respond calmly during the activity. This is a great opportunity to introduce the students to how scientists observe the world: with constant curiosity, wonder, and questioning.

> Scientist glasses are plastic glasses without lens I give to the young students when they are doing observation activities. These glasses give them the power to look at the world like scientists, with wonder and questions, and give greater strength to their eyes and observation skills.

4. What are words that scientists would use to describe the creatures they come across? Where would a scientist look for insects and bugs in the garden?

Activities (15 minutes)

1. In pairs, the students will put on their glasses and explore the school garden with their Life in The Garden worksheets. They will search for slugs, worms, pill bugs, centipedes, and another creature of their choice or discovery, count how many they see, and observe their behavior and habitat.

2. The students will focus on working with their partner to complete their worksheet and should think like scientists by coming up with wonderings and questions about the creatures they encounter. Students who need help writing can seek an adult volunteer to assist them.

Assessment (5 minutes)

1. After the activity is finished and cleaned up, have the students reflect on what they discovered, through a class survey or quick story share.

2. What wonderings and questions did the students come up with during their activity? What new creatures did they discover? How did they treat and behave toward these creatures?

Preparation

1. Life in The Garden worksheets
2. Scientist glasses
3. Clipboards
4. Pencils or crayons
5. Adult volunteers

8 Creating Compost

Time Frame: 30 minutes
Overview: Students will explore the school compost systems and learn about all the ingredients in healthy compost.
Objective: To understand the importance and purpose of the school compost system and the role of decomposers.

Vocabulary

1. Compost
2. Decomposers
3. Brown/Green Materials

Introduction (5–10 minutes)

1. Brainstorm with the students what they know of composting: Do they have a bin at home, in their classroom, or in the cafeteria? What do they put into it? What is a decomposer? What decomposers have the students encountered in the garden before?

2. On a class poster, generate ideas about what materials can be added to a compost bin and work with the students to identify which ones can be considered green and brown composting materials.

Activities (15–20 minutes)

1. The class will read *Compost Stew: An A to Z Recipe for the Earth* by Mary McKenna Siddals and discuss key ingredients in the compost and the importance of creating nutritious soil by limiting food waste. Looking back at the class poster, are there materials that should be added or moved on the list?

2. After the book and discussion, students will each receive an ingredient to add to their own "compost stew" and create compost as a class, by adding living ingredients (soil and decomposers), green, and brown materials. This class-made compost can be added into outdoor or garden compost bins so the students will know which bins to put their compost in during the school day.

Assessment (5 minutes)

1. What ingredients should not be added to the compost bin? Finish the activity with a conversation with the students about fruit stickers in the compost and the dangers of plastic and garbage in the ecosystem.

2. As always, a taste test is a great way to end the activity, especially since it will give the students another chance to practice putting any food waste in the compost bin.

Preparation

1. *Compost Stew: An A to Z Recipe for the Earth* by Mary McKenna Siddals
2. 20–25 compost ingredients, including food waste, decomposing compost, and living decomposers

Resources

- *Compost Stew: An A to Z Recipe for the Earth* by Mary McKenna Siddals.

9 Compost to Soil

Time Frame: 30 minutes
Overview: The students will sift finished compost to put on the garden beds and learn about how compost becomes healthy soil.
Objectives: For the class to continue engaging with the decomposition process, learn how some materials break down faster than others, and how food waste can turn into soil or humus.

Vocabulary

1. Living and Nonliving Materials
2. Nutrients
3. Sift
4. Humus

Introduction (10 minutes)

1. Brainstorm with the students what materials they added to the compost during the previous class. Review what the decomposition process is and why it is important to gardeners. What are living and nonliving elements that are a part of the compost system? How does composting create nutrient-rich humus?

2. Materials in the compost decompose at different rates, and some materials need more time to break down. A good example of this is a sample of nearly finished compost or a sample of the garden soil with many years of mulching materials. In order to remove some of the pieces that need more time to decompose, the students will be sifting the compost, harvesting nutritious soil from the decomposed food waste, and returning the larger pieces back into the compost pile.

3. Demonstrate to the student where the stations are (gathering compost, sifting, compost pile) and how to do each task collaboratively and carefully.

Activities (15 minutes)

1. In small groups or partners, the students will gather and sift completed compost. Every student will use containers (black plastic planting pots) and a small shovel to gather compost from one of the ready compost piles (hot or cold) into their container. Students may need an adult volunteer to help them with the tasks.

2. When students have their samples, they will carry them over to the sifting station. As a group, the students will pour their compost onto a screen or a planting tray with small holes, held over a large bucket or wheelbarrow. Students will gently shake their compost sample back and forth, sifting the finer and decomposed items from the larger and less decomposed pieces with the sifting screen. When their sample has been sifted, students will gather the less decomposed pieces that have fallen down and put them in a bucket to observe later. The students can repeat this process until the activity time is done. The sifted compost will be stored for next week's activity.

3. The students can clean up or return their supplies and tools and return to the gathering space to share their experiences.

Assessment (5 minutes)

1. Student volunteers can pick one item from the less decomposed bucket of materials—the ones that didn't fall through the screen. As a class, the students can try to identify the item and discuss why the material didn't break down. How long do they think it will take for the material to decompose? What insects, bugs, or mammals do they think would like to eat it? Is this material living or nonliving?

2. Pass around a sample of the finished compost. What words can students use to describe the texture and smell of the finished compost?

Preparation

1. Finished compost ready for sifting
2. Adult volunteers
3. 4-inch black planting containers
4. Wheelbarrow
5. Sifting tray
6. Gloves for students to handle compost
7. Trowels

Resources

- "Sifting Compost" in Patrick Lima, *The Natural Food Garden: Growing Vegetables and Fruits Chemical-free.*

10 Soil Health and Roots

Time Frame: 30 minutes

Overview: Students will apply their new knowledge of composting and plant root crop seeds in finished compost.

Objective: To discuss soil health and encourage nutritious soil and humus by applying finished compost in the garden and planting a spring food source.

Vocabulary

1. Root Crop
2. Organic Materials

Introduction (10 minutes)

1. During the last class time, the students harvested healthy soil for the garden by sifting finished compost into a fine soil. This soil is full of nutrients in the form of small pieces of decomposed plants and cafeteria waste that can best be accessed by microbes, decomposers, and plants.

2. By creating soil from organic materials, the students can grow healthy plants for their community to eat. Brainstorm with the students how plants take in these nutrients. What part of a plant helps it in taking up these nutrients? What roles do roots perform for a plant? What kinds of roots do students like to eat? Why do humans eat them?

Activities (15 minutes)

1. Individually, the students will gather cups of finished nutritious compost they gathered and spread it over a garden bed or fill up a deep black planting pot (three- to five-gallon). Students should pay attention to spreading the compost out evenly, not wasting this valuable resource, and doing the careful work of a gardener. They can repeat this process until the compost is gone or the bed/container is adequately filled or covered.

2. When the compost has been spread, the students will receive the seeds of a root crop (carrots, beets, radishes) and plant them in the soil. After they have scattered or sowed the seeds, they can gently put the seeds into contact with the soil by applying soft pressure with the palms of their hands. This will ensure seed-to-soil contact. Students should notice that the seeds are so small that it will be difficult to bury them twice as deep as the size of the seed.

3. Alternatively, the students can fill up a small planting container and accomplish the same activity.

Assessment (5 minutes)

1. When students have completed the activity, they can enjoy a taste test of the same types of root vegetables they planted (carrots, beets, radishes, turnips).

2. For an enhanced activity, have the students compare the taste test between a crop that has been grown in organic soil and another that hasn't, or has been grown with aquaponic liquid fertilizers. Can they experience differences in taste and texture? The student reflections can tie into the previous conversation on the benefits of healthy, nutritious compost.

Preparation

1. 4-inch black planting containers
2. Root crop seeds
3. Prepared taste tests
4. Sifted compost

Winter: Pollination and Seeds

1 Seed Discovery

Time Frame: 30 minutes
Overview: The students will explore different seed shapes and the unique features of each seed variety.
Objectives: To begin understanding the role of seeds and the diversity of their shapes, sizes, and colors.

Vocabulary

1. Diversity
2. Texture

Introduction (5–10 minutes)

1. Brainstorm with the students what they know about seeds. Write these responses on a poster for future reference. What is a seed? How can students tell something is a seed? What are the characteristics of a seed? Why are they so important to gardeners?

2. Students will be introduced to the great diversity of seeds, especially those from the garden. There are many shapes, colors, sizes, and characteristics of seeds that aid them in survival. Visual guides, such as *A Seed Is Sleepy* by Diana Aston, can be provided to help students understand this concept.

Activities & Assessment (20 minutes)

1. Individually, the students will receive cups with mixed seeds and sorting trays (egg cartons). Using any available tools (tweezers, spoons) or their hands, the students will be guided to sort the mixed seeds into cups based on color, texture, and shape.

2. Students who finish their sorting can go a step further by counting the seeds in each variety and recording the number on a board for the class to observe. Then, the students can sort their seeds again based on the next sorting method (color, texture, and shape).

3. Students will share their sorting methods with a partner. As a class, they can also explain their reasoning, observations, and discoveries as they explored the seeds. Were there any seeds that didn't fit into their sorting method? What seeds can students identify by name? How did students sort the seeds in their mixture? What wonderings and questions do students have about seeds?

4. When the students have finished sorting and reflecting, they can clean up their trays and remix their seeds before returning them.

Preparation

1. One egg carton for each student
2. Cups with mixed seeds
3. Tweezers and spoons

Resources

- *Seeds* by Ken Robbins.
- *A Seed Is Sleepy* by Dianna Aston.
- "Seed Sensation: Exploring and Sorting Seeds." (online—see Appendix B).

2 Mini-greenhouses

Time Frame: 30 minutes
Overview: Students will plant seeds for classroom observation and make predictions about plant growth.
Objective: To set up an ongoing experiment in order to observe seed sprouting, growth, and the parts of plants.

Vocabulary

1. Seed Coat
2. Stem
3. Sprout
4. Greenhouse

Introduction (5–10 minutes)

1. What do students know about what seeds need to thrive? What compels a seed to move from a stage of dormancy to sprouting? Brainstorm with the students what they know about seeds and how they grow.

2. Students will be assembling mini-greenhouses in class and watching seeds sprout and grow. As a class, they can prepare for this experiment by brainstorming their "wonderings" and what they think will happen. How does a seed sprout? Discuss with the students about the parts of a seed they will discover during the growth process (seed coat, stem, and root).

Activities (15–20 minutes)

1. Individually, each student will get a clear plastic cup, a napkin, and five seeds for their greenhouse. They will wet their napkins with a spray bottle until that they are damp and gently fold them in the bottom of the plastic cup.

2. Then, they will collect their five pre-soaked seeds and line them up at the bottom of their cup between the napkin and the plastic siding so that the seed is moist but the students can see them. The cups can be placed in a window, on student tables, or in another place where students can observe them growing every day.

3. When the seeds have been "planted," the students will receive a small piece of clear plastic, or another cup, to put over the top to trap moisture and heat. A volunteer can help write student names on the cups.

> Plastic cups are not ideal, but I usually have to use whatever is on hand. Glass jars and other clear reusable containers are better alternatives.

Assessment (5 minutes)

1. When students have finished assembling their greenhouses, they can gather back together and discuss what they discovered about their seeds. Did they see the seed coat splitting or roots emerging on their presoaked seeds?

2. How long do they think it will take the seeds to sprout? Students should understand that not all seeds sprout and not to be disappointed or feel like they failed. Some seeds will grow quickly, others take longer, and some don't grow at all. This is an opportunity to think like scientists and develop wonderings about how and why seeds grow.

> Over the next few weeks, students should be encouraged to observe their seeds sprouting every day and to take the lid off for a short amount of time to let their plants breathe. This is a great chance to encourage scientific thinking in students by creating a class chart documenting seed sprouting and growth.

Preparation

1. Pre-soaked bean seeds
2. Clear containers
3. Napkins
4. Clear lids or plastic wrap
5. Spray bottle
6. Adult volunteer

Resources

• Deborah Stewart. "Creating a Mini-Greenhouse in Preschool." (online—see Appendix B).

3 Seed Survival

Time Frame: 30 minutes
Overview: The class will read *The Tiny Seed* by Eric Carle and learn about what seeds need to survive.
Objective: For students to understand how soil, air, water, and sunlight play a role in the growth and cultivation from a seed to a plant.

Vocabulary

1. Survival
2. Environment

Introduction (5–10 minutes)

1. During the last lesson, students planted seeds in their mini-greenhouses. How will they keep those seed growing and flourishing? What should students do to help the seeds survive? What do seeds grown indoors or outdoors in the garden need to thrive?

2. Before their activity, the students can create hypotheses about seeds and record their thoughts on a classroom chart. Why are soil, water, wind, and sun so important to seeds and plants? What do they each offer for the plant's survival? What environments do different seeds prefer?

Activities (15–20 minutes)

1. As a class, the students will read *The Tiny Seed* by Eric Carle. They should be aware of the differences between the reality of a seed's life and the fiction in the book. What events in the book could happen to a seed in the school garden? What was the right environment for the seed to grow in? Why did the other seeds not survive?

2. The students can observe different soil types in sample containers that are passed around. Each variety of seed needs a certain amount of sunlight, water, soil type, and wind. What differences can students spot in the various soil types? What soils would seeds prefer to grow in and why? Using their hypothesis from the beginning of class, what have students learned about seeds and how they grow? What ideas have changed since the beginning of class?

3. The students can finish the activity by carefully watering the plants in their mini-greenhouses and removing the covers to give them air.

Assessment (5 minutes)

1. Did students note any changes or growth to their seeds in the mini-greenhouses? Are the seeds getting enough water, soil, sun, and air? If not, what changes can the students do to provide these things for the seeds?

2. The students should be prepared to understand that not all seeds survive. Some seeds can have all the right balances of sun, wind, soil, and water but still not grow. That is OK. This happens when gardening, which is why gardeners plant multiple seeds together!

Preparation

1. Soil samples
2. Water for seedlings
3. Classroom poster

Resources

- *The Tiny Seed* by Eric Carle.

4 Sprouts

Time Frame: 30 minutes
Overview: Students will study the growth of their plants in the mini-greenhouses and learn about the nutritional value of sprouts.

Objective: For students to understand the process of seed growth and the different parts of a sprout.

Vocabulary

1. Photosynthesis
2. Cotyledon
3. Seed Coat
4. Roots
5. Germinate
6. Protein

Introduction (5–10 minutes)

1. Brainstorm with the students about what changes they have noticed with the seeds in their mini-greenhouses. Have them reflect on their observations and what parts of the seeds they see emerging.

2. Using a student's sprout as an example, have the class explain the growth of the seed and how the plant is developing. Which part of the plant emerged first? What stories can students tell about plant growth—when did it begin? How fast is it growing?

3. Introduce students to the term *cotyledon*. The cotyledon is the embryonic leaf or leaves, or simply the "baby" leaves of a sprout. As I explain to students, similar to "baby teeth," these leaves will fall off the plant when their adult leaves grow. In the bean seeds that the students planted in their mini-greenhouses, the cotyledon is inside the dormant seed, even when they are dehydrated. These little leaves help the sprouted seed begin harvesting sunlight through photosynthesis.

Activities (15–20 minutes)

1. Individually, the students will carefully handle their mini-greenhouses and observe the changes their seeds have encountered. They will try to identify the cotyledon, roots, stem, and seed coat of their seeds. If possible, they can record their findings on a classroom chart by counting how many seeds sprouted and which have roots or cotyledon (it's never too early to introduce scientific thinking and processes).

2. After students have observed their plant growth, they can reassemble the greenhouses and water their seedlings if needed.

3. The students can enjoy a taste test of pre-sprouted seeds and discuss how nutritious the sprouted seeds are for humans by making proteins more available—for further information, see Resources.

Assessment (5 minutes)

1. After the taste test, the students can gather back together to reflect on their findings and what they recorded on the class chart.

2. Were all the seeds in the mini-greenhouses at the same stage of growth? How many seeds have sprouted? Were there parts of the sprout that they didn't know or couldn't identify?

Preparation

1. Sprouts for students to taste test (alfalfa, mung bean, garbanzo beans)
2. Chart for recording findings

Resources

• J. K. Chavan, et al. "Nutritional Improvement of Cereals by Sprouting." (online—see Appendix B).

NGSS and Activity Extensions

Further in-class studies could include making observations on and describing patterns of what animals and plants need to survive (K-LS1-1: From Molecules to Organisms).

5 The Parts of a Plant

Time Frame: 30 minutes
Overview: Students will explore the basic parts of a bean plant and discuss the role of each part.
Objective: To identify key parts of a plant and make observations about them.

Vocabulary

1. Roots
2. Stem
3. Leaves
4. Flower
5. Cotyledon

Introduction (5–10 minutes)

1. Brainstorm with the class about what observations the students made about the growth of their seeds. What parts of the plant are emerging that they recognize or don't recognize?

2. As a class, the students will help identify at least four different parts of a bean plant: roots, stem, leaves, and flowers, and will define them on a poster for the class to reference later. As each part is listed/drawn, have the students discuss the role of each part to the plant as a whole.

All the parts of a nasturtium plant for the students to observe.

Activities (15–20 minutes)

1. In pairs, students will receive a sprouting seed from their indoor seed experiments and identify the different parts of a plant that they observe. Then, each student will work on The Parts of a Plant worksheet, focusing on correct labeling and spelling the parts of a plant. They can use the class chart as a reference for spelling.

2. When students have finished, they can color or decorate their worksheets.

Assessment (5–10 minutes)

1. Gathering the students back together, have them share the parts of the bean plant they identified with their observation examples. What parts of a plant were they unable to find but was on their worksheets? These observations will be preparation for the next lesson about the process a plant goes through in producing a flower and fruit.

2. Finally, students can disassemble their mini-greenhouses and either compost their seedlings or plant them in pots of soil for further study.

Preparation

1. Sprouting seed examples
2. The Parts of a Plant worksheet
3. Classroom poster

6 The Pollination Game

Time Frame: 30 minutes

Overview: Students will be introduced to the process of pollination and learn about the importance of pollinators in the school garden.

Objective: To value insects in the garden, especially pollinators, and become more comfortable with insects by understanding them as scientists and gardeners.

Vocabulary

1. Pollination
2. Pollinator
3. Nectar
4. Pollen

Introduction (10 minutes)

1. Brainstorm with the class and record their ideas and wonderings on a classroom chart. How and why do some plants produce flowers? What do the students know about pollen, pollination, or pollinators? Have they ever seen any pollinators before? What types of insects visit flowers?

2. Do all pollinators fly? How do they find their way from flower to flower? What would the garden be like if gardeners didn't have pollinators?

> Pollinators play a very important role in gardens. Without pollinators—such as honeybees, mason bees, leafcutter bees, bumblebees, ants, wasps, hummingbirds, butterflies, and moths—we wouldn't enjoy many of the delicious fruits that grow in the garden. Each pollinator has a variety of flower colors and shapes that attract them. That is one reason why there are so many different varieties in flowers. For example, bumblebees prefer small flowers, especially purple and white ones. Hummingbirds tend to go for pink, red, purple, and even blue flowers with a long cone-like shape.

Activities (15 minutes)

1. This game can be performed in a variety of ways, but the students will need space for all of them. For the game, each student will become a pollinator. They will get a picture of a pollinator (hummingbird, honeybee, mason bee, butterfly) taped to the front of their shirt and another picture of a color (red, orange, yellow, white, purple) also taped to their front. The color of the flower should not be one their pollinator prefers. In this way, the student will be both a "flower" and a "pollinator," depending on the rules of the game.

2. The goal of the game is for the students to identify the colors that pollinators prefer and to engage with this knowledge throughout the activity. Again, the flower color on their shirt isn't necessarily the one their pollinator prefers. The games will be tag-like, but the teacher can direct the focus of each round. It can be a free tag game, where each pollinator tries to tag someone who has the flower color their pollinator prefers (hilarious chaos), or a game where only the pollinators that like them tag the red flowers and everyone else freezes for a small amount of time. It can also be as a "hawks and mice" game, where pollinators try to capture certain colors as they run past. Another variation is to have the "flowers" gather together according to color.

> While many pollinators have mechanisms to help defend them, most are harmless for students in the garden. During the games, none of the students should feel the need to "sting" or "bite" as they imagine a certain pollinator might do!

Assessment (5 minutes)

1. At the end of the activity, gather the students back together and have them reflect on the colors their pollinators prefer. Can they remember which flowers the pollinator likes?

2. As a pollinator, is it easier or harder to spot these colors when they are grouped together or spaced apart? How do they imagine gardeners can help out the pollinators in their own pollination race? How can students best plant flowers in the garden while keeping this knowledge in mind?

Preparation

1. Pollinator and flower color identities/stickers

Resources

- "12 Plants to Entice Pollinators to Your Garden." (online—see Appendix B).

NGSS and Activity Extensions

For greater conversations about plant and animal adaptations, the students can extend their studies into learning how plants and animals can change their environment to fulfil their needs (K-ESS2-2: Earth's Systems).

7|8 Butterflies, Birds, and Flower Shapes

Time Frame: 30 minutes for each activity (one hour total)

Overview 7: Students will learn about how butterflies and hummingbirds pollinate and which flower shapes they prefer.

Objective: To identify the unique physical characteristics of hummingbirds and butterflies as pollinators and begin identifying flowers that can support their work.

Vocabulary

1. Antennae
2. Proboscis
3. Nectar

Introduction and Activity 7 (30 minutes)

1. Begin class by brainstorming with the students about the experiences they have had with butterflies and hummingbirds in the garden. How can the students they tell the differences between the two flying pollinators? Record student ideas on a classroom chart.

2. Hummingbirds and butterflies are both pollinators that harvest nectar in similar ways, but with different tools and preferences. Both use a long tongue to drink nectar from flowers. For butterflies, this straw-like tongue is called a *proboscis*. What colors of butterflies and hummingbirds have students seen before? Do they know any specific names of these creatures?

3. Have any students seen the birds and butterflies visit flowers? What shape of flowers did the pollinators visit? Even though hummingbirds and butterflies visit flowers for nectar, how do they help pollinate in the garden?

4. Using books and images, have the students point out key and distinguishing features for hummingbirds and butterflies (feathers, wing shape, colors, antennae, beak, tongue). Discuss how the flexibility of the butterfly's long tongue allows it to drink nectar from a greater diversity of flower shapes, while the hummingbird's rigid beak helps it drink nectar from flowers with long cone shapes. These observations can be illustrated and listed on a classroom chart.

5. If possible, use a model of a cone- or trumpet-shaped flower to demonstrate how butterflies and hummingbirds use their tongues (and hummingbird beak shapes) to drink nectar. Compare the flexibility and length of their tongues to how difficult it would be for a hummingbird to try and gather nectar from a flat or plate-shaped flower, like a sunflower.

Activity 8 Overview 8

Using their knowledge of hummingbirds, butterflies, and the flower shapes they prefer, the students will plant a variety of flowers indoors for future garden pollinators.

Activity 8 Objectives

To learn about garden flower varieties that encourage pollinator activity and provide forage, while also learning how to plant flower seeds in pots.

Activity 8 Introduction (5–10 minutes)

1. Brainstorm with the students about the flower shapes hummingbirds and butterflies prefer. Do students know what garden flowers have these shapes? What colors are the flowers? The class can refer to their previous classroom posters for brainstorm ideas.

2. What garden flowers do the students know about that have colors the two pollinators prefer, such as red, purple, pink, white, and yellow? Honeysuckle, salvia, and bean plants are all options for hummingbirds, while milkweed is a great option for monarch butterflies.

3. For the planting activity, students should be able to recall how deeply to plant their seeds and how high to fill the container with soil. What does good, careful gardening and positive behavior look like in this activity?

Activity 8 (15–20 minutes)

1. Individually or in pairs, the students will plant a variety of flowers for pollinator forage, specifically butterflies and hummingbirds. They can plant multiple types, but they should be in separate containers from each other and make sure they are labeled correctly. The students should focus on carefully planting the seeds and filling up the soil containers to the top with soil.

2. The students can choose from a variety of seed types, especially with the colors and shapes that the pollinators prefer.

3. When the students have planted the seeds and labeled them (with the help of volunteers), they can water them with a watering can.

Activity 8 Overview (5 minutes)

1. After the class has finished planting, they can clean up the workstations and move the seed starts to a greenhouse or indoor growing

A lovely garden pollinator lands on my hand for students to study its anatomy.

area. What flowers did the students plant? What pollinators do they expect will visit them?

2. When the students are done, they can enjoy a taste test of winter garden foods.

NGSS and Activity Extensions

There are so many excellent classroom extensions and resources on hummingbirds and butterflies. These activities merely scratch the surface of potential in-depth studies, reading, and math opportunities about life cycles and pollination. Further in-class studies and storylines could include following bird and butterfly migration routes or developing a model to illustrate the relationship between the needs of various plants and animals within the places they live (K-ESS3-1: Earth and Human Activity). The students could also develop and propose solutions to reduce human impact on natural systems and living things (K-ESS3-2: Earth and Human Activity).

Preparation 8

1. 4-inch pots
2. Soil
3. Flower seed varieties
4. Tape for labeling
5. Markers
6. Watering can
7. Classroom posters

Resources

- "The Bug Chicks: Butterflies & Moths." (online—see Appendix B).
- Everyday Mysteries. "How can you tell the difference between a butterfly and a moth?" (online—see Appendix B).
- *Gotta Go! Gotta Go!* by Sam Swope.
- *The Very Hungry Caterpillar* by Eric Carle.
- *My, Oh My A Butterfly! All About Butterflies* by Tish Rabe.

9 Bees at Work

Time Frame: 30 minutes

Overview: The class will learn about the diversity of pollinating bees and play an identification game.

Objective: To demystify bee pollinators and for students to identify key pollinating players by their different characteristics.

Introduction (10 minutes)

1. Every student will have a story to share about a bee. During the following conversations, the students are encouraged to think like scientists and gardeners and wait to share their stories until the end of gardening class.

2. Brainstorm with the class about what kinds of bees the students know about. What pollinating bees have they seen in the garden before? Why are pollinating bees so important to the garden? Record student ideas on a classroom poster for future reference.

> Young students often want to talk about painful wasp experiences. I prefer to be direct in these conversations by emphasizing repeatedly that wasps and bees are not the same, that wasps and yellow jackets prey upon bees, and that their behaviors are very different.

3. For this activity and all of remaining the gardening lessons, it will be important for the students to become familiar with three active bee pollinators: the honeybee, mason bee, and bumblebee. There are many types of each bee, but most share certain traits that can help in identifying them. If the students can recognize these traits and differences, then they can think and behave like scientists in identifying each species.

4. Using a picture of each bee species (honeybee, bumblebee, and mason bee), have the students identify each one or notice the differences between each type. How can they remember these bees if they were to see them in the garden? What makes each bee memorable and unique? Record these distinguishing differences on a classroom poster.

Activities (15 minutes)

1. For their activity, the students will be thinking and behaving like scientists by using what they know to identify different pictures of the three bees that are hidden across the room or garden space. In pairs, they will first have to find the pollinator picture and then use their Bee Guide worksheets to help them identify what bee species it is. They will make a tally for each species they identify and will report this number back to the class at the end of the activity. If it is possible to go outside and do this activity, I recommend it.

2. Students should focus on teamwork and creative problem-solving. They shouldn't worry about finding all the pollinators, but rather on correctly identifying each bee. This is will also provide practice in looking for live insects in the spring garden, so they can pretend it is real by keeping calm and using a low voice when they do this activity.

Assessment (5 minutes)

1. Bringing the class back together at the end of the activity, have the students share the quantity of pollinators they found and the type of bees they discovered.

2. Were there any pollinators that were difficult to identify? Which bees were easy to discover and name? Use the classroom posters from the beginning of the activity to clarify any misidentified bees.

Preparation

1. Bee Guides worksheets
2. Pictures of a bumblebee, honeybee, and mason bee
3. Classroom posters

10 Planting Seeds for Spring

Time Frame: 30 minutes

Overview: Students will plant vegetable seeds in pots that will later be transplanted into the garden.

Objectives: For students to continue practicing their seed planting and careful gardening, while beginning to imagine what the school garden will look like with their favorite foods growing.

Introduction (5–10 minutes)

1. Now is the time of year to begin planting the seeds that will flourish all summer and feed the students in the fall. Today, they will be planting warm season crops, such as sweet melons, cantaloupe, cucumbers, and tomatoes that will grow in a greenhouse, indoors under a growing light, or near a window until they are ready to be transplanted outside.

2. As a class, have the students brainstorm how deeply to plant seeds. At this point in the year, they have planted many seeds and bulbs of various sizes. All students have different descriptions of how deeply to plant seeds (a finger deep up to a knuckle). I recommend having the seeds twice as deep as the seed is large. This is quite a concept for young students, but they should understand that many seeds need to be completely covered by the soil and not so deep that they are weakened before they reach the soil surface.

Activities (15–20 minutes)

1. Individually, students will pick out which variety of seeds they want to plant and will gather pots, soil, and seeds for their planting activity. Adult volunteers can help the students write their names on tape to put on the containers.

2. Students will need to focus on carefully planting their seeds and how deeply to plant them, as well as how much soil to put into the pots. I encourage students to keep the soil light and fluffy, rather than tamp it down with their hands.

3. The students will rotate through the available stations and work collaboratively and kindly with their classmates to do their best work.

Assessment (5 minutes)

1. When students have finished planting their seeds, they can clean up and enjoy a taste test of foods from the garden (chives, sorrel, kale, cabbage).

2. Students can also check on the growth of the flower seeds they planted weeks before.

Preparation

1. Summer seeds (watermelon, cantaloupe, cucumbers, tomatoes)
2. Soil
3. 4-inch pots
4. Tape and markers
5. Adult volunteers

Spring: Insects and Garden Species

Even slugs this large can provide opportunities for student learning, wonder, and garden math.

1 Slugs and Pests

Time Frame: 30 minutes

Overview: The students will learn how slugs play a role in an ecosystem and go on a slug hunt in the garden.

Objectives: To understand the important role that even perceived pests play in the garden and understand how these creatures are part of a permaculture and ecological garden.

Vocabulary

1. Ecosystem
2. Pest
3. Predator
4. Decomposer
5. Consumer

Introduction (5–10 minutes)

1. What kind of interactions have students had with slugs in the garden? What effects have they noticed in the garden because of slugs? What do they wonder or want to know more about the shape, traits, and behavior of slugs?

2. There are many types of slugs, but three common ones in Pacific Northwest gardens are the Banana Slug, Leopard Slug, and Red Roundback Slug (see Resources for more common varieties in Oregon. If slugs are not a key pest then choose a few that are in your local gardens). Show pictures of three common slugs (of your region) to the class and have the students identify key traits of each variety.

3. What are decomposers and consumers? What kinds of creatures are predators to slugs?

Activities (20 minutes)

1. As a class, the students will go on a slug hunt in the garden and look for evidence of slug activity and the slugs themselves. They will use what they know about the three slugs in the garden to identify them and count what they find.

2. Students are welcome to carefully handle the slugs, as long as they are gentle with them and don't harm the mucus on their skin (picking them up with leaves or dirty hands is preferable). The students should focus on practicing Care for Self, Others, and the Land during this activity.

Assessment (5 minutes)

1. After students have gone on their slug hunt, they can gather back together to share what they found. How many different types of slugs did they find? Do they know which varieties they found? Where did the students find the slugs in the garden? Were they out in the open, feeding on plants, or hiding under logs?

2. Did the students see any evidence of slug activity in the garden, such as slime trails or bite marks? Did they see any evidence of slugs decomposing plants in the garden?

3. What is a pest? Are there such things as pests in an ecosystem?

Preparation

1. Pictures of garden slug varieties

Resources

- Robert Mason. "Snakes slither through the garden." (online—see Appendix B).
- Joshua Vlach. *Slugs and Snails in Oregon: A Guide to Common Land Molluscs and Their Relatives.* (online—see Appendix B).

I add this lesson because I have noticed negative student behavior towards "pests" in the past, such as students poking slugs too hard with sticks or throwing things into the webs of important garden spiders. Much of this behavior is spurred by the language they have heard and by watching adults handle the creatures.

While I understand the damage that slugs and other pests can cause in a garden, from an ecosystem perspective these creatures are still a vital part of the garden system that cannot be qualified by human standards. Slugs feed the snakes and salamanders in the garden, which in turn feed on a diversity of other "pests."

A great seasonal project for a variety of ages can be creating natural, non-harmful slug and pest deterrents around plants as well. When encouraging students to think like scientists and gardeners, observation and wonder are the most powerful tools I can foster in the kids, taking the place of reactionary and thoughtless behavior spurred from bias.

2 Decomposers

Time Frame: 30 minutes

Overview: Students will learn more about decomposers in the garden and use magnifying glasses to help them discover these creatures in the compost ecosystem.

Objectives: To expand student understanding of the garden ecosystem by introducing them to the diversity and value of decomposers.

Vocabulary

1. Ecosystem
2. Decomposers

Introduction (5–10 minutes)

1. What do students know about decomposers? What types of decomposers live in the garden that the students have seen? What role do they play in the garden ecosystem?

> Decomposers play a crucial role in a garden that is essential to creating nutritious foods, particularly by breaking down organic matter into humus. In the learning garden, we practice celebrating our insects, grubs, and bugs!

2. Common garden decomposers that the students may encounter include worms, ants, and fungi. They may also encounter sow bugs, snails, centipedes, millipedes, beetles, and spiders. These are all important members of the soil ecosystem, and the students should focus on their value in the garden, rather than personal feelings toward insects and bugs. Have them imagine what scientists wouldn't have discovered if they let themselves feel uncomfortable when studying the world!

Activities (20 minutes)

1. In pairs, students will explore a soil sample of the school compost pile and search for decomposers at work. They can use magnifying glasses to help make these discoveries. They should thoroughly explore every bit of the soil sample for the decomposers, but also focus on carefully interacting with the living things they encounter. Students can record what they find on the class whiteboard or chart using tally marks or numbers to quantify and observe the populations of each decomposer.

2. Students who finish their exploration early can return the compost pile sample and gather a new sample of garden soil to compare differences in diversity between the two samples.

3. This activity will push some students to expand their comfort zones, so they will need to be reminded to remain calm and quiet during this activity, while also practicing gentleness with the creatures they encounter.

Assessment (5 minutes)

1. When they have finished their exploration and cleaned up their stations by carefully returning the soil samples with the decomposers back to the compost pile, the students can gather back together to share their findings.

2. What kinds of creatures did the students encounter in their explorations? What does the quantity of each decomposer population say about the health of the garden compost or the interactions these creatures are having with each other?

3. What evidence of decomposition did the students find during their exploration?

Preparation

1. Magnifying glasses
2. Observation trays

3. Compost bin sample soil
4. Pictures/examples of garden decomposers

Resources

- James Hoorman and Rafiq Islam. "Understanding Soil Microbes and Nutrient Recycling." (online—see Appendix B).
- "The Science of Composting." (online—see Appendix B).

NGSS and Activity Extensions

Further in-class studies can include making observations about patterns of what plants and animals need to survive and how they can change their environment to meet their needs (K-LS1-1: From Molecules to Organisms, K-ESS2-2: Earth's Systems).

3 Consumers

Time Frame: 30 minutes

Overview: Students will engage in a survey and sketch activity in the garden with older students while identifying consumers and evidence of consumers in the garden.

Objectives: To learn about another key ecosystem relationship in the garden between insects, animals, birds, and humans and observe how they interact together with the plants in the garden, while practicing new science skills.

Vocabulary

1. Consumer
2. Interaction
3. Survey

Introduction (5–10 minutes)

1. Brainstorm with students what they know about consumers in the garden: If a decomposer breaks down materials, what does a consumer do? What elements or creatures do mammals, insects, and birds consume? Where can the students find these creatures in the garden?

2. In order to develop their observation skills and scientific mind-set, the students will be creating a class survey of the consumers in the garden. In a survey, scientists examine and record certain features of what they're studying. For students, this means that they will be searching for consumers in the garden, identifying them, and recording their behavior and activities. If they find evidence of a consumer, they can record this as well and generate questions about the creature.

I like to do this activity with older student partners because it is another opportunity for them to pass on the cultural values of the garden and demonstrate the important science skills of observation, asking questions, illustration, and labeling. It helps if the older students have had a recent lesson that mirrors this one, so the information is fresh in their minds and they can be inspired to share their knowledge by teaching younger students.

Activities (20 minutes)

1. In pairs, students will explore the garden and search for evidence of consumers at work and actual consumers in the garden. The evidence can look like bite marks on leaves, animal prints, insect body parts, feathers, and fur. When they find evidence, the students will sketch what they see and label what they find, writing questions and observations as they go.

2. The students can measure the points of interest with rulers, or use colored pencils after they have done an initial sketch in order to provide more realistic detail.

3. The goal of this activity is to do quality scientific observations about consumers in the garden, so the students should focus on taking their time and doing good work by recording details and generating questions, rather than rushing from one site to another.
4. If the students can identify the name of a consumer at work, they should record this detail, otherwise they can seek help from an adult, another group, or an available reference book.

Assessment (5 minutes)

1. After students have gathered a few, well-detailed observations, they can gather back together and share their findings and sketches with another partner group. They can also share their findings as a class through a quick exchange of unique findings and a survey of the consumers they discovered.
2. What evidence of consumers did students find? What is so interesting and unique about it? What questions did student groups generate about their consumers? What consumers do they think left behind certain tracks and bite marks?

Preparation

1. Sketching paper
2. Pencils (regular and colored)
3. Rulers
4. Reference books

4 Planting Insect Forage

Time Frame: 30 minutes
Overview: Students will plant flower seeds or vegetable plants that pollinators, and kids, prefer.
Objective: To learn how to promote insect forage and habitat in the school garden by practicing direct outdoor planting.

Introduction (5–10 minutes)

1. Gardeners value insects and pollinators that live in and visit the garden. Brainstorm with the class what are some of reasons they are so important? What can students do to help them find food, water, and shelter in the school garden?
2. There are many ways that children can increase and encourage insect and pollinator activity and nesting in the garden. They can plant flowers that insects are attracted to and provide water features. There are garden vegetable plants as well, such as beans and sunflowers, which provide food for pollinators and people.
3. Where are the best places to plant these seeds in the garden? How deeply should students plant them and how close together?

Activities (20 minutes)

1. Individually or in pairs, students will plant flower seeds or bean seeds in the garden. They will focus on careful planting by making sure the soil covers the seeds (but not too deeply) and planting in the best spots for that seed variety in the garden.
2. As they plant, the class should look for pollinators and insects already interacting with flowers in the garden and do their best to observe the flower colors and types of insects that are visiting them.

Assessment (5 minutes)

1. After students have planted the seeds, they can gather back together to share their experiences and their findings on pollinator and insect activity in the garden.

2. Students can finish off the lesson by enjoying a taste test of seasonal foods from the garden.

Preparation

1. Annual flower seeds (sunflowers, calendula)
2. Bean seeds (presoaked)

Resources

- "Gardening for Pollinators." (online—see Appendix B).
- "Invertebrate Conservation Guidelines. Pollinator Conservation: Three Easy Steps to Help Bees and Butterflies." (online—see Appendix B).

NGSS and Activity Extensions

Greater classroom studies can include presenting solutions to decrease the impact that humans have on the local environment and natural resources (K-ESS3-2: Earth and Human Activity).

5 Predators

Time Frame: 30 minutes

Overview: Students will be learning about valuable predators in the garden and build habitat boxes to provide additional space for them.

Objectives: To continue building student knowledge of the garden ecosystem, valuing beneficial insects, and creating solutions for greater habitat.

Vocabulary

1. Predator
2. Habitat
3. Prey

Introduction (5–10 minutes)

1. The students have explored decomposers and consumers in the garden, but what about the insects, birds, and mammals that prey upon these creatures? Brainstorm with the students what kinds of predators live, eat, or travel through the school garden.

2. Predatory insects are valuable to the garden ecosystem because they feed on many insects and bugs that are pests to the garden produce and students. Ladybugs, praying mantis, spiders, and many more insects feed on consumers and decomposers in the school garden. If there is good habitat for them (food, water, shelter, nesting sites), then students can encourage them to live and reproduce in the garden.

3. For the activity, the students will be assembling habitat boxes that will provide nesting cavities for predatory insects. Use a finished and assembled box for students to discuss how

This beautiful black-and-yellow garden spider is a valuable predator in the garden ecosystem.

it was created, what insects will live and nest in it, and the importance of providing habitat for the insects.

One spring there was a gorgeous black and yellow garden spider in the school garden I worked with; she made her large squiggled web in a patch of yarrow. I adored this massive beauty that caught yellow jackets in her trap. I showed her to as many children as I could so we could talk about how, even though she made some students feel uncomfortable, she was incredibly valuable to the garden and the students' experiences in it. And when they learned that she ate yellow jackets, that spider became a superhero to the children.

Unfortunately, not all the children got my message, and we soon found the spider's web doused with bark chips that someone threw at her, without the spider in sight. Even this loss was a great learning experience for many students about how to value garden predators.

Activities (20 minutes)

1. In small groups, students will assemble the habitat boxes with the assistance of an adult volunteer. They should do their best to gather small amounts of the provided materials, or they can go out into the garden and schoolyard to gather needed items. The groups should focus on teamwork, taking turns, and kind language ("Care for Others") during this activity.

2. Adult volunteers can help students tie the different materials into bundles. The students will stack and assemble these bundles into a wooden frame or drawer, to the best of their ability. Material bundles should fill up the entire space. The bundles should be sturdy and not have the materials fall out, but not tied too tightly or loosely.

3. The adult volunteer can also staple on a small section of chicken wire for students to fill with leaves or bark chips for further habitat. When the boxes are assembled, the group will find a safe, partially shaded space to place or hang their habitat box.

I list drawers as potential box structures because I used them and reused them in one school garden for years. They have been easy to find on internet sale sites or free piles.

Assessment (5 minutes)

1. After students have cleaned up and installed the bug boxes, they can enjoy a taste test from the garden and discuss what insects will first visit the habitat. How will the students know they have been there? What evidence could these insects leaves behind?

Preparation

1. Adult volunteers
2. Drawers or wooden boxes
3. Dried leaves
4. Sticks
5. Pieces of wood with drilled holes
6. Hollow stems
7. Precut chicken wire
8. Bark pieces
9. Stapler

Resources

• Robin Plaskooff Horton. "How to Design a Bug Hotel to Attract Beneficial Insects & Bees." (online—see Appendix B).

6 Sowing Seeds

Time Frame: 30 minutes

Overview: Students will sow cover crops in the garden for soil health and insect forage.

Objectives: To learn a new way to plant seeds and to promote soil health and habitat in the garden.

Vocabulary

1. Sow
2. Broadcast
3. Cover Crop
4. Forage

Introduction (5–10 minutes)

1. Today, the students will *sow* seeds in the garden. Another term is *broadcast*, a technique used by some farmers to plant seeds over a large space. Impress on the students the importance of not wasting seeds by scattering them on pathways or living plants (you'll be amazed at where children think are good places to plant), but rather focus on careful, slow garden work. The goal is to scatter the seeds in the exposed soil areas and around/under living plants.

2. The seeds that the students sow will provide important foraging space for the insects in the garden. The flowers will support pollinators and attract beneficial predators. The plants will also serve as a cover crop, which will encourage soil health, suppress weeds, and maintain soil moisture during the summer.

> I've used cover crops like red clover and native ground covers like meadow foam in the garden with great success. I also have used strawberries and sheep's sorrel (a kid favorite) as perennial ground covers.

3. Have students practice the motion of sowing or broadcasting seeds before going out into the garden. They'll need to make sure that their sowing motion is low to the ground so the seeds reach the soil and are not wasted.

Activities (20 minutes)

1. Individually, the students will receive a handful of cover crop seeds (native wildflower mix, clover, legumes, grains) mixed with soil in a bucket and broadcast it in the garden in pre-exposed or designated bare patches. (I will often rake the mulch back prior to this planting activity.)

2. Afterward, they can gently press the seeds into the soil using their hands and return for more seeds to sow until the end of the activity.

3. At the end of the activity, the students will cover the seeds with a little amount of mulch and, if necessary, water them.

Assessment (5 minutes)

1. Gathering back together, the class can discuss the concept of foraging. How do the students forage for food in the garden? What foods do they search for?

2. Individually, the students can forage for taste tests from the garden. Typically, I will announce which plants they can harvest from and how many pieces (leaves, fruits, flowers) they can have. I find regulating this activity leads to less plant damage and enough food for everyone.

Preparation

1. Cover crop seeds
2. Bucket
3. Soil

7 Planting Flower Starts

Time Frame: 30 minutes
Overview: Students will transplant the flower starts they began in the winter for the garden pollinators.
Objectives: To learn how to transplant potted plant starts and provide habitat for pollinators.

Vocabulary

1. Transplant
2. Depth

Introduction (5 minutes)

1. As a class, brainstorm what the students remember about the seeds they planted for pollinators back in the winter. How have the plants been growing and developing? What kinds of pollinators will these flowers support in the garden?

2. Transplanting in the garden means to take a plant from one growing area, like a container, and move it to a new spot. In this case, the students will be taking their flowers starts into the garden and planting them in a flower bed or designated part of the garden. Demonstrate to the students how to carefully remove their plant from the pot, as well as how deeply to plant it.

Activities (10–15 minutes)

1. Individually, or in pairs, the students will transplant their flower starts into the garden. They will need to work as careful gardeners to dig a hole the same depth as the pot and to use digging tools safely. After digging the hole, the students will carefully remove the plant from the container and transplant it. When they are filling up the hole, they should make sure to crumble up any large pieces of surrounding soil and gently fill the hole. They can pat down the soil after they have planted their flowers, but should be aware not to compact the soil too much.

2. After returning the empty planting containers, students can water their flower starts with a watering can and repeat the planting process again; I always keep extra plant starts handy for the students who do good work quickly.

Assessment (5–10 minutes)

1. As a class, explore the garden to make sure all the tools and planting containers are collected and returned and all the flower roots are covered by soil. There are always students who are great at quality control!

2. Finally, the students can enjoy a taste test from the garden or go on an exploration to see if pollinators are already at work.

Preparation

1. Student flower starts
2. Small shovels
3. Watering can

8 Mulching and Weeding

Time Frame: 30 minutes
Overview: Students will help with seasonal garden chores including mulching the garden pathways and weeding.
Objective: For students to participate in the community work by beautifying the garden.

Vocabulary

1. Mulching
2. Organic

Introduction (5–10 minutes)

1. Brainstorm with the students on what they know about weeds. What kinds of plants do they consider weeds in the garden? What is the best way to remove a weed? Demonstrate to the students how to best remove a weed in the garden using plant examples.

I teach students that weeds are not bad plants, they are just plants in places gardeners don't want them. As an organic gardener, if I need to remove plants, then I promote hand-pulling weeds as the most effective and long-term way to remove an unwanted variety.

2. Has any student heard the term mulching or do they recall the activity in the fall? Mulching is the process of layering organic materials on the ground for greater soil health. As gardeners, they can mulch materials in the garden beds for healthier soil and use hardier materials such as wood chips on the pathways.

I typically use burlap covered by wood chips on garden pathways. While both materials break down over time, they allow soil creatures to live comfortably underneath, for mushrooms to grow along the pathway edges, and for students to continue learning about mulching after many years of seasonal work. The materials are also cheap to find in my region. Local coffee companies often donate burlap sacks, and the wood chips can be delivered by local tree trimming services.

Activities (20 minutes)

1. Students will be weeding plants from the pathways or garden beds, as needed. Their goal is to remove the plant by the roots, rather than rip off the leaves or stem only. It's not a competition to see who can get the most pulled, but rather to see who can get whole plants removed—this is a time for high-quality gardening and craftsmanship. Students can work in partners for this activity and put their weeds in a bucket (to dehydrate).

2. When mulching, the students can work in pairs or individually with buckets. They will fill their buckets up at the wood chip pile (by hand or with adult help) and carry it to the designated garden pathway to spread out. An adult volunteer can be there to direct the students and rake out the bark chips. The students will continue this process until the activity time is over.

Assessment (5 minutes)

1. When students have cleaned up and returned all the buckets, they can gather back together to reflect on their activity. Was it difficult for the students to weed? Which weed was the most difficult or easiest to remove?

2. Students can enjoy a taste test from the garden as a reward for their hard work.

Preparation

1. Adult volunteers
2. Buckets
3. Gloves
4. Weed examples
5. Wood chips

9 Spring Harvest

Time Frame: 30 minutes

Overview: The class will learn about seasonal spring foods and learn how to carefully harvest them.

Objectives: For students to practice harvesting techniques and try new foods.

Vocabulary

1. Harvest
2. Seasonal
3. Washing Station
4. Ripe

Introduction (5–10 minutes)

1. Brainstorm with the students: What foods have they noticed growing in the garden? How are these foods different than what was growing over the winter? What plants don't grow well in the spring? What are some of the students' favorite spring foods? Where do these plants grow in the garden? What does a *ripe* vegetable look like?

2. Introduce (or reintroduce) the students to the concept of seasonal foods; some foods grow best in cool spring weather, others like the heat of summer, and others prefer the wet or coolness of winter. In the Pacific Northwest, spring foods like peas, radishes, fava beans, and green garlic taste and grow well during this time of year.

Activities (20 minutes)

1. Demonstrate for the class how to best pick peas and determine if a radish or carrot is ripe. Have them practice the hand motions of harvesting. For peas and leaves (especially chard, lettuce, and kale), I recommend that children use two hands to snap the plant part off, rather than twist or pull. For root vegetables, I recommend that children dig a little space around the root to see if it is big enough. If it isn't, then they can cover the plant back with soil. It is important for students to demonstrate careful harvesting techniques for next week's lesson.

2. Have the students go into the garden to find certain spring food (lettuce, chard, kale, peas, radishes, carrots, fava beans). I typically name and list which plants to harvest and the amount that each student can pick. When they have found the plants, they can carefully harvest them and bring them to the washing station to be rinsed and scrubbed by students with an adult volunteer. Some students may be concerned if they can't find all the plants, but they shouldn't worry because everything that the class gathers will be shared among all the students.

3. When the students have finished their harvest, gather them back together to taste test the seasonal spring foods.

Assessment (5 minutes)

1. As they eat, the students can discuss what tastes and textures they are experiencing. Do these feelings and experiences change as they chew their food?

2. What part of a plant do radishes, peas, and chard come from? What are the students favorite spring plants so far? What flowers and other plants grow well in the spring?

Preparation

1. Examples of spring foods

10 Celebration

Time Frame: 30 minutes

Overview: The students will enjoy a picnic to celebrate the garden work and learning that they have accomplished.

Objectives: For students to identify the plants they enjoy eating and to try new foods as a community.

Introduction (5–10 minutes)

1. Begin class with a discussion of the foods in the garden that students enjoy taste testing the most. What new foods are there to eat in the garden? What seasonal foods will the students be able to eat in the fall?

2. Today, the students will forage for foods they would like to eat for a class garden picnic. Before foraging or harvesting the food, discuss which plants the students would like to eat, what is in season, how to carefully harvest different parts of plants, and how much each student can harvest.

Activities (20 minutes)

1. Individually, the students will go into the garden and gather the available foods, such as lettuce, sorrel, chives, edible flowers, kale, chard, mustard, mint, strawberries, and rhubarb.

2. As students finish gathering their food, they can rinse them in the washing station (with adult help), plate their food, and meet for a class picnic in the garden.

3. Alternatively, schools with nutrition services can arrange with their support staff to process and plate the garden food.

Assessment (5 minutes)

1. As students finish their garden picnic, have them discuss the changes that the garden will go through over the summer. What will the garden look like when the students come back in the fall?

2. What creatures will visit the flowers they planted and what new foods will be growing? What activities did the students most enjoy accomplishing this year?

Preparation

1. Plates
2. Clean washing station
3. Adult volunteers
4. Taste test examples

For those children who experienced the Kindergarten lessons, this year of study in the garden has similar themes and activities. But the slight repetition instills the best garden behaviors, science skills, cultural values, and gardening knowledge after a whirlwind year of sensory overload in Kindergarten. The students may not remember all the vocabulary and processes that go into seasonal garden work, or seed lessons, or about pollinators, but they should remember some of the fun activities and positive behaviors.

The theme of first grade is seeds, and it will color all of the children's garden activities, explorations, and experiments. Beginning with the seasonal work of seed saving, moving to the science and nutrition of different parts of plants, to growing a small indoor garden, the

FALL
The Edible Parts of Plants

1. Garden Expectations
2. Seed Saving
3. Planting Winter Seeds
4. Mulching with Leaves
5. Identifying Edible Flowers

6|7. Pumpkins (Fruit, Seeds)
8|9. Chard (Stems)
10. Carrots and Beets (Roots)

WINTER
Traveling Seeds

1. The Parts of a Seed
2. Seed Sorting
3. Planting for Spring
4. The Traveling Seed
5. Fly, Float, Explode!
6. Seeds in Water

7|8. Make Your Own Seed
9. Seed Investigation
10. Sowing Seeds

SPRING
Sprouting Seeds Experiment

1. Early Spring Starts
2. Growing an Indoor Garden
3. Seed Balls
4. Observing Plant Growth
5. The Carrot Seed
6. Gardener's Math
7. The Seed Bank
8. Final Plant Observation
9. Transplanting Summer Starts
10. Garden Celebration

class will learn about garden plants and how every part of a plant and an ecosystem works in relationship to each other. Students will explore the diversity of plants and their various shapes, soil preferences, root sizes, widths, and flower colors that all promote plant survival. They will learn how seeds have adapted skills to travel and spread their progeny by using a seed model to explore this natural technology.

For their winter science studies, the class will set up experiments for observing plant growth, make scientific observations, and generate questions about how plants thrive. In expanding their growing observational skills, the students will practice measuring, labeling, and using scientific language to construct arguments. Throughout their studies, they will continue exploring new foods and tastes. And the students will especially practice objective observation and systems thinking—essential to a good scientist and gardener—which will help them value and respect the diversity of relationships in the garden.

© contrastwerkstatt / Adobe Stock

Next Generation Science Standards

The students will develop an understanding of how plants and animals use their external parts to help them survive, grow, and meet their needs, as well as how behaviors of parents and offspring help the offspring survive. They will also explore heredity by understanding that young plants and animals are similar to, but not exactly the same as, their parent. The students will expand their science skills as they plan and carry out investigations, analyze and interpret data, construct explanations, and design solutions.

These lessons are extensions or supplemental to in-depth classroom reading, writing, and math units on plant technology and mimicry by humans, daylight variations and the Earth's place in the universe, as well as heredity differences according to environmental changes and needs.

Permaculture Principles

- Observe and Interact
- Obtain a Yield
- Apply Self-regulation and Accept Feedback
- Use and Value Renewable Resources and Services
- Design from Patterns to Details
- Integrate Rather than Segregate
- Use Small and Slow Solutions
- Use and Value Diversity

Fall: The Edible Parts of Plants

1 Garden Expectations

Time Frame: 45 minutes

Overview: As a class, students will set expectations for their behavior in the garden and then will explore the garden and become familiar with its changes, layout, and the plants inside.

Objectives: To establish community behavior expectations and practice them in the garden.

Introduction (10–15 minutes)

1. As a class, brainstorm the behaviors expected of students during garden class and especially in the garden space itself: What behaviors do students appreciate or wish to see? What does it mean to respect or Care for Self, Others, and the Land? These can be listed on a large paper pad so students can visualize this social agreement.

2. Explore what students know about gardens. What kinds of gardens have they seen before? What behavior expectations did they follow in those gardens? What is a garden?

3. Give the students a chance to voice any concerns they may have about being in the garden. Some students carry fears with them about insects or about how other students behave in the garden.

Activities (15–20 minutes)

1. From the garden, choose at least 16 plants that are easily identifiable for students, such as rosemary, apple trees, squash, tomatoes, nasturtiums, beans, and marigolds. Prior to the gardening class, gather images of these plants to help students identify them during the activity.

2. Break students into pairs and hand out plant identification images, one or two images per group. The students will go into the garden and count as many of these plants as they can. They can mark or tally the numbers on the cards or notepaper.

> I like to give the students two cards with one easily recognizable plant and one that they may not know. I also keep extra cards for those students who finish faster than others.

3. After students have completed their survey, or after 15 minutes, gather them back together to share their findings and the quantity of each plant they discovered.

4. Any extra time can be given to taste testing an easy-to-pick plant in the garden, such as kale or lettuce.

Assessment

1. Students should be able to identify these plants by name or show them to another classmate who was not in their group.

2. What other interesting plants or fruits did students come across in their explorations? What does the quantity of the plants they found say about preference in the school garden and the success of different plants? How does size and season play a factor in the current plant populations?

3. What good garden behaviors did the students notice from other groups during the activity?

Preparation

1. Identification cards of garden plants

2 Seed Saving

Time Frame: 45 minutes
Overview: Students will be introduced to the seeds in the garden and participate in dry seed saving from marigolds or available flowers.
Objective: To save and store viable and healthy garden seeds for next year.

Introduction (10 minutes)

1. Introduce the students to the idea of seed saving and show them different types of seeds that gardeners gather in the fall. Brainstorm with the class: What does seed saving mean? Why is it important to save seeds each year? What seeds have students noticed in the garden?

2. How do students identify seeds worth saving for the next year? What kind of colors, textures, and shapes indicate that one plant is healthier than another?

3. What do seeds look like? Where do students find them in the garden?

Young gardeners harvest and save bean seeds for next spring's planting.

Activities (25 minutes)

1. The students will break into small groups and gather, or receive, whole marigold flower heads from which to harvest seeds. They should practice teamwork and careful gardening by using collection cups to store their seeds. When their cups get too full, they can deposit the seeds into a large paper bag.

2. Good student work looks like careful harvesting with limited seeds wasted or on the ground. It will also include considerate teamwork, rather than competitive behavior.

3. When the students have completed the activity, they can put the remaining plant pieces in the compost and gather all the seeds together in a large bag or glass jar to save.

4. Finally, the students can enjoy taste testing edible flowers, describing the flavor and tastes of different varieties, while also reflecting on the discoveries they made and experiences they had during the activity.

Assessment (5–10 minutes)

1. What other creatures save seeds in the fall? What would the garden look like if all these seeds sprouted?

2. What words describe the texture of the seed heads? What wonderings did the students experience or come up with during the activity?

Preparation

1. Collection cups
2. Marker for labeling bags
3. Paper bags for seeds
4. Seed examples for demonstration

3 Planting Winter Seeds

Time Frame: 45 minutes
Overview: Students will plant a crop of fava beans in the school garden.
Objective: To practice how deeply to plant a seed and learn what plants prefer to grow in cold weather.

Vocabulary

1. Dormant
2. Hardy
3. Trellis
4. Hoop House

Introduction (10 minutes)

1. What types of seeds have students planted before? Discuss with the students how deeply and how widely they should plant each seed. Remind them of how gardeners measure in the garden when they don't have tools to help them, such as using fingers to estimate the right depth.

2. So far, the students have saved seeds from plants that die in cold weather, and they have planted seeds that remain dormant over the winter. This week, they will be planting the seeds of plants that prefer to grow in cold conditions.

3. What parts of the bean plant do people usually eat? What seasons would other beans prefer to grow in? What seasons do they not like to grow in? How deeply in the soil should students plant a seed? What plants grow in warm weather and what grow in cold weather? What season is it now?

Activities (30 minutes)

1. In small groups, students will plant their handful of fava bean seeds in the designated garden plot or container. They should focus on sharing their garden space and planting their seeds at the right depth and along a trellis, if it is available. They can also plant their seeds under a cloche or hoop house for a potentially earlier spring harvest.

2. At the end of the planting activity, the students can taste test other foods from the garden with the help of volunteers or pre-picked by an adult.

Assessment (5 minutes)

1. How did the soil feel to the students as they planted? What textures did they encounter? Was it difficult to dig into the soil with their hands?

2. How long do students think it will take for the fava bean seeds to start spouting? What other conditions need to occur for the bean seeds to grow?

Preparation

1. Fava bean seeds
2. Prepared planting areas
3. Adult volunteer

4 Mulching with Leaves

Time Frame: 45 minutes
Overview: Students will learn how tree leaves store energy and nutrients that gardeners can use to make rich soil.
Objectives: To participate in helpful seasonal garden activities and protect the soil health during the winter.

Vocabulary

1. Mulching
2. Dormant
3. Energy
4. Photosynthesis

Introduction (10 minutes)

1. Brainstorm with the students about the importance of leaves on trees and how gardeners can use them: What role do leaves play as a part of a tree? Can trees live without leaves? What happens to the leaves' energy or nutrients when they fall? Where is energy stored in leaves?

2. Demonstrate to the students how to mulch with leaves by being careful and practicing craftsmanship, such as by tucking the leaves around living plants and spreading them evenly over the soil rather than throwing them on top of plants and garden beds. If possible, dig into a garden bed to find decomposing leaves and mulch from previous years. Are there decomposers at work in this organic matter? Is the material dry, wet, rotten, or earth smelling?

The Tree

O tree

Glistening

In the wind

I love you

Tree. It's

So nice to

See you

Glistening

In the

Wind.

— Anonymous
Student Poem

Leaves are a valuable source of nutrients to use as mulch. Leaves also create habitat for migrating amphibians and insects and protect soil during rain, snow, and winter wind.

Activities (30 minutes)

1. Individually, students will take handfuls or buckets of pre-gathered leaves and tuck the garden beds in for the winter. They should focus on mulching one or two garden beds as a class, before moving on to new sites. If there are trees in the garden area, students can rake the leaves prior to this activity or during it while their classmates transport the collected leaves into the garden.

2. If there if time, students can also pour prepared buckets of compost over the leaves to keep the leaves from blowing out of the garden beds in the winter. Chicken wire can also be used to keep the leaves down. This same activity can be done in containers.

Assessment (5–10 minutes)

1. Have students return to the garden and make sure all the leaves are in the garden bed and aren't strewn about the pathways (quality control).

2. Afterward, the students can clean up and enjoy a taste test from the garden and discuss how energy is cycled through the garden each season. What other organic plant materials offer a source of energy and nutrients to the garden ecosystem?

Preparation

1. Leaves
2. Volunteers to help students
3. Rakes

5 Identifying Edible Flowers

Time Frame: 45 minutes
Overview: Students will learn about edible garden flowers and how to engage with their senses, especially taste.
Objectives: To learn new plant names and problem-solve ways to identify key flowers by recognizing colors and patterns.

Vocabulary

1. Edible
2. Pattern
3. Senses
4. Flavor

Introduction (10 minutes)

1. Introduce the students to the word *edible* and the flowers they will be looking for in the garden. What does edible mean? What does it mean if something is not edible? What words do students use to describe taste?

2. Show examples of flowers that are edible. How will the class be able to find these flowers again in the garden? What colors and patterns will help them identify the correct flower? Guide the students to develop a plan on how to find these flowers again by using petal shape or count, distinctive features, size, colors. This activity can be an excellent tie-in to conceptual math ideas in class, especially pattern recognition.

3. Students will be given permission to pick a small sample of the flowers in the garden, such as borage, brassica flowers, nasturtiums, calendula, and fennel flowers.

Love-Lies-Bleeding Amaranth is a unique flowering plant for students during their sensory explorations and identification activity.

Activities (30 minutes)

1. In pairs, students will be given Flower Guide Cards with three common garden flower varieties. In their groups, they will figure out how to identify the three flowers on the sheet and go out to find and pick them.

2. When students have collected their flowers, gather them back together to double-check their identification, with adult volunteer help if needed. When the flowers' identification and edibility have been confirmed, the class can enjoy a taste test of their samples.

3. Have the class think of words to describe the taste of the edible flowers: Are the flavors different from each other? What is the texture of the flowers? Are they spicy, sweet, soft, or crunchy?

4. The students should return the Flower Guide Cards at the end of the activity.

Assessment (5 minutes)

1. Students can share their strategies for identifying flowers and use adjectives to describe the flavor and texture of the edible flowers. Were there flowers in that garden that looked similar but were not the same variety? How could the students tell them apart?

2. The students should also work to remember the names of the edible flowers to the best of their ability.

> If a student asks me to taste test a plant from the garden, I always insist that they recall the common name of the plant if they can, or show me the plant while practicing pronouncing the name. It is important for students to be able to identify a plant based not only on its characteristics but also by name.

Preparation

1. Examples of edible flowers
2. Flower Guide Cards (you can create three identical cards from each worksheet page)
3. Adult volunteer(s)

6|7 Pumpkins: Fruit and Seeds

Time Frame: 1½ hours (2 × 45 minute)
Overview: Students will learn about the parts of a fruit using pumpkin or squash as examples.
Objective: To learn that the fruit is the part of the plant that holds the seeds.

Introduction (10 minutes)

1. Brainstorm with the students what they know about fruits: What is a fruit? What fruits are not edible? What are others foods people eat that hold seeds? What part of a plant is a pumpkin? What are the different layers that students can see in a pumpkin?

2. Demonstrate to the students how they will be harvesting the seeds of the garden pumpkins or squash. They will be saving these seeds for taste testing and for storing to plant next summer.

3. The class should practice slow, careful teamwork and gardening behavior during the activity. What does student behavior look like in accomplishing these goals?

Activity 6 (30 minutes)

1. In small groups, students will remove the seeds from a halved pumpkin or squash, making sure to carefully remove pulp and meat from the seeds.

2. Students will collect the seeds in a cup and then transport them to a sieve by a sink or hose, so an adult volunteer can assist in washing them off.

3. Students should pay attention to the parts of their pumpkins and generate words to describe the texture and smell of the pumpkin. They can share these words during or after the activity.

4. The leftover fruit and seeds can be washed carefully and used for student taste testing.

> In the past, I cut up raw pumpkin or squash pieces for students to experience. It leads to great conversations on how foods tastes and textures change when raw or cooked with spices. I also roast up the seeds without flavoring for the students to taste test during the next class.

Preparation

1. Pre-halved pumpkins or squash
2. Sieve to wash seeds
3. Pumpkin seeds or pumpkin to taste test
4. Collection cup
5. Adult volunteer

Activity 7 and Assessment (35 minutes)

1. In order for students to continue trying new foods, this activity is a cooking extension from the previous lesson (and can be skipped or altered depending on access to a kitchen).

2. Students will continue learning about fruits and seeds by cooking mini-pumpkin pies or a sweet squash curry (slightly spicy or not spicy) as a class.

> I recommend finding a simple and quick mini-pumpkin pie recipe for expediency and simplicity. During past activities, a parent volunteer, or myself, has preroasted the pumpkins as well as the seeds before this activity.

3. This cooking lesson can be performed by students in small groups or as a class. I prefer using "name sticks" where I can draw a name, call on student volunteers, and give everyone a job, no matter how small, from adding spices, mixing different ingredients, and spooning the mixture into cupcake trays.

4. The students can help clean up while the recipe is cooking and then gather back together to review the key parts of the pumpkin and discuss how a fruit can carry seeds. Why would a plant grow a fruit that tastes good to consumers, like humans and other mammals? What other reasons could a plant have to store seeds within a fruit?

5. When the recipe has been completed, the students can enjoy a taste test of their creation. I recommend letting them taste the roasted and unseasoned fruit before mixing it into the recipe as well. The students can discuss the taste test difference, as well as their experience with the roasted seeds.

Preparation

1. Roasted pumpkin or squash
2. Roasted seeds
3. Ingredients for pumpkin pies/sweet squash curry
4. Adult volunteers

Adaptations

This same activity can be accomplished with other fruit surplus from the garden, such as cucumbers, squash, tomatillos, and tomatoes. Of course, the recipe options change with each variation.

Resources

- Jean Richards. *A Fruit Is a Suitcase for Seeds.*
- Inspired by an activity from Eat.Think. Grow. (online—see Appendix B).

8|9 Chard: Stems and Leaves

Time Frame: 1½ hours (2 × 45 minutes)
Overview: Students will harvest their own taste test from the garden and learn about edible stems and leaves.
Objectives: To practice careful harvesting methods and identify new foods.

Vocabulary

1. Photosynthesis
2. Nutrients

Introduction (10 minutes)

1. Name a few stems that the students should be familiar with, such as celery, broccoli, or rhubarb. Brainstorm with the students what

A ground cover of self-sowing chard means more food for students during the fall and winter.

part of a plant these edible foods are from and what they do for the plant itself (hold it up, transfer nutrients, pass along water). What does a stem do to help a plant? What plant parts are connected to it above ground and below ground?

2. Some parts of plants, including stems and leaves, last through cold weather and even get sweeter when it gets colder, as celery, leeks, and parsnips do. Before the winter frost, many cold-loving plants have bitter and strong flavors; over the cool season, root crops increase the amount of sugars in their bodies from starches, and brassica family members increase sugars in their cells which help combat freezing. These physiological changes can be a good connection to weather changes and phenology for further in-class studies.

3. Introduce the students to the stems they will be harvesting from the garden. Demonstrate how to identify these plants and how to carefully pick these plants by using two hands to snap the stem at the base—not twisting or yanking. Some plants have shallow roots, and rough handling can pull them completely out of the ground. Have the students practice the snapping motion before they harvest from the garden.

4. Discuss with the students that not all stems and leaves are edible. Chard is a safe plant for the students to pick, but rhubarb plants should be handled by an adult if students struggle to differentiate it from red chard or dock. Students should not pick rhubarb without an adult at school because the oxalis in the leaves is poisonous in high quantities. A prepared taste test of rhubarb stems can be available to students after they harvest chard.

Activity 8 (15 minutes)

1. As a class, students will go into the garden to identify chard, celery, or other edible stems and then pick a sample, demonstrating that they can harvest them carefully. It is helpful to have the students bring their harvest to another adult to rinse at a washing station.

2. After the produce is washed, students can dissect their foods to study the inside of the stems. Can they see where the water and nutrients are transported in the stem? What other parts of the stem do students see? Are there parts that they can't identify?

3. Afterward, the students can indulge in a taste test of their harvest and any other prepared stems and leaves. They should begin generating words to describe the taste and texture of these foods.

4. If it is possible to have a microscope and a

projector prepared, then an adult can take a cross section of a stem for the students to observe the complexities of the plant stem and compare the view under a microscope to what they can see with their eyes.

Activity 9

Repeat Activity 8 with leaves. Students should understand that some leaves store water and photosynthesize to create energy for the plants that humans and other creatures eat.

1. Students will gather the leaves they want to taste from prepared samples and should be aware of the different flavors. Students can eat the different leaves separately or together.

2. What words would students use to describe the tastes and textures of these parts of plants? If eaten together, how did the flavors change? Are all stems and leaves edible? Could they find any evidence of how stems carry nutrients and water throughout the plant or how leaves photosynthesize and capture energy?

Preparation

1. Adult volunteer
2. Prepared stems/leaves

Resources

- Inspired by an activity from Eat.Think. Grow. (online—see Appendix B).

10 Beets and Turnips: Roots

Time Frame: 45 minutes

Overview: Students will harvest roots from the garden and learn what role roots play as plant parts.

Objectives: To learn how to correctly harvest root crops and identify other edible and non-edible roots in the garden.

Vocabulary

1. Seasonal
2. Edible
3. Nutritious/Nutrients

Introduction (5–10 minutes)

1. Brainstorm with the class what they know about plant roots, edible roots in the garden, and how roots support garden plants: What roots they have seen or eaten in the garden? What kinds of roots do they like to eat? How do they know something is a root? Are all roots edible?

2. How do roots help a plant growing in the garden? Why do people like to eat them? What other creatures like to eat roots?

2. Discuss the role of roots with the students. Roots hold up some plants and absorb nutrients and water from the soil. Some roots are very deep, some are shallow, some grow flat against the ground, and others go straight down.

3. Demonstrate to the students how to identify beets, turnips, and other root vegetables in the garden. If they can't see the root, how else can they recognize the plant? If there are other similar leaf structures in the garden, pick a sample to show the students so that they can determine the differences between the plants.

Activities (10–15 minutes)

1. As a class, go out to the garden and demonstrate how to harvest selected roots. Students should know how to gently brush the soil from around the top of the root in order to check the width of the root before picking. If the width is small, they should spread the soil back over the root and find another plant. This same practice can be done on other roots, such as radishes and parsnips.

2. Individually, students will explore the garden to look for an available root crop. When they have found a plant and checked the width, they can wait for an adult to check their roots before picking them and taking them to get washed by a volunteer.

Assessment (5 minutes)

1. When students are gathered back together, they can taste test of a variety of roots while they discuss their methods of removing the roots from the ground. How can students positively talk about what they taste test? What words do they use to describe taste and texture?

2. How deep were the roots and were they difficult to remove? Did any students find other roots that resembled a carrot? How could they tell the difference between the two?

Preparation

1. Adult volunteer
2. Examples of root crops and look-alikes

Adaptations

This activity can also be done with any root vegetables grown in a deep pot, if they are planted months in advance, or weeks in advance if planting radishes.

Winter: Traveling Seeds

1 The Parts of a Seed

Time Frame: 45 minutes

Overview: The students will learn about three parts of a dicot seed and create a model.

Objective: To recognize and name the parts of certain seeds through hands-on exploration.

Vocabulary

1. Seed Coat
2. Cotyledon
3. Radicle

Introduction (5–10 minutes)

1. Brainstorm with the class what they know about the parts of a plant where there are fruits, seedpods, and/or shells. What is the role and importance of each of these plant parts? With a large dicot seed as an example, have the students try to name the different parts of the seed that they can see. Illustrate these parts of a seed by labeling them on a classroom chart.

2. When working with the seeds, students should focus on teamwork and keen scientific observations.

Activities (30 minutes)

1. Individually, the students will receive the two Parts of a Seed worksheets. They will color and then cut out the pieces, before assembling them together. The order will be the radicle and epicotyl cutout (#1) on top of the cotyledon (#2) on top of the seed coat. (These should be glued together and then folded like a card to create a model of a dicot seed.)

2. Students with extra time can label the parts of the seed, using the classroom chart for reference and correct spelling.

Assessment (5 minutes)

1. Before the activity time is over, the students should go back over their models to make sure they labeled all the parts and have used the correct spelling. At the end of class, each student will have modeled the insides of a dicot seed and learned three new vocabulary words.

Preparation

1. The Parts of a Seed worksheets

2 Seed Sorting

Time Frame: 45 minutes
Overview: Students will handle different seeds and sort them according to size, color, and texture.
Objectives: To explore the diversity of seeds and compare their similarities and differences.

Vocabulary

1. Dormant
2. Diversity

Introduction (5–10 minutes)

1. Brainstorm with the class what they know about seeds: Are they living, nonliving, or in another category? What does it mean when something is dormant? What does diversity mean? What does diversity among seeds look like?
2. How can students generate wonderings during the activity? What are descriptions and observations that student scientists could use during their discoveries?

> Students who engaged in the Kindergarten gardening lessons may remember doing this before. I don't mind repeating this activity because it gives further meaning to student learning, and I can present them with newer, smaller, and more complex seed shapes, such as beet and chard seeds.

Activities (30 minutes)

1. Each student will receive an egg carton and a handful of seeds from different plants (the greater the diversity of seeds, the better). Each student will sort their seeds based on size, color, texture, or another category of their choosing.
2. As they go, the students will create wonderings about the different seeds they encounter. Can they guess which plants the seeds came from? The class should also generate words to describe the texture, weight, and feel of the seeds.
3. If the students finish sorting the seeds by one parameter, have them remix the seeds and sort them in another way.
4. The end of the activity, the students will carefully gather their seeds and return them along with the egg cartons, before meeting back as a class to share their findings.
5. A more intensive math extension for this activity would be to have the students receive the same amount of seeds, sort them according to a parameter (size, shape, color), count the seeds in each parameter grouping, and record these numbers before moving on to another parameter. Afterward, the students can compare the data from the two groups and create a graph or grid of this data as a class. Or, the students could use these numbers as the basis for future math equations.

Assessment (5 minutes)

1. What observations did the students make about their seeds? How would they describe the texture and feel of the seeds? What wonderings did the students come up with during the activity?
2. Were there seeds that were similar to each other, but slightly different? How many ways did the students sort the seeds? Can the

students name any of the plants where the seeds come from?

Preparation

1. Egg cartons
2. Variety of dried seeds

NGSS and Activity Extensions

Further in-class studies could include making observations and formulating arguments on how some offspring are similar but not exactly like their parents (1-LS3-1: Heredity).

3 Planting for Early Spring

Time Frame: 45 minutes

Overview: The class will plant pea seeds for future outdoor transplanting.

Objective: For students to interact with new seeds and prepare starts for the school garden.

Introduction (5–10 minutes)

1. Brainstorm with the class how they have planted seeds in the past: How deeply did the students plant seeds? How far apart should the seeds be from each other? What is the benefit of using presoaked seeds?

2. Introduce the class to the process of planting pre-soaked pea seeds in potting soil and the different stations the students will need to accomplish their planting (gather a container, get masking tape and markers to write names on, fill with soil, acquire seeds, plant, water, and store in planting tray, repeat). Demonstrate the planting process.

> Many young and even experienced students believe that seeds should be planted one finger's depth in the soil. I have no idea where they continuously come up with this idea, but I always address this erroneous belief when I guide planting activities with any age group. This is also a good opportunity to discuss with the students how gardeners measure. Gardeners don't typically carry measuring tapes or rulers around with them to determine how deeply or how far apart to plant seeds or starts. Instead, they estimate using their fingers, hands, and sometimes feet. This intrinsic application of math is an excellent discussion topic for any age.

Activity (30 minutes)

1. Individually or in pairs, the students will plant pea seeds in potting soil and planting containers. They will work as a team to plant the seeds at the right depth and spaced out far enough in the container.

2. The students will make sure to label their planting container with their name, date, and type of seed.

3. They can continue this activity until the seeds have all been planted. Afterward, they can water their seeds with a watering can and then enjoy a taste test from the garden at the end of the activity.

Assessment (5 minutes)

1. How long do students think it will take these seeds to sprout? What are the stages of growth that they should expect to see?

2. How would the students describe the taste,

> It's very important to me that children learn and practice how to talk about food, flavor, and texture beyond the "yuck" or "yum" of a food. Weekly taste testing is an opportunity for students to think critically about raw and cooked foods while exploring the diversity of garden produce.

texture, and experience of eating their taste tests?

Preparation

1. Presoaked pea seeds
2. Potting soil
3. Planting containers
4. Masking tape
5. Markers
6. Watering can

4 The Traveling Seed

Time Frame: 45 minutes

Overview: The class will read *The Seed Is Sleepy* by Dianna Aston and generate questions and wonderings about how seeds travel.

Objective: To identify different ways that seeds travel by creating a class list for future reference.

Introduction (5 minutes)

1. What types of seeds do students like to eat? What seeds have they planted? Are there mammals that like to eat seeds? How do seeds move in order to find the right space to grow?

2. Can students name types of seeds that travel by wind, water, animals, or explode from their seedpods?

Activities (35 minutes)

1. Students will listen to the story of *The Seed Is Sleepy* by Dianna Aston and discuss the lessons of the story, paying attention to the different traveling patterns of seeds. The book has excellent examples of how seeds travel and how their parent plant develops mechanisms to help them fly, float, or explode from a seedpod. As these travel styles are identified, they can be added to a classroom chart as a resource for future activities.

A barren fennel umbel provides evidence of seed travel that has already occurred.

2. The class will explore seed samples that are passed around. If there are seedpod samples, can the students tell which part of the plant is the seed? Students are encouraged to practice careful handling of the samples. What features of each seed help the students discover how they travel (hairs for flying, hooks for sticking)?

Assessment (5 minutes)

1. How do seeds know when to travel and when to sprout?

2. What are the different ways that seeds travel? Why do seeds travel?

Resources

- Dianna Hutts Aston. *A Seed Is Sleepy.*

5 Fly, Float, and Explode!

Time Frame: 45 minutes

Overview: Students will learn about common ways seeds travel and how plants developed these techniques.

Objectives: To recognize the mechanisms that seeds have developed to travel and identify how certain seeds travel.

Introduction (10 minutes)

1. Using various seeds as an example, have the students brainstorm what they know about travelling seeds. They can also use the classroom chart from the previous lesson to help them. How do seeds find new ground to grow in? What tools do these seeds develop to help them travel? As each type of travel is discussed, show the students an example of that type of seed. For example, a dandelion seed uses thin hairs for flight and macuna seeds have a thick hollow shell for water travel.

2. The four types of seed travel the students will learn about are water, animals, bursting, and wind. Seeds that travel by water utilize hard shells to withstand their journey; wind-traveling seeds use long hairs and umbrella shapes to float on a breeze; seeds that travel on animals often use hooks to attach to fur; and seeds that burst from their seedpods are small and hard like little cannon balls and are plentiful. "The Seedy Side of Plants," from the PBS series *Nature*, has video of seeds exploding from seedpods for students to observe.

3. There is an additional seed traveling technique that the class already engaged with when they looked at pumpkins and fruits: seeds that animals eat tend to have sweet, juicy, and nutritious coatings.

Activities (30 minutes)

1. Each student will receive a How Do Seeds Travel worksheet and the corresponding seed picture page. They will carefully cut out each picture, sixteen in total, and arrange them on the worksheet. Using what they know about how seeds travel, the students will act as detectives to analyze the pictures and determine how the seeds travel based on their unique features. They will place these pictures in one of the four columns on the worksheet. Each picture corresponds to a wind, water, animal, or bursting category.

2. When the students feel they have every picture in the right category, they can check it with a teacher or adult volunteer and then will glue the images to their sheet.

Assessment (5 minutes)

1. As a class or in small groups, have the students explain how they determined where the seeds go. What clues did they look for in the pictures that helped them solve the puzzle?

2. Were any seeds in the activity difficult to identify? What methods did the students use to solve these problems?

Preparation

1. How Do Seeds Travel worksheet
2. Seed examples
3. Glue

Resources

• PBS, *Nature*. "The Seedy Side of Plants." (online—see Appendix B).

NGSS and Activity Extensions

Opportunities for greater in-class lessons can include reading and exploring different media to identify patterns among parents and offspring to help them survive, and to compile evidence that offspring are similar but not identical to their parents (1-LS1-2: From Molecules to Organisms, 1-LS3-1: Heredity).

6 Seeds in Water

Time Frame: 45 minutes
Overview: Students will learn how to sprout seeds in water.
Objectives: Students will make predictions about how seeds grow and observe the sprouting process in class.

Introduction (5–10 minutes)

1. Brainstorm with the class: Can seeds sprout without soil? How long can seeds survive in water? What other necessary elements do plants need to survive? What are the different parts of a sprout? Why do humans like to eat them?

2. Introduce the class to the practice of sprouting certain seeds in water for a short amount of time in order to produce a nutritious food source. Demonstrate how to set up a jar for sprouting seeds, as well as what seeds work best and the importance of using fresh water daily.

> Most home sprouting kits involve a plastic jar lid that allows water to drain and air to pass through. There are also larger sprouting kits that can be purchased. I find that a simple jar/clear cup with a coffee filter cover secured by a rubber band works for this activity. If the students are able to carefully remove the cover, rinse the seeds, and provide fresh water daily then the project is most successful and the sprouts are safe to eat.

Activity (30 minutes)

1. In pairs, students will assemble their sprouting systems by putting seeds and water into a container. They will mark their containers with masking tape and markers and then secure a coffee filter lid with rubber bands.

2. The students will be monitoring and caring for these sprouts for the next two weeks or so, depending on the plant variety, so they should incorporate care for the sprouts into a daily routine. When the seeds have sprouted and are ready to eat, then the students will be able to enjoy a taste test of their sprouts. The goal of this activity is to watch how seeds sprout, so the students should add daily or weekly reflections to their schedule about their observations on seed/sprout changes.

3. The class can be guided on how to follow the directions for sprouting seeds on the back of the seed packets or directions online from a reputable resource. Many seeds need to be washed and the water filtered during the first few hours of sprouting. After the first few hours, or day, of soaking, then the sprouting seeds will need to be refreshed with water and then drained. The sprouting jars will need a tray or plate/bowl to capture this excess water.

4. At the end of their activity, the students will enjoy a taste test of presprouted seeds.

Adaptation

This activity can also be done as a whole class using a single quart jar filled with a small amount of seeds.

Assessment (5 minutes)

1. What words do the students have to describe the taste and texture of different sprouting seeds?

2. How long does the class estimate it will be for the seeds to sprout? What changes do students predict will occur to their soaking seeds after having just seen "finished" edible sprouts?

Preparation

1. Small Mason jars/cups
2. Coffee filters
3. Water
4. Sprouting seeds (garbanzo or mung beans)
5. Masking tape
6. Markers
7. Rubber bands
8. Plates/trays for draining water

7|8 Make Your Own Seed

Time Frame: 1½ hours (2 × 45 minutes)
Overview: Students will make their own model of a flying seed by using a simple cutout design or origami, as well as their own imaginations.
Objective: Students will understand the mechanics of seeds that fly on the air.

Introduction (15 minutes)

1. Brainstorm with the class what students know about how seeds travel on the air: What features and characteristics do these seeds (such as dandelions or samaras) have to help them fly or float?

2. What is a model? What are the elements of a flying seed that can be mimicked in a model? Using premade paper examples of flying seeds, let the seed models fly and guide class discussion on the mechanics behind these flight patterns. What observations can the students make about how the seeds move?

3. Demonstrate to the class how to create these models of a flying seed through origami maple seed designs or a simple spinning seed cutout from a long thin strip of paper (see Resources).

Activities (1 hour)

1. Students will be making seeds that travel on the wind using one of the two demonstrated methods: an origami maple seed or a simple spinning seed. Students will need to use good craftsmanship and follow directions if they want to have a seed that flies. If there is enough time, the students can create both designs and compare the flying successes of each.

2. The students can color the seeds after they have created their models, by using available tools. If possible, the students can also decorate their finished flying seed with found materials of various textures that mimic more natural seed parts, such as the feathery extensions of dandelion seeds.

3. Students will be able to fly their seeds in class or off of a high point to test the success of their seeds, troubleshoot problems, and celebrate their creations. When all the students have finished their projects, then they can test them all and make observations as a class. This is also a great opportunity to discuss ideas of success and failure. Students should understand that when things don't go the way they expect then they are given a chance to learn something new. Do all seeds from a maple tree succeed in finding new ground to grow in?

Adaptations

Ideally, I dreamed this activity would have students create seeds from their own imagination, while fulfilling certain guidelines. This version would take longer and be better for an in-class teacher, rather than a once-a-week garden teacher or volunteer, to tackle. In class, the students could accomplish designing and testing their models in a week.

An Extended Version

1. Students will design and build a seed from provided craft materials. The seed must either:

- Float on water
- Travel a distance on the air
- Stick to animals

2. The students will sketch their design and then have it approved by the teacher before building it with the supplies. This is a great opportunity to address craftsmanship with students.

3. The class can watch videos of how seeds travel and discuss what design elements the seeds have to help them travel.

4. The students will use available and gathered materials to build a model of their seed design. During work times, they should be given space to practice their seed's traveling abilities and make changes as needed.

Assessment (15 minutes)

1. What kinds of problems did students face when designing and assembling their seeds? How did they overcome these problems or change their designs?

2. What did they learn or do differently in designing their seed next time? What did they do in designing their seed that they felt was realistic?

Preparation

1. Assorted materials, including waterproof ones, for students to glue or tape together
2. Sample origami or spinning seeds

Resources

- "Seeds on the Move—Seed Dispersal for Kids." (online—see Appendix B).
- "Make a spinning seed." Planet Science. (online—see Appendix B).
- "Origami Maple Seed." (online—see Appendix B).
- "Origami Maple Seed Helicopter." (online—see Appendix B).

NGSS and Activity Extensions

Greater in-class studies can include designing solutions to human problems by mimicking how plants and animals use their design and external features to survive (1-LS1-1: From Molecules to Organisms).

9 Seed Investigation

Time Frame: 45 minutes

Overview: Students will dissect and investigate the parts of seeds that they have sprouted in class.

Objectives: To recognize the different parts of a seed, learn new vocabulary, and review each function.

Vocabulary

1. Cotyledon
2. Epicotyl
3. Seed Coat
4. Radicle
5. Hilum

Introduction (10 minutes)

1. Using a poster or an example, have the students name the parts of a dicot seed that they can recall. Introduce the new vocabulary to students and work together to label these on the classroom chart: *epicotyl*, *cotyledon*, *radicle*, *hilum*, and *seed coat*. What parts of a seed have the students seen emerging in the seeds they sprouted in water?

2. What do the students remember about the purpose or function of each seed part? How will the class carefully open up the seeds to explore them?

Activities (30 minutes)

1. Students will receive one or two soaked bean seeds and slowly break them apart, identifying the hilum, radicle, cotyledon, and epicotyl, and seed coat. Unlike the Kindergarten activity, the students will get to split a bean seed in half by being very careful to remove the seed coat and then finding the split in the seed and separating the two halves. This will require their best work and concentration.

2. The students will explore the anatomy of the seed they dissected and can use magnifying glasses to best see smaller features. As they discover each piece, then they will label their findings on their Seed Investigation worksheet. The students should make sure that the spelling of the vocabulary words is correct by using the classroom chart to check their work.

3. The class can enjoy a taste test of their edible sprouts after the assessment.

Assessment (5 minutes)

1. Did the students notice any part of a plant that was not listed on their worksheet? Was their experience finding each of these parts difficult or easy?

2. What did they observe about each of these parts? For example, what words would they use to describe the cotyledon?

A student carefully plants pea seeds around a homemade trellis.

Preparation

1. Prior to this activity, soak bean seeds in water overnight to soften them and begin the sprouting process.
2. Seed Investigation worksheet
3. Class poster with the parts of seed labeled
4. Magnifying glasses

Resources

• "Parts of a Seed Worksheet." (online—see Appendix B).

10 Sowing Seeds

Time Frame: 45 minutes

Overview: Students will learn how to sow hardy cold weather seeds and cover crops.

Objectives: For students to learn a new way to plant seeds and about plants that can handle the early spring weather.

Introduction (5–10 minutes)

1. Brainstorm with the students what types of seeds they may know or have planted in the winter and spring: What seeds prefer being planted in the cool winter instead of spring weather?

2. What does it mean to sow seeds? What other ways do people sow? Who uses this method to plant seeds? What methods of seed travel does this mimic?

3. Students can practice the motion of sowing seeds before going out to the garden.

Activities (30 minutes)

1. Individually, students will sow seeds in an open space of the garden. The seeds could include clover, rye, vetch, radishes, carrots, turnips, or lettuce. Before the class begins sowing their seeds, they should add potting soil into a bucket and then combine it with the seeds. This will help spread the seeds out and cover them with additional growing medium. The students will use small handfuls of this mixture to spread a variety of seeds or cover crops in the open spaces of the garden.

2. They should practice craftsmanship and slow, careful gardening work. The students will focus on keeping the seeds in the growing area and not on pathways. If need be, they can water the planted areas with watering cans after the activity.

3. If possible, have the students enjoy a taste test of seasonal winter foods after their gardening activity.

Assessment (5 minutes)

1. What was the student's experience sowing the seeds? Was it difficult to spread the seeds?

2. Are there seeds that students would not want to spread this way?

Preparation

1. Hardy winter seeds
2. Bucket
3. Potting soil

Adaptations

This activity can also be accomplished in planting containers.

Resources

• "Seed in the Ground," a song by Connie Kaldor (see Appendix B).

Spring: Sprouting Seeds Experiment

1 Early Spring Starts

Time Frame: 45 minutes

Overview: Students will transplant the pea seeds they started earlier in the year.

Objectives: To learn how to handle young plants and to transplant vegetables.

Introduction (10 minutes)

1. Brainstorm with the class what students know about removing plant starts from growing containers: How can they treat the plants with care and gentleness during the transplanting? How can students Care for Self, Others, and the Land during this activity?

2. How deep should the planting holes be? What other things can the students do to help transition the plant starts into the garden?

Activities (30 minutes)

1. Individually or in pairs, the students will transplant their pea plant starts into the school garden. The students will focus on high-quality work by making sure to dig deep enough holes for the plants, protect the roots, and fill up the holes with soft soil using their hands. This activity can also be accomplished in larger potting containers.

2. The students can also water their starts with a watering can after they have transplanted the plants. They can continue this activity until all the starts are in the garden. If there are planting containers with multiple plant starts, demonstrate a careful way to separate the plants from each other.

3. The class can enjoy a taste test from the garden after their activity.

Assessment (5 minutes)

1. Have the students return to the garden to collect any tools and planting containers left behind.

2. They can also check the quality of their work by making sure that the starts have been watered and are planted deeply enough.

Preparation

1. Plant starts
2. Shovels
2. Watering can

2 Growing an Indoor Garden

Time Frame: 45 minutes

Overview: Students will plant seeds for an indoor soil and seed experiment that they will observe scientifically.

Objectives: Students will experience how seeds grow in soil while practicing introductory science skills.

Vocabulary

1. Sprout
2. Prediction
3. Observation
4. Dormant

Introduction (10 minutes)

1. Brainstorm with the class about what they know about how seeds sprout: What are the difference stages of a sprouting seeds, from dormancy to producing leaves and flowers? What do seeds and sprouts need to survive?

2. What does it mean to make a prediction and an observation? What observations would a scientist and a gardener make about how seeds grow?

Activities (30 minutes)

1. In small groups, students will assemble their supplies from the available materials: a planting container, soil, and name tag, They will work as a team to write their names on the name tag around their container and fill the containers with loose, soft soil.

2. Students will then receive 5 or more presoaked bean seeds and work together to plant them at the right depth in the container (scarlet runner beans are an excellent choice for this activity). The students should be experienced in performing these tasks from their previous activities. Finished students can gently water their plants with a watering can.

4. Students can finish the activity by receiving a raw sprouted bean seed and exploring the seed coat, cotyledon, and texture of the bean seed before taste testing it.

Assessment (5 minutes)

1. Gathering back together, guide the students in making their first observations about the presprouted seeds. Record their observations on a classroom record board. This board will be used for all the class observations, including changes in the size and growth of sprouts.

2. These indoor observations should be paired with wonderings about the seeds the students previously planted in garden.

Preparation

1. Adult volunteer
2. Planting containers
3. Potting soil
4. Pre-soaked bean seeds
5. Masking tape
6. Markers

7. Watering can
8. Class growth/recording poster board

3 Seed Balls

Time Frame: 45 minutes
Overview: Students will create seed balls and learn a new way for seeds to travel.
Objective: For students to learn how they can benefit their communities by spreading seed balls of native flower seeds.

Introduction (10 minutes)

1. Brainstorm with the class what they know about how seeds travel. Introduce the idea of seed balls as a way of spreading seeds without disturbing or tilling the soil—a method that has been used across human agricultural history. Making seed balls requires only a few natural ingredients. And the kids' hands get dirty when they do it, which should always be a goal in gardening class.

2. Demonstrate to the class how to mix and make seed balls. What are the benefits of using native wildflower seeds for this activity? What observations do students have about the process of making seed balls?

3. How can small student groups work as teams and practice cooperation during the activity?

Activity (30 minutes)

1. In small groups, the students will follow directions in mixing the proportions of clay, compost, and wildflower seeds. They will be a team and take turns or work together to mix the ingredients. Then, they will add the appropriate amount of water and mix it gently into the dry ingredients.

2. The students will form seed balls from their mixture and use a sample seed ball as a reference to the ideal size ball for the activity. The groups can lay out these seed balls on trays or plates to dry as they create them. Paper plates can be marked with the name of the small group or students for later use in the garden, or to take home.

A group works together and gets dirty while forming seed balls.

> I usually follow the directions for seed balls described in Resources (below). With students, I alter this process because I prefer to have them shape the seed balls using their hands. The balls are about the size of a quarter in width, or smaller. I use native wildflower seeds in order to limit the spread of potentially invasive plants in any ecosystem. Make sure to check the varieties of plants in wildflower mixtures to limit this possibility.

3. At the end of the activity, the students can wash their hands and enjoy a taste test from the garden.

Assessment (5 minutes)

1. Where would be the best places for students to scatter their seeds? What kind of insects and birds will visit the wildflowers?

2. What were student experiences with texture during this activity? Did they find it easy or difficult to form the seed balls? What other garden seeds could be planting with this method?

Preparation

1. Clay
2. Compost (sifted)
3. Native wildflower seeds
4. Water
5. Bins for mixing
6. Paper plates
7. Markers

Resources

• Andrew Schreiber. "Making Seedballs: An Ancient Method of No-till Agriculture." (online—see Appendix B).

4 Observing Plant Growth

Time Frame: 45 minutes

Overview: Students will measure sprout growth and generate scientific observations about plant growth in the indoor garden.

Objective: To practice and develop science skills by making observations and measurements.

Vocabulary

1. Cotyledon
2. Seed coat
3. Roots/root hairs
4. Germinate

Introduction (5–10 minutes)

1. What do the students know about the different parts of a sprout? What changes have the students observed about their seeds in the indoor garden over the past two weeks?

2. Reintroduce the class to the parts of a sprout, including the *roots*, *root hairs*, *cotyledon*, and *seed coat.* Use a sprouting seed from the indoor garden to demonstrate the different parts of a sprout.

3. Even seeds of the same variety that were planted at the same time often germinate at different rates. The students should understand this going into the activity and should study the growth difference of their seeds as scientists, by generating questions about why and how this happens. The students should be prepared to see substantial growth or no growth with their seeds.

4. Review with the class how to use a ruler to measure and record height and enhance their observations.

Activity (30 minutes)

1. In their small groups, the students will gently explore the sprouting progress of their seeds and make observations about the changes and parts of a sprout that are beginning to emerge.

2. First, they should do this visually by looking at their seeds in the containers. The students can observe the growth of the sprout, emergence of cotyledon, changes of the seed coat, and what part of the sprout is emerging through the soil first. They can count and record these changes as a group to share with the class at the end of their observation time. They can also record the height of the sprouting seeds using a ruler and record the highest and lowest sprout on the classroom chart.

3. After they have visually observed the

sprouting process, the groups can gently remove one sprout from their container, keeping in mind the goal of not disturbing the other plants. What new parts of a sprout can they see? How are these parts, such as root hairs, attaching themselves to the soil? The groups can carefully dissect these sprouts for exploration before putting them in a compost bin at the end of the activity.

4. Gathering back as a class, have the groups share their findings and record any remaining findings on the class record sheet. How many seeds had visible cotyledon emerging or seed coat splitting? How many seeds have not sprouted? What was the tallest sprout the groups recorded? Discuss student wonderings on sprout growth and what they predict will happen next in the plant growth process.

5. At the end of the activity, the student groups can carefully mist or water their seedlings.

Assessment (5 minutes)

1. Were there parts of the sprouts that students couldn't identify? What part did the students believe they were?

2. Looking at the class record sheet and the observations on plant growth and height, how much will the sprouts grow until the next observation?

> Ideally, with in-class teachers, students would be making observations of this caliber every week and recording them on the classroom chart.

Preparation

1. Class record sheet
2. Rulers

5 The Carrot Seed

Time Frame: 45 minutes

Overview: The class will read Ruth Krauss's *The Carrot Seed* and plant seeds in the garden.

Objective: To learn that it takes some seeds a long time to germinate and relate this lesson to the growth of student seed experiments.

Vocabulary

Germination

Introduction (5 minutes)

1. Brainstorm with the class about the growth of their seed sprouting experiments: Are some seeds growing faster than others? Are there seeds that have not germinated at all?

2. Why have these seeds not germinated? Why do some varieties of seeds grow faster than others?

Activities (35 minutes)

1. The class will read *The Carrot Seed* by Ruth Krauss and discuss the character's reaction to the slow growth of the carrot seeds. Be sure that the students understand the fiction of the story. How would a carrot realistically grow in a garden?

2. After a discussion on the slow germination of carrot seeds, the students will sow carrot seeds in the garden. They should be sure to be careful gardeners and gently press the seeds into the soil with their hands after they have spread the seeds over the designated garden beds.

4. Finally, the students will gather in their small groups and observe the changes in their sprouting seeds experiment to share with the class. They can measure the height of the sprouts with a ruler.

Carrot seeds are notoriously slow germinators, and it can be difficult to remember where they were planted as they grow gradually. I learned from my mother to mix radish seeds with carrots when planting. Radishes germinate and grow very quickly, so when they are ready to harvest, the carrot leaves will be popping up and will have grown enough to need more space. You will also more easily recognize where you planted those carrot seeds! When I do this activity with children, I have them mix carrot and radish seeds in a bucket with soil. This mixture helps students spread the seeds over wider spaces.

Carrots and other root vegetables can be tricky plants for young students to correctly identify and carefully harvest. I've had many students show me whole California poppies or parsley plants that they misidentified and picked. There are also many wild members of the carrot family that are not edible or are poisonous. So, I am very cautious about letting children freely pick carrots at their whim if the area is not clearly marked. I usually make them show me the plant, in the ground, before they pick and always before they wash and eat it. I also choose very recognizable carrot varieties so the students can more easily identify them.

Assessment (5 minutes)

1. Gathering back together, have the groups share their new observations about the growth of their seeds and record their findings on the class record board/poster.

2. How many seeds have sprouted? How many sprouts have emerged cotyledon or young leaves?

3. If there are any seeds that have not germinated, have the class brainstorm why the seeds didn't grow. Are there factors other than time that could affect seed growth?

Additional Activity

If there are carrots that have overwintered, demonstrate to the students how to identify a carrot in the garden. If they can't see the root, how else can they recognize a carrot? If there are other carrot-like tops in the garden, pick a sample to show the students so that they can determine the differences between the plants.

Preparation

1. Carrot seeds
2. Radish seeds
3. *The Carrot Seed* by Ruth Krauss
4. Soil
5. Bucket

Resources

- Ruth Krauss. *The Carrot Seed.*

6 Gardener's Math

Time Frame: 45 minutes

Overview: Students will plant tubers in the school garden, learn how deeply to bury them, and how some plants can sprout without seeds.

Objectives: To plant different varieties of potatoes in the garden and practice measuring between plants.

Introduction (5–10 minutes)

1. Begin by leading a brainstorm with the class: Do all plants grow from seeds? What plants do the students know that can grow from stems and roots?

2. Introduce the students to tubers, such as potatoes, and other augmented stems and modified lateral roots. Demonstrate the structure of a potato, its eyes, and the sprouting process of a tuber. Have they even seen a plant sprout even without soil to grow in?

3. Potatoes grow abundantly with a certain amount of space between each plant. How can the students measure like gardeners, with their hands and fingers? How can the class estimate inches and feet with their hands? During the activity, the students will be working with a partner, so they can determine together how to measure 3–12 inches with their bodies.

Activities (30–35 minutes)

1. In pairs, the students will receive potatoes to plant in the garden. Working with their partner, the students will use *gardener's math* to measure the right depth and space between each potato plant. The class can use shovels or their hands to dig into the soil and carefully fill in the holes after they have planted. They can continue this activity until the designated space is full or all the potatoes are planted.

2. After they have finished the planting, they can clean up and then return to their sprouting seed experiment to observe and record the growth and changes. The class can also observe the growth of the pea seeds they planted weeks earlier and compare it to the growth of the indoor seeds.

Assessment (5 minutes)

1. Before sharing their sprout observations, have the class reflect on their experiences during the activity: Did they feel successful in measuring like gardeners? Were there any struggles the groups face when planting? How long will it take for the potatoes to sprout?

2. Have the groups share the progress of their sprouts: Is there any mortality among the seeds? How tall are their sprouts? What parts of the sprouts are emerging? Record their data on the class record poster/board.

Preparation

1. Potatoes (whole or precut)
2. Shovels
3. Rulers

Adaptation

This activity can be performed in large containers instead of garden beds.

Resources

- Bonnie Plants. "Growing Potatoes." (online—see Appendix B).

7 The Seed Bank

Time Frame: 45 minutes

Overview: The students will study the garden soil and search for seeds in the garden seed bank.

Objective: To explore the diversity and vitality of seeds in the soil ecosystem.

Introduction (5–10 minutes)

1. Brainstorm with the class what a garden seed bank is: What do students think when they hear the term *seed bank*? Gardeners can deposit beneficial and desired seeds into the garden soil, other than the seeds and roots of undesirable plants. These banked seeds play a valuable role in the garden ecosystem. The seeds will sprout, grow, and have the potential to provide greater food and habitat for garden

A sprouting seed bank includes an edible ground cover of arugula and other brassicas in spring.

creatures and students, even as they grow along the margins and understory of the garden.

2. What do seeds and sprouting roots or stems look like in the garden soil? How can the students identify what they see? Have students recall their descriptions of plant growth in the indoor garden.

2. Before the activity, review with the class the behavior expectations around handling the tools (magnifying glasses) and any soil creatures and insects they encounter during this activity. How can the students be both scientists and gardeners during this activity?

Activities (30 minutes)

1. In pairs, the students will receive or gather a sample of garden soil on an observation tray and use their magnifying glasses to study and identify different seeds in the garden soil.

2. They can also look for tubers, roots, sprouts, rhizomes, or other evidence of future plant growth. The groups can separate and sort their findings into various groupings.

3. The student groups can share their findings with other student partners and then gather back together with the class to survey and review what they discovered: Were there any seeds or seed bank items that the students couldn't identify? What items were the most prolific in the soil?

4. At the end of the activity, have the students carefully return their samples to the garden and clean up their workstations and tools.

5. Finally, the student groups will return to their sprouting seeds experiment and observe the growth changes of the sprouts.

Assessment (5 minutes)

1. Record the changes in the students' sprouts, including plant growth and emerging parts of the sprouts.

2. Are there new parts emerging which students can't identify? What wonderings and questions do the groups have about the sprout growth?

Preparation

1. Magnifying glasses
2. Observation trays
3. Garden soil samples

8 Final Plant Observation

Time Frame: 45 minutes

Overview: The class will make a final observation of the growth of their sprouts and finish their planting project.

Objective: To practice emerging science skills by making conclusions and generating more questions.

Introduction (5 minutes)

1. Brainstorm with the class: What does it mean to think like a scientist? How can students generate wonderings and draw conclusions from this experiment?

Activities (35 minutes)

1. The student groups will make their final observations about the growth of the sprouts, measuring heights with a ruler, counting seed mortality, and the emergence of leaves and shriveling cotyledon.

2. When the groups have finished their final observations, they will gather as a class and share their findings to write on the class record sheet. What conclusions can the students draw from the data and graphs they have created together? Can the sprout growth continue exponentially or continually, why or why not? What elements are missing for the sprouts to thrive fully? Were there seeds that never sprouted or which died during the growth process?

> There are many excellent math and science connections that can happen during this activity. Using the class record sheet or poster as a visual, the students can be introduced to graphs, percentages, and fractions as ways to understand their data. I enjoy creating a simple line graph on an x/y axis with the height measurements of the tallest sprouts and having the students make predictions about what future height/growth would look like.

3. The students will carefully return their plants to the growing space and clean up the materials from their experiment. They can enjoy a taste test from the garden at the end of the activity.

Assessment (5 minutes)

1. If the students were to do this experiment again, what would they do differently?

2. What questions do the students have about the project and their observations? Were there any unanswered assessments they made?

9 Transplanting Plant Starts

Time Frame: 45 minutes

Overview: The students will review how to transplant starts by moving their experiments into the garden.

Objectives: To practice careful transplanting in the school garden and finish the growth experiment.

Introduction (10 minutes)

1. Brainstorm with the class what they know about transplanting starts into the garden: How deeply should the students dig holes for the plant roots? How can the students gently remove the plant starts from the planting container without damaging the plant stem or roots? What other needs do the plant starts have in order to survive? What gardening techniques have worked for them in past activities?

2. What are the steps the students should perform in order to successfully transplant their starts? What are the behavior expectations for the activity?

3. Demonstrate how to remove plant starts from their planting containers and how deep to dig the holes.

Activities (30 minutes)

1. In their groups, the students will transplant their bean plants into the garden at designated trellis areas. They will practice careful handling of the fragile starts and their roots, making sure to dig deep enough holes, to cover all the roots with soil, and to gently fill up the holes with soft, non-compacted soil.

2. The groups should practice teamwork and collaboration during the activity. They can water their starts deeply with a watering can after they have finished planting. Additionally, the groups can use wooden sticks and markers to label their plants.

3. At the end of the activity, the student can enjoy a taste test from the garden and the edibles they have planted over the past few months, such as fava beans, radishes, or peas.

Assessment (5 minutes)

1. After their taste test, have the student return to the garden and check the quality of the transplanting, as well as gather any remaining tools and planting pots.

2. Are all the plants watered? Are their roots covered by the soil or are they planted deep enough?

Preparation

1. Shovels
2. Watering can
3. Wooden sticks
4. Markers

10 Garden Celebration

Time Frame: 45 minutes

Overview: The class will enjoy a taste test experience and garden celebration.

Objective: To celebrate student learning and accomplishments over the year.

Introduction (10 minutes)

1. During the activity, the students will have the option of harvesting different foods from the garden. What are the best ways to pick the produce while being gentle with the plants? What foods do students want to eat?

2. Make a class list of the types of food and the amount each student can harvest, such as strawberries, lettuce, peas/pea leaves, edible flowers, and radishes.

Activities (35 minutes)

1. Individually or in pairs, the students will embark on a large taste test in the garden and gather the foods that they would like to eat for the celebration. When they have gathered these foods, they can wash them at a washing station, plate their food, and gather back together to eat.

2. Guide the class through a taste test meal in which students can describe the flavors and texture of individual plants and then create combinations of plants in order to explore the varying plant tastes. Are there combinations of plants, or recipes, that students enjoy? Allow students to share their favorite recipes of garden plants with their classmates.

3. As students enjoy their foods, lead a class discussion on their accomplishments this year. Have the class reflect on their favorite or most memorable lessons of the year. What do they know now that they didn't know at the beginning of the year? What surprised them about the school year in the garden? What did they do that they had never done before? What do they want to do again in the next school year?

Preparation

1. Adult volunteer
2. Washing station
3. Example taste tests

SECOND GRADE—GRADE TWO

POLLINATORS AND CYCLES

By the end of this year, even the most squeamish students will love to look for bugs, insects, and pollinators in the garden. This is a year for valuing the vital and diverse living organisms in the garden ecosystem and engaging in conservation and action to build student science and advocacy skills.

After engaging in the joys of fall garden activities, the class will begin their winter studies by delving into focused observations. The students will learn how scientists label sketches and ask questions, while ecologically minded gardeners observe the system as a whole with the plants, insects, animals, and soils in relationship. Then, they will identify and interact with beneficial and non-beneficial insects in

FALL
Seasons and Garden Changes

1. Garden Behavior
2. Garden Safari
3. Harvesting
4. Dry Seed Saving
5. Cover Crops
6. Mulching and Compost
7. Focused Plant Observations
8. Identifying Medicinal Plants
9. Fascinating Fungi
10. Spore Prints

WINTER
Beneficial Insects and Birds

1. Predators and Prey
2. Praying Mantis
3. Garden Spiders
4. Ladybugs
5. Hummingbirds
6. Planting Hummingbird Seeds
7. Building Habitat
8. Planting Forage
9. Creating Watering Basins
10. Beneficial Insect Release

SPRING
Observing Pollinators

1. Pollinators: Myth and Fact

2|3. Butterflies and Moths

4|5. The Colors of Food

6. Citizen Scientists
7. Planting Colors for Pollinators
8. The Insect Hotel
9. Honey Bees
10. Garden Celebration

the garden before moving onto exploring valuable pollinators.

In mimicking the pollinating methods of hummingbirds and by actively building pollinator habitat in the garden, the seasonal garden chores and studies of the class will enhance the students' science skills and gardening knowledge. At the end of the year, the students will have become citizen scientists and learned how to promote healthy soil systems, value the marginal, plant beneficial habitat, and create pollinating opportunities.

Next Generation Science Standards

During this year in the garden, the students will enhance their already budding science skills through deeply focused observations, sketching, and labeling of the garden ecosystem. They will explore the relationships between plants, insects, and human interpretations of whether they are beneficial or nonbeneficial. The class will map out the diversity of life, from the soil to the sky, and study the populations of diverse insects. And they will work hard to classify and identify different key garden players and record and analyze the data collected through surveys.

Further in-depth classroom studies and readings can follow themes or units on: diversity and populations of life in comparative ecosystems and habitats, long-term or seasonal land changes, and pollinators and the process of pollination.

Permaculture Principles

- Observe and Interact
- Obtain A Yield
- Use and Value Renewable Resources and Services
- Produce No Waste
- Design from Patterns to Details
- Integrate Rather than Segregate
- Use Small and Slow Solutions
- Use and Value Diversity
- Use Edges and Value the Marginal
- Creatively Use and Respond to Change

Fall: Seasons and Garden Changes

1 Garden Behavior

Time Frame: 45 minutes
Overview: Students will learn how to carefully harvest food from the garden.
Objective: To build empathy and consideration when interacting with plants and to encourage students to eat healthy food.

Introduction (5–10 minutes)

1. Review the behavior expectations for the students and discuss why they are important in the garden. How do students practice Care for Self, Others, and the Land while they are in the garden?

2. Introduce the students to a few plants that they will help harvest from the garden so they can identify them in the garden, such as tomatoes, cucumbers, kale, or lettuce. Choose plants with similar methods for harvesting. This will help students practice good picking habits. How can students carefully gather plants from the garden? What language can they use to talk about how food tastes?

3. Have the students practice the hand motion of snapping stems with two hands, rather than pulling with one hand.

Activities (30 minutes)

1. Demonstrate how to pick a variety of vegetables to the students. Try to choose different flavors so every student can try something they like. Have them pick a plant themselves, demonstrating their harvesting skills, and then taste test their choice. The students can continue taste testing from the garden or partner with another student as they explore the garden for its seasonal changes and fascinating finds.

2. Review garden expectations with the students so they can follow them if they want to eat from the garden during other times of the day. Have the students gather back together to share their discoveries and experiences.

Assessment (5 minutes)

1. What new or favorite flavors did the students experience in the garden? Were there flavor combinations that they enjoyed? Was it difficult or easy to harvest plants?

2. What discoveries, new features, or creatures did the students encounter?

Preparation

1. Sample plants for harvesting

2 Garden Safari

Time Frame: 45 minutes
Overview: Students will go on a safari in the garden to look and listen for animal activity.
Objectives: For students to look for signs of life in the garden, develop observation skills using the senses, and begin fostering an understanding of the garden ecosystem.

Introduction (5–10 minutes)

1. Brainstorm what types of evidence the students might see that was left behind by animals and insects in the garden. Encourage the students to pretend that the garden fences are a part of a new ecosystem and the students are discovering amazing creatures that have never been seen before. What creatures migrate through the garden? What insects and animals live in the garden? How can student behavior impact the creatures that live in the garden?

2. Places to look for evidence or wildlife include under leaves, bark, on shrubs, in the cracks of the sidewalk, on blades of grass, in the soil around plants, along the edges of fences, and on fence posts. Evidence and signs could be egg masses, spider webs, feathers, nests, animal tracks, or holes nibbled on leaves.

3. Before students go on a safari, they need to practice *deer ears*, *owl eyes*, and *fox feet*, which will help them detect a larger number of animal signs and insect activity. Students can create deer ears by cupping their hands around their ears, mimicking a deer's powerful and large ears. Similarly, students can make circles with their hands to put over their eyes and focus their gaze, like owl eyes. Feet like a fox require quiet, careful steps and that the students be aware of what surfaces they are walking on (see Resources).

4. Remind the students that if they find something exciting, they can whisper to a friend or teacher what they have found. They can also remember what they discovered and share their findings with everyone at the end of class.

Activities (30 minutes)

1. Individually or in pairs, the students will go on a quiet safari in the garden equipped with their owl eyes, deer ears, and fox feet. They will practice garden behavior expectations and positively interact with the garden ecosystem. They should mentally record what evidence they see and hear on animal activity in and around the garden.

2. Finally, the students can enjoy a guided taste test of fall harvest food in the garden. After their exploration, gather the students back together to share findings.

Assessment (5 minutes)

1. What animal and insect signs did the student groups find in the garden? What creatures or insects left these signs?

2. What do these signs say about the creatures living in the garden? Were there moments where student behavior helped them interact with an insect or animal? How can students continue to practice Care for Self, Others, and the Land within the garden?

Resources

• Jon Young et al. *Coyote's Guide to Connecting with Nature*, 2nd ed.

3 Harvesting

Time Frame: 45 minutes

Overview: Students will gather ripe fruit from the garden for eating and sharing.

Objectives: To learn how to identify the qualities of a ripe fruit and properly pick them without harming the plant.

Introduction (10 minutes)

1. Have the students brainstorm about the seasons and the physical changes they can see occurring in the garden. Discuss what plants have ripe fruit in the fall and which ones should be harvested before the frost can freeze burn the food. What season is it? How can students tell from their surroundings that it is fall? What foods are seasonally available to eat? What previous harvesting methods have the students practiced before?

2. The class will be harvesting fruits from

the garden from whatever foods are available and ripe, such as tomatillos and tomatoes. The students will use two hands to carefully twist fruits off the vines or stems. Using sample fruits to demonstrate, discuss with the class how to tell if a fruit is not ready or is ripe to harvest. What does a ripe fruit look like?

Activities (30 minutes)

1. Students will work in pairs or individually to harvest as many ripe fruits as they can. The goal is to be careful and gentle, rather than quick and competitive. Have them bring their harvest to an adult or teacher with a harvest basket. This will give the adult time to double-check the ripeness of the fruit.

2. The groups can be split into food themes for harvesting. For example, one group can harvest tomatoes, another squash, and another cucumbers.

> In the past, I have used this harvest to share at school parties, feed the community at seasonal events, or process in the school cafeteria.

3. At the end of the harvest time, gather the class together and have students look over their combined harvest and double-check for unripe fruits or poorly picked produce.

4. Students can enjoy a taste test of their ripe harvests, or compare the flavor of a ripe fruit with an unripe one.

Assessment (5 minutes)

1. How else could students have harvested the foods so that they didn't pick unripe fruits? Was it difficult to find the fruits?

2. What are the flavor differences between a ripe fruit and an unripe one?

A student discovers the delicate leaf structure of late fall tomatillos.

Preparation

1. Pre-picked ripe fruits
2. Pre-washed foods for the taste test

4 Dry Seed Saving

Time Frame: 45 minutes

Overview: Students will harvest seeds from dry seedpods and practice saving them for springtime.

Objectives: To sort and categorize healthy seeds and review how seeds are both living and dormant.

Vocabulary

Dormant

Introduction (10 minutes)

1. Reintroduce the students to the word *dormant* as something that is alive but not actively growing—as if it were sleeping. Explain that the seeds they will be collecting today are dormant and waiting for the right growing conditions; to be gathered and saved in a dry, cool, and dark place to plant for the spring.

2. Brainstorm with the class what other seeds they have saved in past gardening activities: What techniques helped them remove the seeds? How could they tell which seeds were ready to store away and which weren't? What dry seeds can they find in the garden? Demonstrate how to remove seedpods or husks to reveal and gather the seeds.

Activities (30 minutes)

1. In groups, students will break open seedpods, or separate bulbs, and gather all the seeds they can from each plant part. Garlic or radish seedpods are good options for the students to work with. Have the students sort the seeds into collection cups and the husks into another pile for composting.

2. Students will need to work carefully and slowly in order to properly categorize seeds and not waste them. The students can store their seeds in a large paper bag or glass jar and deposit the husks and organic materials into the school compost bins.

3. When students are done collecting, have them taste test from the same plants from which they saved seed. For example, if they harvested chard or bok choi seeds, have them taste test chard and bok choi leaves. Other potential seeds and bulbs to be saved include: garlic, radishes, sunflowers, or calendula.

Assessment (5 minutes)

1. What did students notice about the seeds that they never noticed before? What were the textures the class encountered during the activity?

2. What kind of seeds did they save today? Was it difficult to harvest the seeds?

Preparation

1. Seedpods or garlic bulbs
2. Paper bags or glass jar
3. Collection cups

5 Cover Crops

Time Frame: 45 minutes

Overview: Students will plant cover crops and edible winter starts in the school garden.

Objective: To explore new plants that like to grow in the cold and can provide food throughout the winter.

Vocabulary

1. Seasonal
2. Cold Hardy
3. Cover Crops

Introduction (10 minutes)

1. Brainstorm with the students what plants like to grow in cooler or cold weather: What plants can the school grow so students can eat food throughout the winter and spring? What is a cover crop? How can these plants also help nourish the garden soil and future plant growth?

2. Demonstrate to the students how to plant their starts, bulbs, or seeds. Seeds should be planted at least twice as deep as they are large. Without measuring tools, what can students use to determine the right planting depth?

3. Where are good places to plant these seeds in the garden? How will students make sure to have seed-to-soil contact during the activity? How closely together should students plant the seeds?

Activities (30 minutes)

1. In pairs or individually, students will get a handful of seeds or bulbs to plant together. Students should do this work carefully and slowly, like good gardeners do, by paying attention to the details and working in the designated space.

> Alternatively, the students can explore the garden and determine the best place to plant the seeds on their own. In this way, they will apply lessons on what growing conditions their plant prefers and identify these spaces in the garden. This makes for a wild jungle of planting possibilities, rather than straight rows. Personally, I enjoy this wild aesthetic, but at the same time I make sure to suggest student planting in certain areas so they can guarantee a better harvest.

2. The groups can continue this activity until all the seeds are planted. The groups should work in one garden bed at a time, rather than jump from bed to bed. The pairs can water their seeds with a watering can after they have planted.

3. When students are done, they can taste test from picked or prepared garden foods.

Assessment (5 minutes)

1. Did the groups see any other seeds sprouting as they planted? Did they encounter any living things? What plants will be growing alongside their cover crops? Will these plants survive the cool weather?

2. What strategies did students have for practicing collaboration and teamwork during this activity?

Preparation

1. Prepared foods for taste testing
2. Cover crops such as oats, peas, barley, clover, or other legumes
3. Watering can

6 Mulching and Compost

Time Frame: 45 minutes

Overview: Students will use organic matter to mulch the garden beds.

Objective: To learn that mulching is similar to tucking the garden in for winter, by keeping the soil protected from the seasonal weather conditions.

Vocabulary

1. Mulch
2. Compost

Introduction (10 minutes)

1. Brainstorm with the class what they know about mulching, or laying organic materials down on the soil. How can straw or fallen leaves still contain nutrients that can make the garden soil healthier? What habitat for insects and worms does it provide during the winter seasonal weather? What living creatures will enjoy this cover of mulch? What living things won't thrive under it? Why is this an important seasonal practice for gardeners?

2. Some students will have performed this activity in previous gardening activities. What lessons and techniques did they learn to help them? Demonstrate to students the ideal amount of the material to hold in their hands—just one handful at a time so as not to waste the valuable material—and how to carefully tuck the organic matter around the living plants. Where are places in the garden that need mulch? What does careful gardening look like during this activity?

Activities (30 minutes)

1. Have students break into groups, or work individually, and begin to mulch in the garden beds. One or two student volunteers can rake fallen material into piles for other students to gather. This is a great activity for students who are uncomfortable or itchy with the material.

2. Students should focus on mulching one garden bed at a time and doing quality work in each space.

3. When students have finished mulching, gather them back together and have them do one more check over their work to make sure there aren't mulching materials discarded in pathways or thrown over established plants. Afterward, the class can enjoy a taste test of fall foods from the garden.

Assessment (5 minutes)

1. How would the students grade the quality of their gardening work? Did they feel like they did the best work they could have done?

2. Was it easy or difficult to work with the mulching materials?

Birds are squeaking
while the air around
them is whistling and the
wind is silent and the bees
are humming while airplanes
are soaring and flies are buzzing
wildly and I think it is quite peaceful.

— Anonymous Student Poem

Preparation

1. Straw bale or gathered leaves
2. Gloves

7 Focused Plant Observations

Time Frame: 45 minutes

Overview: Students will begin their own nature journals of beneficial and useful garden plants.

Objectives: For students to identify plants and their uses, while also learning how to scale their drawings realistically.

Vocabulary

1. Medicinal
2. Native Plants

Introduction (10 minutes)

1. Students will be engaging in nature journaling for the next few sessions as a way to learn about different plants in the garden, especially native and medicinal plants, and to develop strong observation and science skills.

2. Brainstorm with the class what the difference is between realistic and fictional drawing: How can students focus on realistic drawings? What are important features to add to realistic drawings? What methods do scientists use if they are struggling to illustrate what they are seeing? See Resources for links and images to references and discuss realistic drawings with the class.

3. For the first part of class, all the students will pick the same plant and focus on drawing it accurately and to scale (to the best of their abilities). They will also be encouraged to use descriptive words or questions in the drawing's margins to help record what they see. Demonstrate what the plant looks like and its key features. What are the medicinal, edible, and desirable qualities of this plant?

Activities (30 minutes)

1. Individually, the students will go into the garden to identify and sketch the plant, using their observation skills to analyze the features of the flower and whole plant structure.

2. As students finish their sketches, they will display them for the class to see. The students can explore the differences between each of the drawings and observations their classmates made.

3. When students have finished their sharing of their sketches with their classmates, they can return to their drawings to make any final adjustments and then engage in a taste test from the garden.

Assessment (5 minutes)

1. Encourage the class to reflect on their own observations and what they saw in their classmates' work. What were key features that students noticed about the plant? What took them by surprise or inspired them to ask questions?

2. What did students feel they did well with their sketches? What do they want to improve on? What details did they admire in the work of other students?

Preparation

1. Notebooks or journal paper
2. Pencils
3. Pencil sharpener (for the field)

Resources

- Dorothy Hinshaw Patent. *Plants on The Trails With Lewis & Clark.*
- John Muir Laws et al. *Opening the World Through Nature Journaling: Integrating Art, Sciences, and Language Arts*, 2nd ed.
- University of Nebraska Lincoln. "Images of Plants and Maps." *Journals of the Lewis and Clark Expedition* (online—see Appendix B).

NGSS and Activity Extensions

Extensions for in-class studies can include investigating, describing, and classifying different types of matter and observing their properties (2-PS1-1: Matter and Its Interactions).

8 Identifying Medicinal Plants

Time Frame: 45 minutes

Overview: Students will learn about medicinal garden plants and sketch the leaves of each plant.

Objectives: To draw and observe more complex plants, while generating questions about medicinal plants.

Vocabulary

Medicinal Plant

Introduction (10 minutes)

1. Brainstorm with the students about how to generate scientific questions on their drawing subjects: What questions do scientists come up with about what they are studying? What detailed observations would a scientist make about new things they encounter?

2. Introduce the class to medicinal plants in the garden and their uses: What is a *medicinal plant*? What types of plants do students know have medicinal uses for humans? Medicinal garden plants could include yarrow, calendula, angelica, catnip, coneflower, mint, or comfrey.

3. Pass around leaf examples from two or three varieties of medicinal plants. How can students identify these plants in the garden? What are key features that they can remember?

Activities (30 minutes)

1. With their nature journals, the students will enter the garden and identify two or three medicinal plants. They can pick a leaf from these plants to sketch and make observations about their texture, shape, insect bite marks, or interesting features.

2. Students who finish their sketching early can go back to the medicinal plants and continue sketching the other parts of the plant, including stems, flowers, and seeds.

3. After sketching, the students will share their observations with a partner and work on creating more questions and observations about their subject.

4. Finally, the class can enjoy a taste test of plants in the garden, including the medicinal plants they studied, if possible. They can add their observations about the taste and texture of these plants in their journals.

Assessment (5 minutes)

1. Gather the class together to reflect on their experiences with these plants. What key observations did the students make about their plant samples?

2. What differences did they notice between the medicinal plant leaves? How would they describe these differences; for example, were some leaves serrated or smooth?

The *Amanita muscaria* is a recognizable mushroom for most students and makes for easy identification.

Preparation

1. Medicinal plant leaves

Resources

- Lesley Tierra et al. *A Kid's Herb Book.*

9 Fascinating Fungi

Time Frame: 45 minutes

Overview: Students will practice quick sketches of a mushroom by drawing it from three different angles.

Objective: Students will learn about the shape and properties of fungi.

Introduction (10 minutes)

1. Brainstorm with the class what they know about mushrooms and fungi: What concerns do they have about them? Where can students find mushrooms and fungi in the school garden? Why are they important decomposers? What shapes of mushrooms have students seen? What creatures or insects eat mushrooms?

2. During the activity, the students will be using magnifying glasses to enhance their observations. What are the behavior expectations while using these tools?

Activities (30 minutes)

1. Each student will receive a pre-picked mushroom and observe and sketch its details in their nature journal. They will do three quick sketches from different angles, such as from the top looking down, from the side, from the bottom.

2. Students who finish their in-depth study can journey in the garden to find another mushroom and finish one more sketch.

3. Finally, the students can enjoy a taste test from the garden before gathering back together to share their work. A taste test of edible mushrooms would be ideal—raw or cooked.

Assessment (5 minutes)

1. What scientific observations did students make about their mushrooms? What valuable questions did students generate?

2. Were the students able to notice some features of the mushrooms at one angle that they weren't able to see at another angle? What were some of the differences between their sample mushroom and the one in the garden? What observations could students make about mushroom shapes and features with the magnifying glasses?

Preparation

1. Pre-picked mushrooms

10 Spore Prints

Time Frame: 45 minutes

Overview: Students will continue learning about fungi by creating an art and science piece with mushroom spores.

Objectives: To learn key terms and understand more fully the life cycle of a mushroom and its role in an ecosystem.

Vocabulary

1. Mycelium
2. Volva
3. Sporulation
4. Gills/Cap
5. Spores

A late fall fungi uses water propulsion to emerge from pine needle duff.

Introduction (10 minutes)

1. Have the students brainstorm what they know about mushrooms and fungi: What do fungi feed on? What would the world look like without decomposers? What are the parts of a mushroom? What role do spores play in a mushroom's life cycle?

2. Using a diagram, introduce the students to the parts of mushrooms, explaining the significant differences in their anatomy.

Activities (30 minutes)

1. Each student will get one large, pre-picked mushroom as a sample. If time and weather permit, have them go into the garden or a natural space and look for additional mushrooms.

2. The students will carefully remove the stem from their sample and lay them out on ½ of a white sheet and ½ of a dark sheet of paper.

Spores can be a variety of colors, depending on the mushroom, so two differently colored sheets will reveal which color spores the mushroom releases. Students should make sure their names are written on their papers.

3. Spores prints are best captured during this process if a bowl or container is placed over the mushroom cap so that air currents don't affect the fall of spores and if the caps are left alone for 24 hours. The following day, have the students carefully remove and compost the mushroom caps, revealing their spore prints below.

4. An adult can adhere these prints with a spray fixative so the students can preserve them in their nature journals.

Assessment (5 minutes)

1. When the spore prints have been preserved and are dry, have the students reflect on the product of their activity: What is surprising about the spore prints? Were the students expecting them to look differently? What are the size of the spores?

2. What are the different colors of spores represented? Can students see the shape of mushroom gills in the spores or evidence of airflow? How do mushrooms pass on their spores in nature?

Preparation

1. Pre-picked mushrooms for student use
2. White and dark paper
3. Fixative spray
4. Volunteers

Additional Activities & Resources

- Go Science Girls. "Mushroom Spore Prints." (online—see Resources).

Winter: Beneficial Insects and Birds

1 Predators and Prey

Time Frame: 45 minutes

Overview: The class will build a garden food web together and learn about beneficial predators and their prey.

Objective: For students to understand the balance and value of predators and prey in the school garden and the role of adaptation in these relationships.

Vocabulary

1. Beneficial
2. Non-beneficial
3. Predator and Prey

Introduction (5–10 minutes)

1. Brainstorm with the class what they know about the vocabulary terms and types of predators and their prey in the garden. What spiders, insects, bugs, and birds have the students seen hunting in the garden?

2. The students should be encouraged to try and avoid using the words *good* and *bad* but rather *beneficial* and *non-beneficial* when talking about insects and bugs in the garden.

Activities (30 minutes)

1. As a class, the students will build a garden food web together on a classroom chart. Each student, or a pair of students, will receive a Predator and Prey card and will determine whether their item is predator or prey. With a small piece of tape, they will add their creature to the class poster, starting with prey and then predators.

2. The class will discuss whether all the cards are in the right space on the food web poster. Is it better to have more predators or prey in the garden? What about beneficial and non-beneficial insects? What category are the students as members of this food web? What can students do to encourage beneficial predators in the garden?

> I have a collection of preserved insects, from beetles to yellow jackets, in glass jars that I gathered in various gardens. These are great to pass around in a class circle and discuss what role as predator, prey, or both, each creature plays.

3. There are many tag-based games that are extensions of predator and prey relationships, so these are options for further engagement with the theme. While I usually prioritize physical activity over time spent sitting, most of these tag games aren't accurate representations of the dynamic balance between predators and prey in the garden.

Assessment (5 minutes)

1. Are there predators and prey missing from the food web of the school garden? What creatures are both predators and prey?

2. This is the opportunity to acquire praying mantis egg sacs and a habitat frame for the next in-class student study. These are easy and inexpensive to find online, and the egg sacs can be spotted by inquiring students in the garden, on walls, leaves, or twigs.

Preparation

1. Predator and Prey Cards
2. Classroom food web chart

2 Praying Mantis

Time Frame: 45 minutes

Overview: The students will learn about the life cycle of a praying mantis and be introduced to the new class pet: a praying mantis egg sac.

Objective: For the class to value beneficial predatory insects through engagement and observation.

Vocabulary

1. Beneficial
2. Predatory
3. Ootheca

Introduction (10 minutes)

1. Brainstorm with the class what they know about praying mantises and record their thoughts on a classroom chart: What do these

A praying mantis hunts through raised beds.

insects eat and how do they hunt? What do students know about the life cycle of a praying mantis?

2. What is a *predatory* insect? How can they be *beneficial*? Why are they so valuable to a garden?

Activities (30 minutes)

1. Introduce the praying mantis egg sac, or *ootheca*, to the class to gently pass around. Record the group observations on a classroom record chart. What are questions that students have about the shape, structure, texture, and formation of the egg sac? How was it formed? What is it made of? How many praying mantis eggs are in it? What scientific questions do students have about the egg sac and the life cycle of a praying mantis?

2. Have the students act out the behavior of a hunting praying mantis. How do its head and arms move? How does it hunt its prey? Use the videos and pictures (in Resources) as inspiration for student action. The class can all become praying mantises and explore the classroom hunting and safely acting out how they would capture and eat their prey.

Assessment (5 minutes)

1. What changes do students hope to see with the egg sac over the next few weeks? What conditions will the eggs need to experience in order to emerge?

> The class will check on the egg sac for changes over the next few weeks. There are many extension activities and even units that can follow this, if a classroom teacher reserves time every week for students to observe or discuss the changes of the egg sac.

Preparation

1. Praying mantis egg sac
2. Habitat case
3. Classroom record charts
4. Illustrations, book, or online images for reference

Resources

- National Geographic. "Praying Mantis." (online—see Appendix B).
- Bug Chicks. "Order Mantodea, Mantid." (online video—see Appendix B).
- "A Long Tradition of Imitation" in Jon Young et al. *Coyote's Guide to Connecting with Nature.*

3 Garden Spiders

Time Frame: 45 minutes

Overview: Students will learn about common garden spiders, how they hunt, and label their anatomy.

Objective: For the students, as scientists and gardeners, to value and identify predatory spiders.

Vocabulary

1. (Cephalo)thorax
2. Fangs
3. Exoskeleton
4. Abdomen
5. Spinnerets

Introduction (10 minutes)

1. How do scientists and gardeners talk about spiders, insects, and bugs—even if they are uncomfortable? How can a spider be both a pest and a beneficial creature?

2. Introduce the class to a few common garden spiders, such as the goldenrod crab spider, black and yellow garden spider, and wolf spider. Why are these spiders so important to the

I understand that both adults and children can be uneasy around spiders. I don't share this experience; as a child, I actually had a nice collection that I kept as pets. I encourage students to not react fearfully in response to things that make them uncomfortable (spiders, yellow jackets, bees), but rather to calmly move their bodies away from the source and then act as gardeners and scientists to make observations and engage in wonder about these creatures. I model this behavior for students because they will react in the ways they see that adults do.

garden? What do they eat? What are the features and colors of each variety that can help students identify them in the garden?

3. What are the different parts of a spider's body that students can observe in pictures of the garden spiders? Label these parts on a classroom chart showing a simple spider illustration. Add other body parts that students may not know and include terminology/vocabulary to the chart.

Activities (30 minutes)

1. The class will read *The Very Busy Spider* by Eric Carle for its illustrations on a spider building a web, how it begins the process, and this method of hunting. Discuss the hunting techniques of different spiders (see Resources).

2. Afterward, the class will label and color their The Parts of a Spider worksheet, using the classroom chart as a reference. The students will color their spiders realistically in the patterns and colors of common garden spiders. If there is extra time, the students can add illustrations of the landscape where the spider lives and what it eats.

Assessment (5 minutes)

1. Knowing what they do about spiders now, how will students react when they see spiders in the garden or at home?

2. What did they not know about spiders that took them by surprise?

3. Be sure to check on the status of the praying mantis egg sac and record any changes in its development.

Preparation

1. The Parts of a Spider worksheets
2. *The Very Busy Spider* by Eric Carle
3. Projector and computer

Resources

- Eric Carle. *The Very Busy Spider.*
- American Museum of Natural History. "How Do Spiders Hunt?" (online—see Appendix B).
- BBC Earth. "Spider With Three Super Powers." (online—see Appendix B).
- KidZone. "Spider Facts: The Body of a Spider." (online—see Appendix B).
- Spider ID. "Spiders in Oregon." (online—see Appendix B).
- Spider ID. "*Argiope aurantia* (Black and Yellow Garden Spider)." (online—see Appendix B).

4 Ladybugs

Time Frame: 45 minutes

Overview: The students will explore the role of ladybugs as predators to aphids by transplanting plants to the garden that encourage ladybug activity and habitat.

Objectives: To understand the role of ladybugs as predators in the school garden and to promote their population growth.

Introduction (10 minutes)

1. Brainstorm with the class what they know about ladybugs: What do they eat? Why are they important to an organic garden? How are they predatory insects?

2. Introduce the class to the diversity of ladybugs and beetles that look similar (see Resources). What are the different colors and spots of similar beetles that students have seen? Why are ladybugs red, white, and black, while other beetles and insects are camouflaged?

3. Now is a great time to discuss different feelings and opinions about ladybugs and spiders—a conversation that started in the previous class. Many students are excited about ladybugs and cringe at spiders. Why do the students have different feelings about each creature? Would their perspective or experience change if they were an aphid, or a broccoli plant, or a bird?

4. Demonstrate the activity for the day, showing students how to build habitat for the predatory ladybugs by transplanting yarrow starts. What are other uses for yarrow? Why is it a valuable garden plant?

Activities (30 minutes)

1. In pairs, students will transplant yarrow starts into large potting containers or directly into the school garden beds. They will be careful gardeners by making sure the roots of the plants are buried down into the soil, rather than folded up. They can water their starts thoroughly after they have planted.

2. After they have planted the ladybug habitat zones, the students can check in on the development of the praying mantis egg sac and make observations about its status.

> Yarrow is a wonderful pollinator plant, as well as habitat for ladybugs during the summer. Planting it alongside aphid-attracting crops can encourage ladybug predation. Yarrow is also easy to transplant. In the winter and spring, the new rhizomes and shoots can be separated from the main plant and transplanted it to new spaces. Yarrow seeds are also prolific, so if they were saved in the fall, then students can spread the seeds around during this activity time, especially in flower beds and wildflower spaces.

Assessment (5 minutes)

1. How long have the mantis eggs been dormant? What environmental conditions could progress their development?

2. How will ladybugs and praying mantises interact in the garden?

Preparation

1. Potting soil
2. Yarrow starts
3. Planting containers
4. Watering can

Resources

- J. Loomis and H. Stone. "Lady Beetle (*Hippodamia convergens*)." (online—see Appendix B).

5 Hummingbirds

Time Frame: 45 minutes

Overview: The class will transition from learning about beneficial insects to beneficial pollinators by studying two hummingbird varieties and making a hummingbird mask.

Objectives: For students to value these fascinating pollinators and begin understanding the process of pollination.

Vocabulary

1. Beneficial
2. Pollination

Introduction (10 minutes)

1. Brainstorm with the class what they know about hummingbirds: What colors of hummingbirds have students seen? Where have they seen them and what were the birds doing? Do students know what food hummingbirds eat and drink from? How big are these birds?

2. Using illustrations and book resources, introduce the class to two local hummingbird species. In the Pacific Northwest, this is typically the Rufous and Annas hummingbird. What are the different colors and features of each bird? What are characteristics of each bird that the class notices which will help students identify these birds in the garden?

3. How do hummingbirds pollinate? What is it about their physical shape that indicates how they pollinate? What is pollination? Demonstrate the activity and how to make a hummingbird mask with the provided materials. What realistic colors can students use to decorate their masks?

Activities (30 minutes)

1. Each student will receive a Hummingbird Mask template to color with realistic markings of either variety of hummingbird they learned about. Rather than scribbling, they should practice craftsmanship to create work they are proud of. When the masks are decorated, the students will carefully cut the beak hole from the mask and then cut out the entire mask from the template. I intentionally left these masks simple so that the students can create the coloring and markings of the hummingbird species in their own region.

A student practices drinking nectar from a hellebore flower.

2. The students can collect a beak made from rolled construction paper, prepared in advance. They may require adult help to cut the beaks into right length for their masks. Carefully, they will fold and tape these beaks into the beak hole of the mask. With adult help, each mask will have two holes punched into either side of the mask and string tied to the ends. The completed mask can be fitted to each child's head.

3. When the masks are completed, the new flock of "hummingbird students" will explore the large flower examples set up around the learning space, or go out into the garden to visit any flowers in bloom. They can pretend to be hummingbirds visiting flowers for food and drinking the nectar from the flowers, as well as pollinating as they go. After each student has visited each flower, they can fly their way back together to reflect on their play.

4. An addition to this activity would be to have the students practice flying in the manner of hummingbirds, who don't flap but rather swing their wings in a figure-eight pattern. How much energy does it take the students to "fly" like this? What is the purpose and benefit of flying this way?

Assessment (5 minutes)

1. What were students' experiences "drinking" from the large flower examples? Was there a shape of flower that was easier for the hummingbirds/students to drink from and access? What do these experiences tell the class about what flowers the students should grow in the garden to attract hummingbirds?

2. Student volunteers can make observations about the praying mantis egg sac and report their findings to the class.

Preparation

1. Hummingbird Mask template
2. String, hole punch, and clear tape
3. Coloring pencils/markers/crayons
4. Beaks made in advance: construction paper rolled into a point and taped to hold its shape
5. Large flower examples (flat faced and trumpet shaped)
6. Adult volunteers

Resources

- Donald and Lillian Stokes. *The Hummingbird Book; The Complete Guide to Attracting, Identifying, and Enjoying Hummingbirds.*
- Sally Roth. *Attracting Butterflies & Hummingbirds to Your Backyard.*

NGSS and Activity Extensions

For further in-class extensions, the students can continue investigating and building models that mimic the function of an animal in pollinating or spreading seeds (2-LS2-2: Ecosystems).

6 Planting Hummingbird Food

Time Frame: 45 minutes

Overview: Students will plant seeds that will feed both pollinators and students.

Objective: To plant beneficial seeds and cultivate pollinator habitat in the school garden.

Introduction (10 minutes)

1. What is a pollinator? What birds pollinate in the garden? Why are these pollinators so important?

2. One way that students can build habitat for pollinators is to provide their preferred food sources or flower shapes and colors. What colors and flowers do hummingbirds prefer? Hummingbirds have long beaks and tongues; this indicates that they are adapted to drink nectar from flowers with long, cone-like shapes. Some plants that produce flowers with this shape (bell/trumpet) include honeysuckle, beans, and salvia.

3. How deeply should students plant seeds? How far apart should the seeds be from each other? What elements, other than soil, will the seeds need to thrive?

Activities (30 minutes)

1. Individually or in pairs, the students will plant varieties of seeds that will grow inside or in a greenhouse until late spring planting. These seeds will provide nectar for hummingbirds and food or medicine for the children.

2. Students should focus on carefully planting

the seeds at the best depth and labeling their containers. They should also practice teamwork and collaboration in planting each variety of seed. The groups can water their seeds once they have been planted and labeled.

3. After the activity, the students can enjoy a taste test from the garden's early spring plants.

Assessment (5 minutes)

1. Gathering back together, have the students discuss what they think will happen as the plants grow or when the hummingbirds come visit the flowers.

2. What other plants and flowers do hummingbirds visit?

3. If there are any changes with the praying mantis egg sac, be sure to discuss and record these observations as a class.

Preparation

1. Potting soil
2. Planting containers
3. Labeling materials
4. Flower or bean seeds (scarlet runner beans are a good variety for this activity)

7 Building Habitat

Time Frame: 45 minutes

Overview: Students will build wreaths to hang in the garden and provide nesting material for local hummingbirds.

Objective: To engage in constructive activities that encourage hummingbird nesting and habitat building.

Introduction (10 minutes)

1. Discuss with the class how hummingbirds make their nests and what materials they gather, using pictures and illustrations as a guide. In the pictures, what materials can students identify in hummingbird nests? How big are the nests and how many eggs do they hold?

2. Demonstrate the activity for the day and how to assemble the nesting wreaths with available materials. How can student groups work together collaboratively and as partners during this activity?

Activities (30 minutes)

1. In pairs, the students will build wreaths from pre-gathered lichens and mosses. First, the students will gather a flexible stick (a grape vine, willow stick) to create a frame for their wreath. They will tie both ends of the stick together with string and form a loop.

2. They will use the provided materials to tie or wrap their wreaths with the moss and lichen, bunching it together piece by piece until the whole wreath is covered. The students will use creative problem-solving to create a wreath that is held tightly together so the materials won't fall out, but loose enough for bird to remove the fibers and moss. The groups should do their best to follow instructions and work as a team.

3. When the wreaths are finished, the class will hang their wreaths in the garden or throughout the community and return to clean their workstations. Afterward, they can enjoy a small taste test from the garden.

Assessment (10 minutes)

1. Have the pairs check the stability of their wreaths and make any necessary adjustments. Do the students' wreaths hold together strongly?

2. How will student know if their nesting materials are being used? Other than hummingbirds, what other creatures could use the nesting materials?

3. What changes have occurred with the praying mantis egg sac?

Preparation

1. Moss, lichen, organic nesting materials
2. String or yarn
3. Flexible sticks (willow, hazelnut suckers, grape vines)
4. Scissors

Resources

• *The Hummingbird Book* by Lillian and Donald Stokes.
• *Attracting Butterflies & Hummingbirds to Your Backyard* by Sally Roth.

8 Planting Forage

Time Frame: 45 minutes
Overview: Students will sow wildflower and herbs seeds in the garden.
Objective: To provide food and forage habitat for pollinators and beneficial predators.

Vocabulary

1. Forage
2. Native Seeds
3. Germinate

Introduction (10 minutes)

1. Brainstorm with the class what they know about food sources for beneficial predators, insects, and pollinators: What can students do to provide even more habitat and food for these creatures in the garden? Why is it important to attract these creatures? What effect will their increased presence have on the garden?
2. Introduce the students to the varieties of seeds that they will be planting during the activity. What are native plants and wildflower seeds? What does it mean to sow or broadcast seeds? What techniques have students used successfully in the past?

> To spread seeds evenly across the garden soil, I mix available soil with the seeds in a bucket. During the activity, I hold onto the bucket and the students return to me for additional handfuls of this soil/seed mixture. Student volunteers can help thoroughly mix the seeds into the bucket with soil at the beginning of the activity.

Activities (30 minutes)

1. Individually, students will receive a handful of the wildflower mixture to spread in designated garden beds, making sure to do careful gardening work in sowing the seeds. They can repeat this activity until the seeds are all spread.
2. Then, the class can gently press the seeds into the soil using their hands and making sure there is soil-to-seed contact for better germination. Student volunteers can water these seeds and the garden plots afterward.
3. The class can enjoy a taste test at the end of class and be introduced to new early spring plants.
4. Have the students check on their praying mantis egg sac and observe any changes in their emergence. If the eggs hatched into nymphs, spend at least ten minutes reflecting on student observations and recording them and their questions on the classroom observation chart.

Assessment (5 minutes)

1. Are all the seeds connected to the soil? When do the students think the flowers will start to grow? What colors will they be?

2. What beneficial insects and pollinators will visit these flowers and plants?

Preparation

1. Native wildflowers seeds (yarrow, hyssop, poppies, goldenrod)
2. Potting soil
3. Bucket

9 Creating Watering Basins

Time Frame: 45 minutes
Overview: Using recycled materials, the class will create watering basins for insects and pollinators.
Objective: Students learn another valuable way to provide habitat for garden pollinators.

Introduction (10 minutes)

1. Brainstorm with the class on the importance of providing water for pollinators and insects, as well as how to create a watering system that the insects can easily drink from: How do pollinators find water where it is safe for them to drink? How can students build a watering hole that insects won't fall into? What types of pollinators will drink from a garden watering basin?
2. Introduce and demonstrate the activity to the class, from painting or coloring the outside of the basin, to how high to fill it up with gravel/marbles and how much water to pour into the basin.

Activities (35 minutes)

1. In pairs, the students will select a basin to decorate with a quick-drying, waterproof paint. Markers that can be covered with a strong sealant can work too. The students should make sure to only paint or color the outside of the basins and not the inside where the pollinators will be drinking water. The pairs can color their basin with images from the garden or in colors that will attract pollinators.

> If it is difficult to find recycled or reusable shallow basins, terracotta drainage trays work well too.

2. After the students have cleaned up, they can enjoy a taste test from the garden while waiting for their basins to dry.
3. When the paint or sealant has dried, the pairs will fill up their basin with provided gravel, marbles, or river rocks and then place the basins around the school garden. Under hoses is a good location, where water can leak from faucets. The students can fill up the basins with water until the gravel/marbles are still dry on top but surrounded by water on all other sides.

Preparation

1. Variety of bowls, basins, and shallow cups
2. Marbles, pea gravel, river rocks
3. Water/watering can
4. Quick-drying, waterproof outdoor paint or markers
5. Clear liquid sealant

10 Beneficial Insect Release

Time Frame: 45 minutes
Overview: The class will release the praying mantis nymphs that they raised in their classroom.
Objectives: To make observations about the nymphs and what they'll need to survive to adulthood.

Vocabulary 2 and 3

1. Diurnal
2. Nocturnal
3. Antennae
4. Proboscis

Introduction 2 (5–10 minutes)

1. What experiences do students have with moths and butterflies in the garden? How can they tell the differences between the two?

> I have specimen examples of butterflies and moths that I collected in glass jars which I share with the students so they can more closely study the differences (they were dead at the time I gathered them or were given to me). Picture references also work well.

2. Moths and butterflies are both pollinators that consume nectar in the garden in similar ways, but are easily identified if closely observed. Moths typically have fringed antennae, are nocturnal (not all), and have wings that lay flat when they are at rest. Butterflies are diurnal, have smooth antennae, and wings that fold up when they are at rest. A remarkable thing about both species is the long tongue (proboscis) that they use like a straw to consume nectar.

3. Using a large classroom chart, illustrate the key differences between moths and butterflies. Students can also use pictures and what they know of the insects' distinct body features to help identify each type.

Activity 2 (15–20 minutes)

1. For the activity, the students will make a three-dimensional moth or butterfly. They can choose a two-dimensional picture of either a moth or butterfly (see Resources) and determine which realistic colors and markings they would like to use on their insect.

2. When students have finished coloring and adding the realistic traits, they can cut out their insects and fasten the paper to a long straw using a hot glue gun (with adult help).

3. Students can clean up after their work is done and work on folding the wings of their insects (flat if a moth, upright if a butterfly).

4. Finally, students will receive a small cup of "nectar" or fruit juice of choice to drink from using their proboscis/straw. In this way, the students can imagine they are the pollinator drinking nectar from a flower.

Assessment 2 (5 minutes)

1. What details did the students pay attention to when coloring their moth or butterfly?

2. What types of flowers would their pollinator visit and what colors would they prefer?

Preparation 2

1. Tape
2. Scissors
3. Crayons/Colored Pencils/Markers
4. Hot glue gun
5. Long straws
6. Small cups
7. Fruit juice
8. Butterfly and Moth coloring sheets

Overview 3: Students will practice identifying moths and butterflies in the garden according to their differences and unique features.

Objective 3: To study the anatomy of key pollinators and identify them in the school garden.

Introduction 3 (10 minutes)

1. Brainstorm with the class what they know and recall about the features, similarities, and differences between moths and butterflies: How do these insects pollinate the garden? What flowers do they prefer?

2. Review with the class what the physical and behavioral differences are between moths and butterflies.

Activity 3 (30 minutes)

1. In pairs, or with older student partners, the class will use the Butterfly and Moth Guide to identify and count the two pollinators they are studying. If possible, older student partners can assist in this activity and exemplify scientific engagement and learning to the younger students.

2. Student should focus on the four key differences between butterflies and moths to guide their investigation.

3. After they have finished their survey, they can share their findings and reflect on the activity before enjoying a taste test in the garden.

Assessment 3 (5 minutes)

1. Gathering back together, have the students participate in a short survey about how many butterflies and moths they found.

2. Where did they find the most specimens? What colors did they notice on the butterflies and moths? How many different kinds of butterflies and moths did they find?

3. What behaviors did they notice the creatures were doing? Were any of the insects pollinating?

Preparation 3

1. Butterfly and Moth Guide
2. Older student partners

Resources 2 and 3

- "The Bug Chicks: Butterflies & Moths." (online—see Appendix B).
- Library of Congress. "How can you tell the difference between a butterfly and a moth?" (online—see Appendix B).

4|5 The Colors of Food

Time Frame: 1½ hours (2 × 45 minutes)
Overview 4: Students will learn which flower colors attract certain pollinators.
Objective 4: To understand pollinator sight and how flowers interact with and attract insect pollinators.

Introduction 4 (10 minutes)

1. As a class, students will engage in a presentation about pollinator sight, with various pictures of flowers, but as bees, butterflies, and birds might see them (see Resources).

2. What differences do students see in the human and insect ways of seeing? Brainstorm the various qualities and techniques that flowers use to attract pollinators.

3. How do bees navigate through a garden? What parts of their body do they use?

Activity 4 (30 minutes)

1. Individually, students will be creating a version of pollinator sight, using small handmade kaleidoscopes binoculars. They will construct these by placing small square pieces of white tissue paper onto contact paper and attaching it to a paper roll, as a way to mimic the two large compound eyes that honeybees have with their thousands of facets.

2. The student should practice craftsmanship during this activity and perform it twice so that each eye can have a viewpoint.

3. When they are done, they students will look at solid blocks of color and at mixed colors, either in the classroom or in the garden. They will be encouraged to think like pollinators by analyzing the difference between looking at large blocks of color and scattered or mixed colors.

Assessment 4 (5 minutes)

1. After students have made observations with their new tools, gather them back together to discuss what student experiences were while looking through their models of bee eyes.

2. If the students were to imagine they were an insect pollinator, how would they be able to find the colors they are attracted to most in the kaleidoscope of other colors? Was it easier for students to see color when it was blocked together or when it was scattered? What can gardeners do to make it easier for pollinators to find garden flowers?

Preparation 4

1. Pictures and background information for pollinator sight presentation
2. Toilet paper rolls
3. Contact paper
4. Tissue paper

Overview 5: The class will explore what colors certain pollinators are attracted to and play a pollen game.

Objective 5: For students to understand what flower qualities draw in pollinators.

Introduction 5 (10 minutes)

1. Brainstorm with the class what flowers they know are visited by garden pollinators: What types and colors of flowers are visited by bumblebees, hummingbirds, mason bees, and honeybees? Are there certain shapes that they prefer?

2. How do these pollinators spread pollen when visiting flowers? What methods are accidental or intentional?

Activity and Assessment 5 (35 minutes)

1. Students will play a game where they are each given small pieces of crumpled yellow paper and asked to put their "pollen" on different flower pictures around the room as they visit them. They can choose whatever flowers they want to pick to put some or all of their pollen pieces on.

2. When every student has "awarded" their favorite flowers with pollen, the class will gather back together and see which flowers have the most pollen or have won the game.

3. Have the class explore why these flowers were the students' favorites. What characteristics did the flowers have that attracted students to them? Discuss why the flowers were chosen, comparing their reasoning to why pollinators prefer certain flowers in a garden according to color and shape.

4. At the end of the activity, the students should understand that pollinators are attracted to flowers for similar reasons as the children, such as color, shape, and size. And now that the students know what colors and shapes certain garden pollinators prefer, they can plant flowers with these qualities in the garden.

Preparation 5

1. Yellow paper
2. Flower pictures

Resources

- Michael Harrap. "The diversity of floral temperature patterns, and their use by pollinators." (online—see Appendix B).
- T. Kleist. "Bee Navigation: The Eyes Have It." (online—see Appendix B).
- S. Papiorek et al. "Bees, Birds and Yellow Flowers: Pollinator-Dependent Convergent Evolution of UV Patterns." (online—see Appendix B).
- PBS. "How Bees Can See the Invisible." (online—see Appendix B).
- Klaus Schmitt. "Photography of the Invisible World." (online—see Appendix B).
- Rose-Lynn Fisher. *Bee.*

6 Citizen Scientists

Time Frame: 45 minutes
Overview: Students will observe the pollinators that visit certain flowers in the garden.
Objective: To use scientific observations to discover which color flowers different pollinators prefer.

Vocabulary

1. Citizen
2. Scientist
3. Survey

I recommend using the Xerces Society Bee Monitoring Tools for this activity and any other extensions from it. The resources they provide are accessible to many ecosystems and provide a basic structure for identifying pollinators, as well as flying patterns and differences in evolutionary traits. While the pollinators of each ecosystem will differ, the techniques of identifying and monitoring them are common. This activity is just a small taste of potentially larger units and studies that can occur in every class and age group that wants to pursue the science behind insect monitoring.

Introduction (10 minutes)

1. Brainstorm with the class what kinds of pollinators and flowers they expect to see growing and active in the garden: What does counting and observing their numbers tell students about the health of the garden and pollinator habitat at school? What pollinators can students identify, and what features do they use to discern them from others?

2. What does it mean to be a citizen and a scientist? How does a scientist take a survey? What behaviors can students use to make quality observations in the garden?

3. What environmental factors, such as weather and season, could affect pollinator activity? What student activity or behaviors could also impact pollinator activity?

Activities (30 minutes)

1. Before heading out to the garden, have the class collect their worksheets, pencils, and clipboards and then come back together to fill out the first section together. The students will record the weather conditions, temperature, date, and site locations while discussing what impact these variations can have on pollinator species. Each group will be assigned a pollinator species to monitor and should review the worksheet before engaging in the activity.

2. Individually or in pairs, the students will explore the garden and survey pollinators they see at work. They will visit different flower colors and count how many of their pollinators visit the flowers for a certain amount of time. Then, they will record their data on the Citizen Scientists: Pollinator Monitoring worksheets.

3. A teacher or adult volunteer can be the timekeeper for this activity and let the students know when to move to a new flower. Students should be sure to visit different colors of flower, but if there is enough time, they can revisit each flower color again to compare their population counts.

4. The class should practice teamwork in making their observations and act with respectful garden behavior when interacting with one another and observing the insects.

5. After they have accomplished their survey, the students will gather back together to share their findings and discoveries.

Assessment (5 minutes)

1. What type of pollinator did the students see that were most active? Which flowers or flowers colors were they visiting? What flowers were the most popular among all the pollinators?

2. Were there pollinators that the students weren't familiar with in the garden? How many of each pollinator did they find?

Preparation

1. Citizen Scientist: Pollinator Monitoring worksheets
2. Clipboards
3. Pencils

Resources

• Xerces Society. *Xerces Bee Monitoring Tools.* (online—see Appendix B).

NGSS and Activity Extensions

In-class studies can extend into research and observations on plants and animals that compare to the diversity of life and relationships in various ecosystems (2-LS4-1: Biological Evolution).

7 Planting Colors for Pollinators

Time Frame: 45 minutes

Overview: Using what students know about pollinators and their preferred flower colors, the class will plant a variety of flowers for the garden pollinators.

Objectives: To learn about garden flower varieties that encourage pollinator activity and provide forage.

Introduction (10 minutes)

1. Brainstorm with the students about the flower colors their pollinators preferred during the games and activities of the previous lessons. Is it easier for pollinators to find the flower colors they prefer if they are grouped together or spread apart?

2. What garden flower colors do the students know pollinators like (red, purple, pink, white, yellow)? Sunflowers, salvia, marigolds, calendula, and bean plants can provide flowers in this color range. For the planting activity, students should be able to recall how deeply to plant the seeds before they engage in the activity.

Activities (30 minutes)

1. Individually or in pairs, the students will plant a variety of flowers for pollinator forage in designated spaces of the garden or in containers prefilled with soil. The students should focus on carefully planting the seeds and making sure that there is soil-to-seed contact.

2. The students can choose from a variety of seed types, especially with the colors that their favorite pollinator prefers. When the students have planted their seeds and pressed them into the soil, they can water them with a watering can.

Assessment (5 minutes)

1. What flowers did the students plant? What pollinators do they expect will visit them? When the students have finished their activity and reflection, then they can enjoy a taste test of garden foods.

Preparation

1. A diversity of flower seed varieties
2. Watering can

8 The Insect Hotel

Time Frame: 45 minutes

Overview: Students will help assemble an insect hotel by gathering different natural materials that provide diverse pollinator habitat.

Objective: To explore practical ways to build

pollinator habitat and learn what certain pollinators prefer to nest in or lay eggs inside.

Introduction (10 minutes)

1. Brainstorm with the class what they know about pollinator and beneficial insect habitat: What do beneficial insects need to thrive in the school garden? Why is it important to encourage their populations and health?

2. What is a hotel? How can students build a hotel for insects? What will a habitat space look like considering the diversity of pollinator and beneficial insect species?

3. Introduce the students to the activity of the day and the materials they will be working with. Demonstrate the best way to bundle the materials and stack them in the insect hotels. What does teamwork look like during this activity?

Activities (30 minutes)

1. In small groups, the students will gather materials such as bamboo sticks, dried leaves, bark pieces, pine needles, dried seed heads, and pine cones to bundle and stack in their insect hotels. These materials can be gathered around the schoolyard or from provided materials. The students should be sure to harvest only dead, dried, or fallen materials, rather than pick fresh greens, sticks, or leaves that may rot.

2. The groups should focus their work as a team, taking turns as they construct insect hotels, and focusing on a collaboration and craftsmanship to fill up their whole box with insect habitat and nesting materials.

3. When the groups have finished assembling their insect hotels, the class can install them at the designated areas in the school garden or campus.

One group's insect hotel uses pine cones, grasses, and a simple wood frame.

This reusable insect hotel is well-crafted and mounted in a school garden.

4. At the end of the activity, the students can explore the garden and harvest a seasonal spring taste test.

Assessment (5 minutes)

1. Encourage the students to double-check their work. Are the nesting sites secure in the box and not bundled too tightly or too loosely?

2. What types of beneficial insects will use this habitat? What will they eat in the garden?

> There are a variety of ways to accomplish this activity. Many people use pallets or larger wooden structures to build a large insect hotel. The method I describe here is one that has worked well for me in the past—using drawers, approximately 8 × 14 inches, which I secure on fence posts around the garden. A staff member once made similar-sized boxes from old fencing and put a shelf in each to provide layers for the insect hotels. It worked very well too, as has hanging old coffee cans.

Preparation

1. Wooden boxes
2. Bamboo sticks cut to size
3. Dried leaves
4. Bark pieces
5. Pine needles/cones
6. Dried seed heads (umbels)
7. Wide wooden sticks with holes drilled in
8. String
9. Scissors

Resources

- Bob Borren. "Building an Insect Hotel." (online—see Appendix B).

9 Honeybees

Time Frame: 45 minutes

Overview: The class will explore honeybee social structure, behavior, and learn from a beekeeper what it means to care for bees.

Objective: For students to value honeybees and their importance to humans and gardens.

Introduction (10 minutes)

1. Brainstorm with the class what they know about honeybees and how they have interacted with them in the garden: What flowers and colors do honeybees prefer? Where do they live?

2. Using pictures or posters of honeybee hives, introduce the students to the social dynamics of bees, the different roles bees play in the hive (queen, worker, drone), and how the bees gather pollen and make honey (see Resources).

Activities (30 minutes)

For taste testing, the students can compare the different flavors of honeys, such as wildflower or clover (not artificially flavored). Are there some types of honey that are darker or sweeter than others? How many worker bees does it take to make a jar of honey? How is honey considered medicine as well as food?

Assessment (5 minutes)

1. What questions do students have about bees and beekeeping?

2. How can students care for honeybees and other bee varieties by providing habitat and forage for them?

Preparation

1. Beekeeper

There are many local beekeeping organizations that can support honeybee education in schools—reach out to your local chapter or community member and ask if the class can interview a beekeeper and ask them to bring in some supplies, such as suit, gloves, mesh hat, pry tool, smoker, burlap, old combs, dead drones, dead workers, and raw honey.

I enjoy adding this activity into pollinator studies because most students are familiar with this type of bee, or its product, and honeybees have had a vital role in human history and agriculture. My goal with this activity is to impart the wonder and incredible value of these insects.

2. Beekeeping supplies
3. Honey samples
4. Spoons or tasting sticks

Resources

• Girl Next Door Honey. "Educational Bee Posters." (online—see Appendix B).
• The Honeybee Conservancy. (online—see Appendix B).

10 Garden Celebration

Time Frame: 45 minutes
Overview: As a culmination of their garden activities, the students will harvest foods from the garden and eat the produce together.
Objective: To celebrate all the students have learned and accomplished in the garden.

Introduction (5–10 minutes)

1. For the garden celebration, the class will be harvesting certain ripe garden foods for taste testing. Review the methods and careful harvesting techniques that the students know about while introducing the types of plants they will be harvesting.

2. Where in the garden do these foods grow? What steps must be done to ensure plant safety in harvesting and human safety before eating?

Activities (30 minutes)

1. In pairs, the students will be given an identification card of a certain edible food in the garden. They will have to find this item and pick the amount that their card indicates. They will bring these pieces to the washing station, or school cafeteria, and wash with adult supervision.

2. When all the food items have been collected and washed, the class will be able to pick from the buffet table of taste tests. They will be encouraged to discuss the food and what they have learned about seeds, pollinators, and taste testing from throughout the year.

Assessment (5 minutes)

1. What new facts and activities did students learn and do this year that they hadn't done or learned before?

2. What were their favorite memories in the garden? What are their favorite garden foods that they taste tested?

THIRD GRADE—GRADE THREE

BECOMING SOIL SCIENTISTS

This year, the students will delve deeply into the heart of the school garden. With soils as their focus of study, the class will explore living and nonliving components that work together to create a healthy and thriving soil system, from the valuable decomposers to the minerals and nutrients that feed plant roots. Students will map out the soil ecosystem and learn to value soil for the natural resource that it is.

> "The real work of creating soil, structure and habitat is done by the organisms that live in the soil. Our job is to support them, not to put them out of work."[1]

In the fall, the class will participate in seasonal garden activities and keep in mind how their work contributes to the health of the garden soil. They will identify the decomposers,

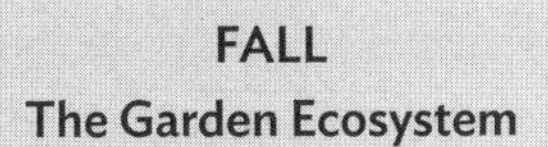

1. Garden Exploration
2. Sunflower Seeds
3. Mulching and Compost
4. Living and Nonliving Relationships
5. Exploring Insect Life Cycles
6. Insect Traps
7. Insect Exploration

8|9. Decomposers, Consumers, and Producers

10. Valuing the Garden Ecosystem

WINTER
Soil Vitality Experiment

1. The Soil Under Our Feet
2. Percolation Tests
3. Potting Soil for Seeds
4. Soil Vitality Experiment
5. Soil Colors as Scientists
6. Soil Colors as Artists
7. pH Testing
8. Organic Fertilizers
9. Plant Growth
10. Transplanting Starts

SPRING
Fostering Healthy Soils

1. Creating Compost
2. Green Manure
3. Manure: Full Cycle Gardening
4. Hügelkultur Soils
5. Planting Seeds for Summer
6. Foraging for Wild Edibles
7. Planting in Layers
8. Companion Planting
9. Nutritious Soil and Plants
10. Garden Journals

consumers, and predators at work in the garden before transitioning to an in-depth soil study during the winter months. Some of the activities will be repeated when they study composting in the fifth grade, but this repetition will reinforce the value of these skills and offer more opportunities to engage with complex ideas. In the spring, the students will return with their newly developed ideas and apply them in the garden through creative and organic gardening methods. They will catch and store nutritional energy in the garden soils and use small solutions to make the garden healthier. Through their hard work and scientific study, the garden will be prepared for the heat of summer, retain greater moisture, and grow nutritious plants to feed the community. After this year, the students will know what it means to be a steward of the soil.

This is the first year that the students will create a Garden Journal, as the final activity. They will continue this tradition into the sixth grade. In preparation for this end-of-the-year activity, I like to prepare designated folders that each student can use to collect their work. The children will be amazed to look back on their progress and to see how much they've learned.

© LIGHTFIELD STUDIOS / Adobe Stock

Next Generation Science Standards

What an exciting year for students to have quality scientific experiences and develop their science skills and mind-set in the garden. In their studies, the class will engage in a detailed exploration of the living aspects in the soil that will expand their understanding of the garden as a system of thriving, living creatures inhabiting the nonliving components that support life; develop arguments using evidence from their experiments and making recommendations as knowledgeable scientists and gardeners on how to create and nourish healthy soil systems; and apply organic solutions for gardening issues, such as soil leaching and erosion.

Themes for classroom extensions in reading, writing, storylines, and math include life cycles, changing ecosystems and diversity, living and nonliving relationships, Earth sciences in soil structure (Earth's crust), soil types in ecosystems across the world, and seasonal weather events and changes in the local landscape.

Permaculture Principles

- Observe and Interact
- Catch and Store Energy
- Use and Value Renewable Resources and Services
- Integrate Rather than Segregate
- Use Small and Slow Solutions
- Use and Value Diversity
- Use Edges and Value the Marginal
- Creatively Use and Respond to Change

Fall: The Garden Ecosystem

1 Garden Exploration

Time Frame: 45 minutes
Overview: Students will embark on a scavenger hunt in the garden and review their past lessons on soil, living organisms, and plants.
Objective: To observe the changes in the garden and discover interesting plants and insects.

Introduction (5–10 minutes)

1. Review expectations for behavior in the garden with the students before they head out into the garden. How do students treat each other, themselves, and the plants and animals they encounter?

2. What are some of the flower seeds and vegetable starts that students planted in the garden during the spring? Where did they plant these, and what types of pollinators should be visiting them?

Activities (30 minutes)

1. Each student will get a copy of the Garden Exploration sheet and find a classmate with which to explore the garden. These partner groups can work on sharing the parts of the garden they like and know, as well as checking each other's spelling and growth of the plants they began in the springtime.

2. When they have completed their exploration, the students can enjoy a taste test from seasonal garden foods before gathering back together to share their findings.

Assessment (5 minutes)

1. Students can share what they discovered in the garden and review the names of insects and plants that they learned in the previous year.

2. What plants or discoveries did the groups have that amazed or excited them? How well did the flowers and plants they sowed in the spring grow?

Preparation

1. Garden Exploration worksheet
2. Clipboards
3. Pencils

2 Sunflower Seeds

Time Frame: 45 minutes
Overview: Students will learn about edible sunflower seeds by harvesting them from the garden.
Objective: To make observations about the structure of sunflower seeds and seed heads while discussing what they provide for the garden ecosystem.

Introduction (10 minutes)

1. Brainstorm with students what they know about sunflowers and their seeds: Is a sunflower seed a living thing? How does a sunflower seed form? What other creatures feed off sunflowers? What other seeds do students find in the garden? Why are these plants called sunflowers?

2. Demonstrate to the students how they will be harvesting the seeds from the sunflower heads and where to store the seeds. Why is it important to harvest seeds and save them for spring planting?

Activities (30 minutes)

1. In small groups, students will receive a sunflower head with ripe seeds and carefully remove all the seeds. They will sort the seeds into a bowl and clean out any plant debris or insects that they may have collected. Later, the class can combine their seeds in a paper bag or glass jar that is labeled to save for the spring and summer.

2. The groups should take their time and work as a team to accomplish their goal.

3. When all the seeds have been collected, the groups will compost any remaining plant matter. They can also dissect the remaining pieces and explore the textures and parts of a sunflower.

4. The class will finish with a taste test from the garden, as well as a sample of raw sunflower seeds and roasted sunflowers seeds.

Assessment (5 minutes)

1. Gather the students back together and let them enjoy a taste test of sunflower seeds. Have them focus on the differences between the shell and seed. How would they describe a sunflower seed's taste and texture? What are the flavor differences between raw seeds and roasted ones?

2. What were student experiences as they removed the sunflower seeds from the plant head? Was it difficult or easy? Did they notice patterns or parts of the plant that they couldn't identify? What was the texture of the seed heads?

Sunflower seeds offer a beautiful visual of the Fibonacci sequence.

Preparation

1. Clippers to chop down sunflower heads
2. Pre-picked sunflower heads
3. Prepared sunflower seeds for taste testing

3 Mulching and Compost

Time Frame: 45 minutes

Overview: Students will use green and brown materials to add long-term compostable materials to the garden beds.

Objective: To apply mulch directly to the garden beds and learn how to encourage a healthy soil system.

Introduction (10 minutes)

1. Brainstorm with the class what they know of mulching: What kinds of mulch have students laid in the garden before? What does it mean to mulch or compost?

2. What are brown or green materials in the garden? How do dry or brown materials like straw or leaves help the garden soil and ecosystem? Review with the class the importance of brown and green materials in creating garden soil, by providing healthy habitat for decomposers, nutrients to the plants, and suppressing weeds.

3. If possible, have students dig in the garden beds for evidence of past mulching activities. What materials did they find, and how well decomposed were the pieces?

4. Before beginning the activity, establish with the class what behaviors the students should demonstrate during the activity. How can they Care for Self, Others, and the Land while completing their task?

Activities (30 minutes)

1. Individually, students will gather the dry or brown ingredients provided, lay them out over the garden soil or designated beds, and pay attention to any living plants and sprouts in that space. The activity can also be accomplished in containers. Encourage students to take small handfuls in order to practice good craftsmanship and to work on one garden bed at a time.

2. The class can also mix green materials into the garden beds, such as sunflower stalks and fresh organic materials that have been chopped down. Then, the students can gather prescreened compost and lay approximately 1 inch of compost across the organic matter they laid out, in order to keep the material from being disturbed over the winter and to increase soil health.

3. After the activity is completed, have students go back through the garden individually to fix any messy work, such as mulch on living plants or pieces that have fallen into the pathways.

4. The class can enjoy a seasonal taste test from the garden after they finish mulching and their quality control inspection.

Assessment (5 minutes)

1. When students were mulching, did they see evidence of other past mulching activities? What materials were being used?

2. What garden plants are ready to be mulched into the beds, and what plants are still growing and thriving?

Preparation

1. Large piles of leaves or other brown organic matter
2. Prescreened compost

Resources

• Toby Hemenway. "The Ultimate Bombproof Sheet Mulch" in *Gaia's Garden: A Guide to Home-scale Permaculture*, 2nd ed.

4 Living and Nonliving Relationships

Time Frame: 45 minutes

Overview: Students will explore the soil ecosystem, practice their observation skills, and identify the nonliving and living items they find in the garden.

Objective: To learn that soil is a living thing comprised of living and nonliving elements and to understand this vital balance in the garden.

Vocabulary

1. Ecosystem
2. Microorganism

Introduction (10 minutes)

1. Have the students review what types of nonliving, living, decomposing items they might find in the garden. What are scientific and student definitions of living and nonliving? What are microorganisms?

2. What is a soil ecosystem? Why is it so important to have all of these elements interacting, or in relationship, with each other in the garden ecosystem?

4. How does a soil scientist behave? What behaviors will students exhibit when interacting with the living things in the school garden? How will they treat the tools they use?

> "In a single teaspoon of fertile soil... live 4 billion bacteria, up to 325 feet of mold or fungus filaments, 144 million actinomycetes."[2]

Activities (30 minutes)

1. In pairs, the students will use magnifying glasses to explore a selected spot of the garden and record the living and nonliving things they find on the Life in Our Garden's Soil worksheet. In their plot, the groups will use keen observation skills to identify and record what is living and nonliving in, on, and under the soil, rocks, logs, and mulch.

2. The students can also dig their hands into the soil to look for living and nonliving items. They can interact with the plants and materials in their plot by picking up leaves, sticks, clumps of soil.

> I have noticed that students get very excited about the living things in the soil, but forget to record the nonliving items. They may need time to go back over their space to record these items.

3. Before the activity time is over, the groups can wash their hands and harvest seasonal taste tests from the garden. Then they can gather together to share their findings.

Assessment (10 minutes)

1. What kinds of living things did students find, and how were they interacting with nonliving items?

2. What types of living things did the students discover? Were there creatures that the groups couldn't identify? Did any groups find seeds in their plot? Is a seed living or nonliving?

4. What other creatures and microorganisms could the students have interacted with but not seen? Why are these relationships between living and nonliving elements important to an ecosystem? Do ecosystems ever stay stagnant or are they dynamic and changing? What roles do balance and resiliency play in ecosystems?

Preparation

1. Life in Our Garden's Soil worksheet
2. Magnifying glasses
3. Clipboards
4. Pencils

Resources

- James Hoorman and Rafiq Islam. "Understanding Soil Microbes and Nutrient Recycling." (online—see Appendix B).
- "Field Guide to Soil Food Webs." Soil Science Society of America. (online—see Appendix B).

Activity Extensions

Further in-class studies can include Earth sciences, geology, and the origins of human agricultural history.

5 Exploring Insect Life Cycles

Time Frame: 45 minutes

Overview: Students will search for insects in different stages of life using a garden map.

Objectives: To practice reading a map and to correctly identify the life stages of different insects.

Introduction (10 minutes)

1. Review with the students what insect life cycle stages they are familiar with or remember from past lessons, such as egg sac, chrysalis, adult, juvenile, and cocoon. What do these life stages look like for common garden insects?

2. What behavior expectations will students

practice during an insect hunt? Where have they experienced or found a lot of insects in the garden? Introduce the students to the map of the garden, asking them to point out the tools on the map that will help guide their exploration or points of interest they see.

> In the past, I have used a garden map that I sketched for students to use. Satellite photos can also work if the details are clear. For greater in-class studies, the students can create their own maps of the garden to use for this activity.

Activities (30 minutes)

1. In pairs, the students will look for insects in the garden and identify and label the type of insect and its life stage on their garden map, creating a colorful guide for insect exploration. The groups will travel through the entire garden and mark points of interests and life stages as they go. They will practice collaboration and make sure the map can be read by other students. They can use provided resources and guides to assist in identifying the insect species.

2. Before they gather back together to share their findings, the groups can pick a taste test from the garden. Then, they can share their maps with another group and, as a class, discuss the insects they encountered by engaging in a short survey and share time.

Assessment (5 minutes)

1. Did students have difficulty identifying an insect's life stage? Were there more of one life stage in the garden than another? What strategies did the groups have for finding their way around the garden and correctly identifying where they found the insects?

2. How many different insects did the class find? How many of each type did they count? What does this small survey of population and diversity say about the health of the garden?

Preparation

1. Maps of the school garden
2. Clipboards, pencils or markers

Resources

- Ross H. Arnett and Richard L. Jacques. *Simon & Schuster's Guide to Insects.*
- Whitney Cranshaw and David J. Shetlar. *Garden Insects of North America: The Ultimate Guide to Backyard Bugs*, 2nd ed.
- Eric Eaton and Kenn Kaufman. *Kaufman Field Guide to Insects of North America.*
- Arthur V. Evans. *National Wildlife Federation Field Guide to Insects and Spiders & Related Species of North America.*

NGSS and Activity Extensions

Greater opportunities for student engagement can include developing models of how organisms have varying and unique life cycles, but all reproduce, grow, and die (3-LS1-1: From Molecules to Organisms).

6 Insect Traps

Time Frame: 45 minutes

Overview: Students will create insect traps for a closeup study of the living creatures in the garden.

Objectives: To continue building student knowledge of the garden soil web and encourage appreciation for the creatures that live in the soil.

Introduction (10 minutes)

1. As a class, students will review the parts of the soil ecosystem using a poster diagram or classroom chart. They should be able to

A student shows off their thoughtfully crafted insect trap.

brainstorm the nonliving parts of the soil ecosystem that they have already learned about and identify the living components as well.

2. What is the soil ecosystem? What key players have students learned about that impact the health of the garden? What are nonliving elements that can be found in the soil? How are they important for the living creatures? What are living elements that can be found in garden soil? What happens to the health of the garden when some of these creatures are removed?

3. In preparation of the activity, discuss the ethics of handling and capturing insects and set standards for student interaction (nonharmful practices). What does it mean to observe?

4. Have the students hypothesize what types of insects they expect to find in their traps. Record their hypotheses on a classroom chart.

Activities (30 minutes)

1. In pairs, students will design a method for trapping insects in a nonharmful way, using the provided materials and what they find available in the garden. The traps should fulfill simple guidelines and pass inspection before they are finished.

Guidelines:

a. Insect traps should not harm the insects in any way. The goal is to temporarily trap them for observation.

b. Traps should include habitat, shelter, and food.

2. In the final five to ten minutes of the activity, the students will secure their traps in the garden and make any final additions and changes.

Assessment (5 minutes)

1. How did the students work together to build their trap? How did they share their ideas and build something constructive?

2. Why types of insects and bugs do the students expect to find and catch?

Preparation

1. Yogurt cups
2. Scissors
3. Assorted recycled materials
4. Markers
5. Cotton balls
6. Sugar water

Resources

• "Safe Bug Trap." Easy Outdoor Activities for Kids, How Stuff Works. (online—see Appendix B).

7 Insect Exploration

Time Frame: 45 minutes

Overview: Students will check on their insect traps and record their findings, doing their best to identify the creatures they have captured or found near their traps.

Objectives: To value and observe the living components of the soil web by identifying common garden insects.

Introduction (5–10 minutes)

1. Before checking on their traps, the class will review the expectations in handling the creatures: What behavior will the students have when interacting with insects or bugs? How will they touch, treat, and respect them?

2. How will the students gently return the insects to the garden at the end of the activity?

Activities (30 minutes)

1. Using the Insect Exploration worksheet as a guide, the students will check their insect traps and record, describe, and identify the bugs and insects they find. The goal should be to gather detailed observations from the insects in the trap, rather than count the quantity of insects and bugs. Students are encouraged to look for small details that distinguish some of the creatures from others. How many legs or eyes do they have? How are they similar or different? Do they have antennae? How many legs or wings do they have?

2. If the students struggle to identify the insects they have trapped, then they can use the available resources and references to assist them. When students have finished their worksheet and exploration, then they can release the insects they trapped and dissemble their traps as necessary, making sure to reuse or recycling as much as possible.

Assessment (5 minutes)

1. Gather the students back together and have them share their findings with other students. Did the groups find similar creatures? What was the most exciting insect discovery they made?

2. Conduct a quick survey on the different insects and bugs the students found and how many they caught. What surprising details did students notice about these insects? Are these insects and bugs the ones that the class hypothesized they would encounter?

Preparation

1. Insect Exploration worksheets
2. Magnifying glasses

Resources

- Ross H. Arnett and Richard L. Jacques. *Simon & Schuster's Guide to Insects.*
- Whitney Cranshaw and David J. Shetlar. *Garden Insects of North America: The Ultimate Guide to Backyard Bugs*, 2nd ed.
- Eric Eaton and Kenn Kaufman. *Kaufman Field Guide to Insects of North America.*
- Arthur V. Evans. *National Wildlife Federation Field Guide to Insects and Spiders & Related Species of North America.*

8|9 Decomposers, Consumers, and Producers

Time Frame: 1½ hours (2 × 45 minutes)

Overview: Students will learn new terminology in identifying and understanding key garden characters and relationships.

Objective: To explore the soil ecosystem relationships and create portraits of different living creatures.

Introduction

1. As a class, create a diagram of the garden community, including examples of producers, consumers, and decomposers. Have the

A banana slug feeds on another decomposer.

students brainstorm what roles the decomposers, consumers, and producers play in the garden ecosystem. What creatures and insects in each category have the students seen in the garden during their recent activities?

2. How can students identify decomposers, consumers, and producers in the garden? What behaviors or habitats can they look for to help them identify these creatures?

3. Discuss the importance of doing careful drawings and during these activities what scientific observations look like. How does a scientist record data and make observations?

Activities

1. To the best of the student's ability, have them use the provided worksheet to name and illustrate a decomposer, a consumer, and a producer from the garden by paying special attention to the subject's anatomical details and behavior. They can also make observations about the environment around the creature and label the parts of the insect or animal that they are drawing.

2. If students have extra time, they can write a short poem about the daily life of one of these creatures.

Assessment

1. Review student drawings and provide time for students to go back over their drawings if craftsmanship is needed.

2. Have students share their findings and drawings with another student, while focusing on active listening and asking questions about other student work.

3. Did any students see their producer, decomposer, and consumer in action? Were there any creatures that played more than one role?

Preparation

1. Prepared poster of the three different categories within the garden community
2. Decomposer, Consumer, Producer worksheets

Resources

- "Producers, Consumers, Decomposers." PBS Learning Media. (online—see Appendix B).
- USDA Natural Resources Conservation Service. "Soil Biology." (online—see Appendix B).
- "Nature's Recyclers" in Project Learning Tree. *PreK-8 Environmental Education Activity Guide*, 11th ed.

NGSS and Activity Extensions

For greater in-class immersion and activity extensions, the students can develop an argument that some animals create groups to aid in survival (3-LS2-1: Ecosystems); gather and analyze data on how offspring inherit traits from their parents and that variations of these traits exists within a group of organisms

(3-LS3-3: Heredity); explain how genetic traits can be impacted and altered by environments (3-LS3-1: Heredity); and develop an argument, with evidence, that in certain habitats, some organisms can survive, some survive less well, and others struggle to survive (3-LS4-3: Biological Evolution).

10 Valuing the Garden Ecosystem

Time Frame: 45 minutes
Overview: Students will play a game that explores the relationships, balance, and resiliency within the garden ecosystem.
Objective: To learn that each element of the garden ecosystem has a valuable role in the garden.

Introduction/Activities (35 minutes)

1. Each student will be given a Garden Ecosystem card with an organism or item on it. Each card shows an element or member of the garden ecosystem which relates to other cards in the group. The class should sit in a circle and keep their card a secret until each clue is presented.

2. Read each clue aloud and have the class deduce the answer, one at a time. The answer to each question is one of the ecosystem cards in the group. When a clue is answered, the person with the answer card will get to hold onto a piece of a yarn. As each clue is discovered, these pieces eventually create a web in the circle of students.

3. As the game goes on, brainstorm with the students what they notice and have discovered about ecosystems so far. What elements in the web are missing? Are there patterns that students notice in the relationship between the ecosystem members?

4. When the web is complete, introduce environmental changes into the balance of the garden ecosystem web and allow the web to become uneven. For example, if pesticides are sprayed on the garden and the insects die, then what will happen to the web elements that rely on insects? Do all the insects die? Allow for student reflections on the visual effect of these changes and what they can predict about the actual effect of these changes on the garden ecosystem, in terms of balance, stability, and resiliency.

Assessment (10 minutes)

1. What creatures benefit the most when certain ecological changes occur? What happened to the garden ecosystem when one species was eliminated? When would this happen in a garden?

2. What connections can students make to the garden ecosystem web and the soil ecosystem? Are there more relationships that have not been mentioned during this activity?

> **"Soil is the habitat living organisms have synergistically created for themselves."[3]**

3. What role do humans play in the ecosystem? What were students' experiences during the activity? What changes did their characters undergo?

Preparation

1. Yarn
2. Garden Ecosystem cards

Resources

- "Field Guide to Soil Food Webs." Soil Science Society of America. (online—see Appendix B).

• Inspired by the "Family Activity: Web of Life." Project Learning Tree. (online—see Appendix B).
• Heide Hermary. "Soil in an ecosystem" in *Working with Nature: Shifting Paradigms: The Science and Practice of Organic Horticulture.*

NGSS and Activity Extensions
Further in-class studies could include research into the origins of a problem caused by environmental changes that affect the types of plants and animals that live there, as well as supporting a solution to this problem (3-LS4-4).

Winter: Soil Vitality Experiment

1 The Soil Under Our Feet

Time Frame: 45 minutes
Overview: The students will observe and examine soil samples for various colors and their corresponding mineral or material content.
Objectives: To explore the soil composition in the school garden and develop the students' observation and science skills.

Vocabulary
Particle Size

Students work close to the soil to record their observations on soil composition.

Introduction (10 minutes)
1. Brainstorm with the students about what they know about soil and what types of soil, or soil materials, they have heard about: What living and nonliving items make up soil? How is soil made? What binds soil particles together?
2. What soil colors have students seen in natural spaces such as swamps, riverbanks, and gardens? What does nutrient-rich garden soil look like?
3. When talking about soil, what does *particle size* mean? What do different soil colors say about where the soil came from or its composition? How would a scientist explore soil samples and ask questions about where it came from and what it contains? How do all the pieces of the soil ecosystem support the whole system?

> "Soil is the result of interaction between three equal partners: the a-biotic components of the soil, living organisms, and environmental components (temperature, water, air)."[4]

Activities (30 minutes)
1. In pairs, students will work with a sample of soil that they have brought to share or have

gathered from the garden. They will explore the particle size, colors, and recognizable substances in the soil, recording their data on the Soil Investigations worksheets. While filling out their worksheets, the students should focus on making scientific observations, use clear handwriting, and write complete sentences.

2. After 10–15 minutes, have the students analyze a new soil sample and continue listing and observing recognizable materials in the soil, as well as color and particle size.

3. At the end of the activity, the student groups can write their final observations of the differences between their two samples on their worksheets.

Assessment (5 minutes)

1. Gathering back together, the students should be able to thoroughly explain how they investigated the soil sample. What clues did they look for and how detailed were they in their descriptions? What recognizable non-living materials did they find (moss, sand, clay, pine needles)? What do these materials say about where the soil was found?

2. What were materials that students could not identify or had questions about? How would a soil scientist attempt to answer these questions? What living elements or evidence of once living things did they found in the soil?

Preparation

1. Soil samples
2. Soil Investigations worksheets
3. Magnifying glasses
4. Observation trays

Resources

- Dig Deeper. "Soil Experiments and Hands-On Projects." The Soil Science Society of America. (online—see Appendix B).
- "The Color of Soil." University of Illinois Extension. (online—see Appendix B).

2 Percolation Tests

Time Frame: 45 minutes

Overview: Students will learn about the drainage properties of different soil textures and particle sizes.

Objective: To explore how different soils and particles sizes change the way water drains through the soil.

Vocabulary

1. Percolation
2. Drainage

Introduction (10 minutes)

1. Before the activity, lead the class in a group brainstorm: Why would gardeners and scientists want to know how water moves through the soil? What hypothesis do students have about how fast water will move through fine particles and large particles? What observations have students made about how water moves across or into soil, such as on rainy days?

2. Introduce the students to their activity and demonstrate the process for testing water percolation in soils. Discuss the goal of the activity in determining the rates that various soils percolate water.

3. What does positive group work and teamwork look like during this activity? How can every group member participate and play an important role?

Activities (25 minutes)

1. In small groups, students will conduct percolation experiments using four different soil samples in cups with holes in the bottom or in upside-down soda bottles (see Resources). Carefully, and working as a team, they will pour a preset amount of water in the soil sample and use a stopwatch or clock to measure the amount of time it takes for the water to pass through the soil. They will use a large classroom poster to record the amount of time, in minutes and seconds, for each soil sample test.

2. Students will practice patience, teamwork, and communication during this activity.

3. If there is extra activity time, have the students repeat the experiment with one or two of the soil samples, just as a scientist would do, to compare their data.

4. When the groups have finished their recordings, they can clean up their stations and return the tools that they used.

Assessment (10 minutes)

1. Bring the students back together and have them analyze and make observations on the data they gathered. What story does the data tell the class about how water moves through different soils? Are percolation times similar between groups? What numbers stand out to students?

2. For the groups who had extra activity time and repeated the experiment with the soil samples, were the times recorded similar or different from each other? What factors and variables could contribute to differing times?

3. What does the data tell the class about the soil samples themselves? What would a gardener want to know using this data? Are there ideal particle sizes for garden soil?

4. What would students do differently if they were to design their own project to study water percolation and drainage?

Preparation

1. Four different soil samples for small student groups
2. Timers/stopwatches/clock
3. Poster for recording data
4. Cups for water with premade line of measurement
5. Bowls or plates to catch the draining water

Resources

- "Soil Percolation Test" in Project Learning Tree. *PreK-8 Environmental Education Activity Guide*, 11th ed.
- "Perkin' Through the Pores—Slip Slidin' Away." Agriculture in the Classroom. Utah State University Cooperative Extension. (online—see Appendix B).

3 Potting Soil for Seeds

Time Frame: 45 minutes

Overview: Students will mix available organic materials to create their own potting soil.

Objective: To illustrate student knowledge of how large and small soil particles help drain water at the right speed for optimal plant growth.

Introduction (10 minutes)

1. Brainstorm with the students what they recall about the last lesson on soil particle size and drainage: What do they know about large and small particles? How does water move through sand or clay? Which soil samples had faster or slower drainage?

2. The class will be working in small groups, so students should remember to use kind words and teamwork to accomplish their task.

> With a little preparation, potting soil mixtures can easily be created at school or home. I have used various mixes of screened compost, a leftover bag of perlite, bonemeal, blood meal, and screened topsoil and compost to accomplish the task.

Everyone should participate: dirty hands are a goal of the day!

3. What key things can students look for to tell them when their potting soil is ready to plant in?

Activity (30 minutes)

1. In small groups, students will be mixing together their own potting soil by using different-sized materials and their own knowledge of which soil particles are best for good drainage. Each group will have a bucket to fill and should be able to go from station to station filling up the bucket with materials as they go. One group member will record the amounts of each material on a Potting Soil Recipe worksheet.

2. Students can be encouraged to take one scoop of the materials at a time, mix, and then repeat if they believe it is necessary. They will need to break up any large pieces of soil and remove any large particles that can't be broken down.

> Gardeners want a good balance of large, medium, and small particles in order to allow water to move through a container, but slowly enough that a plant's roots have time to absorb it. If the water drains too slowly, then the plant roots risk being oversaturated, weak, and potentially rotting. Soils that drain too quickly cause unnecessary stress and dehydration to the plant and can damage its growth.

3. The students will determine the best potting soil consistency by examining the feel of the soil. Then, they can take samples of their mixture and pour water through it (in the cups or containers they used for the previous activity) to analyze its drainage quality and adjust the mixture as needed. As a group, the students can discuss: Is the water sitting on top of the soil without draining? Is the water flushing through the soil too quickly?

4. Students should write their names on their potting soil container for future use and clean up their tools and stations when finished.

Assessment (5 minutes)

1. What decisions did student groups make when mixing their recipes? How did they troubleshoot drainage problems or concerns?

2. What part of the activity do the groups feel they did very well (teamwork, collaboration, recipe ingredients)?

Preparation

1. 10–12 empty buckets
2. Materials to make potting soil
3. Tape
4. Markers
5. Potting Soil Recipe worksheets
6. Drainage cups
7. Watering can

4 Soil Vitality Experiment

Time Frame: 45 minutes

Overview: Students will begin their experiment in growing healthy plants by sowing cool-season seeds in their potting soil.

Objective: Students will set up their science experiments and begin to think of variables, a care schedule, and determine flexible roles in their small groups.

Introduction (10 minutes)

1. Brainstorm with the class about what seeds need to thrive and survive: What variables could affect seed germination and plant growth?

2. The students will be planting seeds at the end of the activity, so the class should review how deeply to plant small seeds. What are cool-season crops? What do these plants need in order to thrive, and what makes them unique?

Activities (30 minutes)

1. In their small work groups, students will fill planting containers with their potting soil mixture and make sure to keep the soil fluffy and non-compacted. Then, they will receive cool-season seeds to plant in the containers. Students should write their names and the seed varieties on the containers and water the seeds, as necessary. They can continue planting until their potting soil is used. Some of these seeds will be for student experiments, and the others are for the community and school garden.

2. All the groups should plant the same varieties of seeds and sow two or three different types (cabbage, kale, onions, lettuce). They can record the first stage of the experiment on their Plant Growth Chart worksheets and then clean up their work areas and water the seeds.

3. Before the end of the activity, have the students discuss a care schedule for watering the seeds and record the roles and responsibilities of each group member.

Assessment (5 minutes)

1. How did group collaboration and teamwork help during the activity?

2. How quickly will it take some plant varieties to sprout? What are the care schedules and roles in each group?

Preparation

1. Cool-season seeds (cabbage, kale, onions, lettuce)
2. Small planting containers
3. Tape and markers
4. Plant Growth Chart worksheets

5 Soil Colors as Scientists

Time Frame: 45 minutes

Overview: Students will delve in the mineral composition of different soils and create a visual representation of what they have learned.

Objective: To identify nutrient composition and minerals associated with different soil colors.

Introduction (10 minutes)

1. What do students remember about their previous exploration into soil colors? What soil colors do gardeners strive to create and how would these soil colors vary with regional location? Why would scientists or gardeners want to know about soil color?

2. Introduce the students to a soil color chart. Discuss how various soil colors occur in different places and are composed of a diversity of rocks, minerals, and organic matter. For example, these colors indicate that they contain these minerals and elements:

- Dark brown colors often mean more organic matter
- Red and yellow soil could indicate lots of iron
- White shows calcium in the soil
- Black indicates manganese, sulfur, and nitrogen
- Grey is usually found in acidic, water-logged soils

3. Demonstrate to the students how to fill out the worksheet graphs as they explore their soil samples. How can students use magnifying glasses as scientists would during the activity?

Activities (25 minutes)

1. In their small groups, students will explore three different soil samples and identify the minerals and substances in each. Using a Soil

Spectrum worksheet and a magnifying glass, each student will fill in the colors they see in each sample. They can use colored pencils to fill in the graphs and magnifying glasses to identify colors and minerals.

2. The teams can work alongside each other to fill out the sheet, but should produce their own worksheet by the end of class. This is a small and simplified piece of a more complex scientific system for recording soil colors. The goal of this activity is to simply engage the students into thinking about soil colors, as well as work with a graph form.

3. When the groups are finished, they can check on the development of their growing experiments and record plant or seed changes on the Plant Growth Chart worksheets, before returning to discuss their findings with the class. The small groups can also analyze the color compositions in their potting soil, if they have extra activity time.

Assessment (10 minutes)

1. What soil colors did the students discover in their exploration? Were there colors that they were surprised by or ones they couldn't identify?

2. Using the soil color chart as a reference, what do the soil colors say about the composition of minerals and rocks in each sample?

3. What changes do students notice with their plant experiments? Were any seeds sprouting? Was the soil moist, dry, or wet? What alterations should be made to the small group care schedules?

Preparation

1. Magnifying glasses
2. The Soil Spectrum worksheets
3. Soil samples
4. Colored pencils

Resources

- University of Illinois Extension. "The Color of Soil." (online—see Appendix B).
- Fred Magdoff and Harold van Es. *Building Soils for Better Crops*, 3rd ed. USDA Sustainable Agriculture Research and Education. (online—see Appendix B).

6 Soil Colors as Artists

Time Frame: 45 minutes

Overview: Students will engage with soil color and structure by making paintings from soil pigments.

Objective: To enhance learning about soil colors and composition using art as a medium for exploration.

Introduction (10 minutes)

1. Review with the students what they know about soil colors and what minerals and compositions are indicated by the various shades and hues. What soil colors have students interacted with so far?

2. Introduce the students to soil paintings as an artistic and cultural practice. What colors do students see in these paintings? How did the artists use soil to create a story? Present students with images on the work of soil artists across the world. Emphasize soil as one of the most vital resources in the world, not only for growing food and sustaining life but also as a building material, part of cultural identity, and artistically as a painting medium.

Activities (30 minutes)

1. Student table groups will share a collection of prepared soil pigments from the samples they brought in and others that have been previously prepared or collected. With these pigments, the students will use watercolor paper and paintbrushes to create their own soil paintings.

> Many students are most successful when they have a focus for their painting. They can paint a favorite garden plant or use still life items from the classroom. They can also gather leaves to rub and make green or purple colors. Other students enjoy the chance to paint from their imaginations.

2. Students who finish early will check on the development of their experiments and record plant or seed changes on the Plant Growth Chart worksheets.

Assessment (5 minutes)

1. Before class is over, let the students go through a gallery experience of the class's work and see what other students accomplished with soil pigments.

2. Were there some soil colors and pigments that were easier to paint with than others? What did students notice about the texture and feel of the paints? Did the textures change with the soil colors?

Preparation

1. Soil samples for pigments
2. Water cups
3. Presentation on soil artists
4. Paintbrushes
5. Watercolor paper

Resources

- Soil Science Society of America. "Paint with Soil!" (online—see Appendix B).
- USDA Natural Resources Conservation Service. "Painting with Soil." (online—see Appendix B).
- UN Food and Agriculture Organization. "World Soil Day: Soil Painting Competition." (online—see Appendix B).
- USDA Natural Resources Conservation Service. "Painting with Soil: Jan Lang's Images of the Lewis and Clark Expedition." (online—see Appendix B).

7 pH Testing

Time Frame: 45 minutes

Overview: Students will test the pH, nitrogen, phosphorus, and potassium levels of the garden soil and their group's potting soil.

Objectives: To foster student science skills while learning how soil feeds garden plants in four different ways.

Introduction (10 minutes)

1. Have students ever heard the term pH before? Where have they heard it? What do students know about nitrogen, potassium, and phosphorus?

2. Gardeners care a lot about the pH and NPK levels in the garden. They look at soil as scientists do in order to learn about its quality and condition. Gardeners use this information to inform what plants they should grow in their garden and to determine what types of materials they should add to their soil in order to encourage the levels to be at a more preferable balance.

- pH tells how acidic or alkaline the soil is. Every plant prefers different ranges to grow successfully, and you can change these levels by adding different organic materials.
- (N) Nitrogen impacts how plant leaves grow and how tall the plant is (tomatoes with high nitrogen in their soil will grow bushy and thick with leaves).

- (P) Phosphorus affects the genetic health of the seeds within the fruit and variations in vitamin content.
- (K) Potassium levels often affect the taste and color of a plant's fruit.

3. Demonstrate the activity and testing procedures. Why is it important for scientists to perform multiple tests on a single sample of soil?

Activities (30 minutes)

1. The students will be performing pH and NPK tests using soil they have collected from the garden and samples of their potting soil. They will use a ruler and go out in the garden to collect a small soil sample from three inches below the soil surface.

2. Using this soil sample, the groups will fill the small vial of their test kit and carefully remove the covering of the indicator pill, pouring the contents in the vial as well. They will fill it with water to the designated line and shake for a few seconds. Then, they will let the vial sit for two minutes and observe color changes as the soil settles. This activity provides a great opportunity for students to practice measurements, counting, and timekeeping with a clock.

3. When the time is done, each student will record their findings for both the garden soil and potting soil on a chart for the class to later analyze.

4. Finished groups can check on the development of their growth experiments and record plant/seed changes on the Plant Growth Chart worksheets.

Assessment (5–10 minutes)

1. Gathering back together as a class, encourage the students to share their recommendations and analyze their records. What was their experience in collecting this data? Was it difficult or easy to read the tests? What does the data tell them about the pH and NPK levels of their potting soil and the growing conditions for their sprouts?

2. As gardeners, what do they conclude about the nutrient balance in the garden soil? What plants would they recommend grow in the school garden?

Preparation

1. Approximately two Luster Leaf Rapitest (#1605) Soil Testing kits (see Resources)
2. Chart for recording data
3. Garden soil (if you don't want the students to collect it themselves)
4. Watering can
5. Stopwatches

Resources

- Luster Leaf Rapitest (#1605) Soil Testing kit (online—see Appendix B).
- "Natural Fertilizer" in Patrick Lima. *The Natural Food Garden: Growing Vegetables and Fruits Chemical-free.*
- Fred Magdoff and Harold van Es. *Building Soils for Better Crops*, 3rd ed. USDA Sustainable Agriculture Research and Education. (online—see Appendix B).

8 Organic Fertilizers

Time Frame: One hour

Overview: The class will create organic fertilizers that provide the necessary nutrients for healthy growth of their plant starts.

Objective: For the students to learn simple and effective ways to feed young plants and apply the lessons they learned into real world solutions.

Vocabulary

1. Leaching
2. Fertilizer
3. Organic

Introduction (10 minutes)

1. Brainstorm with the class what role liquid fertilizers have in an organic garden: Why is it important to monitor the fertility of growing mediums and plant starts? What is the long-term effect of water passing through soil and nutrients leaching?

2. Review with the class what the pH and NPK findings were during the last activity. What did their testing say about the growing potential of their potting soil? What organic materials can be added or utilized to enhance the growth of the plant starts and create a better balance of nutrients in the soil?

3. On a classroom chart, make a list of different organic materials that can be made into a fertilizer and the benefits of each (see Resources). The small groups will use this chart as a reference for what they should add to their mixture later.

Activity #1 (30 minutes)

1. In their small work groups, the students will mix the provided organic items to create a liquid fertilizer that will best work for their plant starts. The groups will keep in mind the measurements and data they recorded in the last activity, as well as the reference chart they created during the brainstorm and introduction.

2. When the groups have gathered their materials, they will assemble them into a jar and fill it with water. They can cover their jar with a permeable cover, such as coffee filters, and place it in a sunny location. The groups should label their containers with their group name(s).

3. Afterward, the groups will check in with their plant starts and record the growth and changes on their worksheets.

Activity #2 (15 minutes)

1. In the morning, three days after the groups have made their mixtures, have the students use water to lightly hydrate their plants. This will help open up the plant roots to receiving the fertilizers and keep them from being burned.

2. In the afternoon, the small groups can remove and compost the large material pieces in their mixtures and dilute their fertilizers with water. Then, they will pour it onto their plant starts, being sure to get as little as possible on the young plant leaves. The groups can repeat this process over the next week or until their fertilizer is completely used.

Preparation

1. Grass clippings
2. Banana peels
3. Comfrey leaves
4. Coffee grounds
5. Compost
6. Watering can
7. Jars with lids
8. Tape and markers

Resources

- "Natural Alternatives" in Patrick Lima. *The Natural Food Garden: Growing Vegetables and Fruits Chemical-Free.*
- Barbara Pleasant. "Free, Homemade Liquid Fertilizers—Organic Gardening." *Mother Earth News.* (online—see Appendix B).
- Barbara Pleasant. "Homemade Fertilizer Tea Recipes—Organic Gardening." *Mother Earth News.* (online—see Appendix B).
- Jonathon Engels. "4 All-Natural Liquid Fer-

tilizers and How to Make Them." One Green Planet. (online—see Appendix B).
• "Here's the scoop on chemical and organic fertilizers." Oregon State University Extension Service. (online—see Appendix B).
• "Organic fertilizers" in *Working with Nature: Shifting Paradigms: The Science and Practice of Organic Horticulture.*

9 Plant Growth

Time Frame: 45 minutes
Overview: Students will use their NPK and pH data to make amendments to the garden's soil.
Objectives: To provide students the opportunity to use scientific data to create positive organic gardening solutions.

Introduction (10 minutes)

1. Have the student groups review their data/graphs from the pH testing activity. What did the experiments indicate about the NPK and pH balance in the school garden soil? What were some of the nutrients that could be added to the garden soil to make it healthier or more balanced for plant growth?
2. Some of the materials students could use include:

- Nitrogen: Coffee grounds, composted manure, fresh-cut organic greens
- Potassium: Banana peels, citrus rinds, wood ash
- Phosphorus: Fresh compost, liquid fish fertilizer, bone meal

3. Students can now move from the mind-set of scientists to that of gardeners. Using their data, the students will address key growth concerns and supplement any deficiencies in the garden soil. They will mulch and add organic materials to the garden beds in order to increase soil vitality and health. It will be important that they practice slow, careful gardening during this activity.
4. Previously, the students made a homemade liquid fertilizer from organic materials. During this activity, they will be burying or mulching the organic materials into the garden beds. By doing this in the winter and cutting or breaking the organic matter into small pieces, the materials should have enough time to begin decomposing in the soil before late spring planting.

Activities (30 minutes)

1. In small groups, students will amend the garden soil in designated spots using the various provided materials. They are expected to practice safety with the gardening tools and return all of their materials at the end of the activity. This activity can also be accomplished in container gardens.
2. If they are using fresh compost or partially decomposed food scraps, the students will have to bury them carefully in the garden bed. Each group will use the provided containers to put an equal amount of the necessary materials and spread them equally throughout their garden bed/working space. Finer materials, like grass clippings and coffee grounds can be gently spread, sprinkled across the garden bed, and tucked around existing plants.
3. At the end of the activity, the students should clean up their materials and put their supplies away. Then, they can check on the growth of their starts and record the data on their worksheets.

Assessment (5 minutes)

1. Were there any gardening difficulties they discovered as they worked? Did they find evidence of other organic materials in the garden beds? What decomposers do the students think will feast on the added materials?

2. What experiences did they have with the different materials and the tools? Were the students able to use kind words and teamwork to accomplish their tasks?

Preparation

1. Mulching materials
2. Gloves
3. Buckets
4. Trowels/shovels

Resources

• Steve Albert. "Organic Fertilizers and Soil Amendments." Harvest to Table. (online—see Appendix B).

10 Transplanting Starts

Time Frame: 45 minutes
Overview: Students will plant spring vegetable starts in garden plots full of healthy soil.
Objectives: For students to engage with healthy soil and its connection to growing nutritious plants.

Introduction (5–10 minutes)

1. Brainstorm with the students how the nutrients they have added to the garden impact the growth and health of the vegetables they will plant: What types of nutrients have the students added to the garden soil over the past few weeks? How will these fertilizers and nutrients impact plant growth? How will the class know that the nutrients are making a positive change in the soil?

2. The class should understand that the taste, color, shape, seed health, and nutrition of garden foods are positively affected by the work the students have done.

3. Discuss with the class how to perform the activity and how to safely remove a young plant start from its growing container. How deep should the planting hole be? What else does the plant start need to survive the transition? Gardeners use their hands to measure. What kinds of measurements can students use with their hands in order to best plant the starts?

Activities (30 minutes)

1. Before planting, the small groups will gather together to provide a final recording of plant growth on their worksheets. They should work together to answer these questions:

- How successful was the small group growth experiments?
- What plants starts were the most successful?
- What variables did the groups encounter during the experiment?

A healthy cabbage is transplanted from the student plant growth experiments.

- What would they do differently in the future?

2. In their small groups, students will transplant their cool season plant starts from the growth experiment into the school garden. They should pay attention to the spacing needs of their starts and the depth at which they plant them.

3. The groups can deeply water their starts after they have transplanted and use wooden sticks and markers to record the variety of each start.

Assessment (5 minutes)

1. How did students use their hands to measure planting distance between their plant starts? What were the distance requirements for their plant variety?

2. In regards to their plant growth experiments, which experiments were the most successful? What did each group do that was most successful during the experiment or least successful? How can students determine the health of a plant?

3. What recommendations would each group make to future experiments and growing conditions in the school garden? Record student recommendations and advice on a classroom chart for future reference and use.

Preparation

1. Watering can
2. Markers
3. Wooden sticks
4. Classroom chart

Spring: Fostering Healthy Soils

1 Creating Compost

Time Frame: 45 minutes

Overview: In small groups, students will explore the different stages of decomposition and create compost with raw materials.

Objective: For students to explore the compost ecosystem and learn another organic gardening technique that benefits the garden soil.

Vocabulary

1. Organic Matter
2. Humus

Introduction (10 minutes)

1. Brainstorm with the class: What is compost? How does it benefit the garden soil? What types of materials go into the compost and what doesn't go into the compost? What are some living creatures that students might find in a compost pile? How does finished or healthy compost look and smell?

2. What does the compost life cycle entail? Encourage the class to image each step of decomposition and the role of decomposers in it. Illustrate student ideas and the process of composting on a classroom chart.

> In a learning garden, compost is an important recycling system for food waste and surplus organic materials, where the students participate in nurturing the soil without commercial products and fertilizers.

3. Students might approach the activity using words like "gross" or exaggerating their reactions, but they should be encouraged to be scientists during the activity by asking questions and making objective observations. How does a scientist interact with compost texture, smell, and ecosystems?

4. Demonstrate the activities at each workstation and have the class discuss the behaviors expected at each place.

> The terms "humus" and "organic matter" are often used interchangeably, but organic matter is the raw material...and humus is the rich, dark, crumbly finished product.[5]

Activities (30 minutes)

1. In small groups, the students will rotate between stations representing different compost stages. They will work at each station for about 10 minutes, before moving to the next.

- **Station #1:** Creating compost with raw materials. Students will learn about the dry/wet and brown/green material ratios needed for healthy composting. They will use measuring cups and provided materials to mix the best ratio together for raw compost that will decompose well.
- **Station #2:** Identifying half-composted material. Using magnifying glasses and gloves, the students will explore the creatures that are essential to decomposition. They will use their senses to determine smell, texture, and color of the compost and decide how healthy the compost is and what should be added to make it more productive.
- **Station #3:** Filtering compost. Students will use a screen and buckets to filter healthy compost and spread it onto the designated garden beds or planting containers. They should pay attention to what materials are not decomposed and discuss their questions as to why these materials have not decomposed as quickly as the others.

A student adds organic materials to the school's cold compost pile.

Assessment (5 minutes)

1. What did the student-scientists discover during their activities? What questions did they generate at each station?

2. How healthy is the compost created from the school compost system? What changes should the class make to create healthier compost?

Preparation

1. Adult or older student volunteers for each station
2. Screen, shovels, and buckets
3. Magnifying glasses
4. Gloves
5. Green/brown materials for composting

Resources

• "Sifting Compost" in Patrick Lima. *The Natural Food Garden: Growing Vegetables and Fruits Chemical-free.*

2 Green Manure

Time Frame: 45 minutes
Overview: Students will plant legume seeds in the garden after studying their ability to fix nitrogen in the soil.
Objective: Students will learn a simple way to capture nitrogen from the atmosphere and distribute it into the soil.

Introduction (10 minutes)

1. Lead the class in brainstorming the key ideas for the activity: What impact does nitrogen have on the garden plants? Why is it an importance resource in the garden and what other organic materials have nitrogen in them? What is a legume? What are other types of *green manure*?

> Legumes have a relationship with bacteria that can benefit garden soil by capturing nitrogen. When alive, but also as the legumes die or their plant bodies are turned into the soil, the nitrogen captured through this relationship releases into the soil during decomposition and is made available to other garden plants.

2. Many legumes can handle cooler spring temperatures in the garden. Students will be planting a variety of legumes to capture the nitrogen for the soil. How deeply should the students plant the seeds? Where should they plant them?

Activities (30 minutes)

1. Individually or in pairs, the students will plant legumes in designated garden beds and practice the gardening techniques they have learned. They should work as careful, considerate gardeners and slowly plant their seeds. They, or student volunteers, can water the seeds after they have been planted. Some possible legumes to plant include: fava bean, cowpeas (depending on climate), and any variety of garden beans, clovers, and lupines.

2. At the end of their planting time, have students clean up and return any seeds they haven't planted. Then they can wash their hands and enjoy a taste test from the garden. They can also check on the growth and health of their transplanted starts.

Assessment (5 minutes)

1. How healthy are the students' transplants? What else do they need to thrive?

Preparation

1. Legume seeds

3 Manure: Full Cycle Gardening

Time Frame: 45 minutes
Overview: Students will carefully spread composted animal manure in designated garden beds.
Objectives: Students will understand the importance of nitrogen in the garden and how they can use waste to benefit soil health.

Introduction (10 minutes)

1. Have any students used manure to fertilize gardens before? What types of manure can gardeners use to add nitrogen to their soil? Recalling the pH and NPK lesson, which garden beds need additional nitrogen?

The manure in the garden should be from herbivores and some omnivores rather than predators. For this activity, it is best to use composted herbivore manure. This material is often available from local zoos, dairies, and stables, but you should check to make sure the feed given the animals is pesticide-free.

Composting with manure is an environmentally conscious method to turn a problem into a solution. Not only is it great for building healthy soil, it makes a positive impact on a larger waste issue.

2. What language and behavior would scientists or gardeners use during this activity?

Activity (30 minutes)

1. In pairs or individually, students will fill buckets with composted manure and spread it out as a thin layer of mulch over designated garden beds. Students are encouraged to use the language and behavior of scientists and gardeners during this activity. As they work, students can look for evidence of partially decomposed organic matter or make observations about living creatures they spot in soil ecosystem.

2. Student volunteers can help shovel the manure into the buckets or rake it out over the garden beds.

3. After the beds have been layered with manure, the students can put away their tools and gloves and then wash their hands before taste testing in the garden.

"Feed the soil, not the plants"[6]

Assessment (5 minutes)

1. What were student experiences handling the manure? Was it what they expected (texture, smell, consistency)?

2. What effect can non-composted or raw manure have on a garden bed?

Preparation

1. Composted manure (horse, chicken, rabbit)
2. Gloves
3. Buckets and shovels

Important Advice to Teachers

"Avoid too much of a good thing"[7]

"Whenever we have more nutrients available than can be utilized within the ecosystem they will be lost to the ecosystem through leaching or volatization."[8]

Keeping this advice in mind, I caution on using excessive compost in the garden and would not recommend performing all these mulching and composting activities in the same garden beds or spaces. Rotate the activities in different beds every year, but monitor the health of each bed as you go. Students can perform some of these activities in pots as well, or compare the effect of these different organic materials on various garden beds.

4 Hügelkultur Soils

Time Frame: 45 minutes

Overview: The class will work together to build a Hügelkultur bed in a new part of the garden.

Objective: To expand the garden's growing potential by creating a long-term, experimental garden bed.

Introduction (10 minutes)

1. Brainstorm with the class about what they know about building healthy soils: What mulch and materials have they added to the school garden soil to improve its health and vitality?

What are essential ingredients to healthy soil? What are brown and green materials?

2. Introduce the class to the Hügelkultur method of building garden beds (see Resources). By building piles of layered brown and green materials, with a woody base as a foundation, gardeners can potentially create compost systems that generate heat for a longer growing season and retain water during dry conditions.

3. What does class-wide teamwork look like during this activity? What does high-quality work look like?

Activities (30 minutes)

1. The students will be building Hügelkultur beds in a new area of the garden, in an existing garden bed, or a large planting container. I recommend edge areas as a great place to situate Hügelkultur beds. As a class or in teams, the students will layer logs and branches in a previously made trench or shallow pit. Then, they can add wood chips, finished compost, leaves or green manure, straw, and topsoil on top of their foundation.

2. As the materials as being layered, an adult or student volunteer can water the pile gently with a hose.

3. The class will practice group work and craftsmanship during the activity. Afterward, they can clean up any tools and materials they used, before enjoying a taste test from the garden. They can also install a sign describing the layers in the bed and its purpose.

Assessment (5 minutes)

1. How long will it take for the Hügelkultur bed to begin decomposing? What ingredients and living things will help the decomposition process?

Preparation

1. Logs and sticks
2. Wood mulch/chips
3. Leaves/green manure
4. Compost/raw and finished
5. Topsoil
6. Straw
7. Hose/watering can
8. Sign materials

Resources

- "The Many Benefits of Hugelkultur." *Permaculture Magazine.* (online—see Appendix B).
- "Hugelkultur: The Ultimate Raised Garden Beds." Richsoil website. (online—see Appendix B).
- "Hugelkultur Bed Construction." Deep Green Permaculture website. (online—see Appendix B).
- Christine Zieglar Ulsh and Paul Hepperly. "Good Compost Made Better." Rodale Institute. (online—see Appendix B).

5 Planting Seeds for Summer

Time Frame: 45 minutes

Overview: The class will plant corn and sunflower seeds in the beds they built.

Objective: Students will begin growing two culturally and agriculturally valuable crops.

Introduction (5–10 minutes)

1. During the activity, students will be planting sunflowers and corn. What recipes can corn be used in? What stories do students tell about corn and sunflowers?

2. Introduce the class to the beneficial relationships between corn (maize), sunflowers, beans, and squash. Like other plants originating in North America, corn has an inspiring and valuable relationship with specific Indigenous Peoples, through thousands of years of

Educators: If necessary, take this opportunity to go a step further than this activity outlines. Please make space and time to discuss the cultural importance of corn, beans, and squash through multi-cultural engagement, storytelling, and sharing. My concern is that if non-Indigenous educators teach simple and short lessons about the Traditional Ecological Knowledge (TEK) of certain Indigenous Peoples, such as in lessons about The Three Sisters, we risk misrepresenting and appropriating TEK and Oral Traditions. Please make space and time to discuss the continuing history and cultural importance of North American cultivated plants and TEK through thoughtful engagement, community storytelling, and relationship-building. This opportunity continues in Lesson #7: Planting in Layers.

Spring is almost over I hear the sparrows
say time for seeds to bloom, time for
gardens to grow huge. Time for rabbits
to rustle in and take the food.
Time for hot hot sun. Time for
wonderous birds to come.

— Anonymous Student Poem

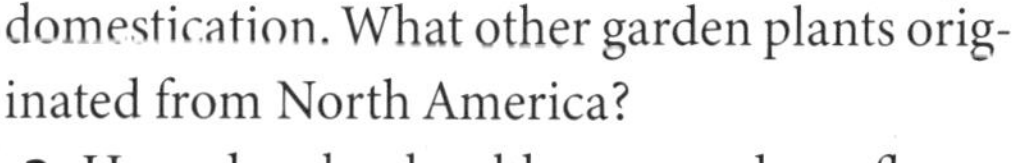

domestication. What other garden plants originated from North America?

3. How deeply should corn and sunflower seeds be planted and how far apart? How long will it take for them to sprout and grow?

Activities (30 minutes)

1. In pairs or small groups, the students will plant corn and sunflower seeds in the school garden, including the Hügelkultur beds they built, in full sun with level soil. They can plant these in a pattern appropriate to the space and school needs.

2. After they have planted the seeds, the groups can water the seeds with a watering can or hose. Then, they can enjoy a taste test from the garden or participate in taste testing a recipe of prepared corn.

Assessment (5 minutes)

1. How tall will the corn and sunflowers grow? What crops is would benefit from this group planting? Why should students wait to plant other cro

From past experiences in school gardens, I have found that corn grows better if it is started before squash leaves can shade the corn plants.

Preparation

1. Corn seeds (presoaked)
2. Watering can
3. Bean seeds

Resources

- Waheenee and Gilbert L. Wilson. *Buffalo Bird Woman's Garden.*
- Marlinda White-Kaulaity. "The Voices of Power and the Power of Voices: Teaching with Native American Literature." (online—see Appendix B).

6 Foraging for Wild Edibles

Time Frame: 45 minutes
Overview: The class will discover healing and delicious properties of wild plants.
Objective: Students will learn to identify and use wild edible plants.

Introduction (10 minutes)

1. Are there any plants that students love to eat when they are in a forest or a wild space? What plants can they identify?
2. What are medicinal plants and native plants? What does it mean to forage?
3. Introduce the students to three to five wild edible or medicinal plants, such as dandelion, mint, lambs' ears, plantain, yarrow, comfrey, western dock, and willow. How can the students identify these plants in the garden or a wild space?

> I like to choose plants that students can safely use to help heal wounds. Most injuries young students get in the garden are painful mentally more than physically, or they get small thorn marks and light scrapes. I find that giving children medicinal plant options is extremely effective in encouraging their own agency to solve their problems and mental concerns.

Activities (30 minutes)

1. As a class, read a story from *The Kid's Herb Book* by Lesley Tierra, such as the story of yarrow, comfrey, or lemon balm. Discuss the realism and fiction of the story, as well as what valuable lessons the students can learn from it about medicinal plants.
2. After the story, the students will embark in the garden to identify and gather small amounts of medicinal plant leaves, flowers, or stems that they have learned about. Before they eat or interact with them, they should have an adult or older student volunteer confirm their identification. Consulting an adult, expert, or book is a valuable lesson for any student who wants to forage.
3. The students can eat some of these plants or steep them in hot water for tea.

Assessment (5 minutes)

1. What plants heal cuts, mend bones, or reduce fever? What plants remove splinters and can be natural bandages?
2. Are there other medicinal plants that students know about in the garden? What about plants that they eat as food, such as garlic, honey, and sage?

Preparation

1. Examples of medicinal and wild edible plants

Resources

- Lesley Tierra et al. *A Kid's Herb Book.*
- Scott Kloos. *Pacific Northwest Medicinal Plants.*
- Jim Pojar and Andy MacKinnon. *Plants of the Pacific Northwest.*

7 Planting in Layers

Time Frame: 45 minutes
Overview: The class will plant bean and squash seeds in the garden.
Objective: The class will utilize horizontal space by planting squash seeds and vertical space by planting beans.

Introduction (5–10 minutes)

1. Brainstorm with the class how squash grows (tall or spreading)? What are some key characteristics of a squash plant?

2. How will corn, sunflowers, beans, and squash grow together and support each other? How will the bean plants grow?

Activities (30 minutes)

1. In their planting groups or pairs, students will plant squash and bean seeds in the garden, among the corn and sunflowers they planted weeks before. The students will make sure to sow seeds carefully around other living plants and use tools gently around their roots.

2. After students have planted their seeds, they can deeply water the seeds and starts, making sure that the seeds are covered by the soil. Then, the students can wash their hands and explore the garden to forage for their favorite taste tests.

Assessment (5 minutes)

1. What other plant groups do students see emerging in the garden (tomatoes and marigolds, dill and cucumbers)?

2. How well are the corn and sunflower seeds growing? With what students know about soil health, what recommendations would they make to increase plant vitality?

Preparation

1. Squash and bean seeds
2. Watering can

Resources

- Waheenee and Gilbert L. Wilson. *Buffalo Bird Woman's Garden.*

8 Companion Planting

Time Frame: 45 minutes

Overview: Students will plant a variety of seeds or starts to promote the presence of beneficial insects and deter non-beneficial ones.

Objective: To layer garden plants and herbs that support each other for optimal growth and health.

Introduction (10 minutes)

1. Brainstorm with the class what plants they know grow well next to each other: What plant groupings have students seen or planted together in the garden? How can plants work to support or benefit each other? What does it mean to be a companion?

2. What kinds of insects are beneficial to the school garden? Which insects do students want to attract or deter? How could these variables or goals affect how the students will plant?

3. What does good gardening work look like during this activity? How will the students Care for Self, Others, and the Land during this activity?

Activities (30 minutes)

1. In small groups, the students will plant starts and seeds for additional edible plants that are beneficial to other favorite foods. For example, the marigold group will plant flower starts around the tomatoes in the garden. The nasturtium group can plant seeds under and around the bean poles and plants. The onion group will plant starts carefully among the brassicas. The dill group can plant seeds around the cucumber plants. For additional groupings, see Resources.

2. The groups will use teamwork and collaboration to carefully plant all their starts or seeds. They will also make sure that their starts are gently removed from their containers and that the seeds have contact with the soil. The groups that plant seeds will make sure to put them in the soil at the right depth, and the groups that plant starts will gently bury the plant roots facing down.

3. After the groups have planted all their seeds and starts, they can water them deeply with watering cans. Then, they can gather taste tests from the garden to enjoy.

Assessment (5 minutes)

1. Will any of these plants compete with each other for nutrients and space?
2. What other plant partnerships did students notice in the garden during the activity?

Preparation

1. Marigold, basil, cilantro, and onion starts
2. Dill and nasturtium seeds
3. Shovels
4. Watering can

Resources

- Louise Riotte. *Carrots Love Tomatoes.*
- Sarah Israel. "An In-Depth Companion Planting Guide." (online—see Appendix B).

9 Nutritious Soil and Plants

Time Frame: 45 minutes
Overview: Students will harvest ripe fruits and vegetables from the garden and celebrate all of their hard work.
Objective: Review how living and nonliving elements work together to create nutritious foods.

Introduction (10 minutes)

1. Brainstorm with the students what they know and have learned about soil: What are living and nonliving things students have found in the soil? What creatures live in the garden ecosystem, eat the soil, or live off the soil? How do soils differ from each other? What relationships exist between the living and non-living elements of the garden ecosystem? How do these relationships contribute to the health of the garden soil and the plants?
2. Review with the students how to properly pick different plants from the garden using two hands and careful harvesting techniques. What plants are ready to harvest? What is in season?

Activities (30 minutes)

1. As a class, the students will go out to the garden to carefully harvest selected seasonal foods. Have a designated area where they can bring produce when their hands are full and wash what they have gathered.
2. Gather the students back together to discuss their experiences harvesting. Share the taste tests with the students and encourage them to focus on the flavor and texture of certain foods. As they eat, have the class discuss their tasting experiences.

Assessment (5 minutes)

1. What words would students use to describe the experience of taste testing garden foods? Does the flavor change after chewing for some time?
2. What connections can students make between the flavor of the food and the soil it grew in? What does flavor tell you about how the plant was grown? What nutrients are essential for healthy, nutrient-rich foods?

Preparation

1. One or two volunteers to wash and prepare taste tests

10 Garden Journals

Time Frame: 45 minutes
Overview: Each student will compile a garden journal from their worksheets and activities.

Objective: For students to create a memento of their garden work and celebrate their achievements.

Introduction (10 minutes)

1. In this last activity of the year, the students will be making garden journals from the work, activities, and experiments they have completed. Not every student will have all the worksheets, but the focus of the activity is to celebrate student achievement, so each child can be pleased with whatever they learned and created this year.

2. Demonstrate how to make a garden journal and what tools are available for student use.

Activities (30 minutes)

1. Each student will gather their worksheet papers and activities together, organize them, and hole punch the sheets (potentially with the help of an adult or volunteer). The Soil Scientist cover page should be the top page for this year's unit of study.

2. Then, the students can carefully weave the pages together with colored yarn and tie it off to bind their book. Using colored pencils, the students will decorate their cover pages.

3. After they have finished making their journals, the students can enjoy a seasonal taste test from the garden.

Assessment (5 minutes)

1. What were the students' favorite memories from this year in the garden? What did they build, create, and learn?

2. What are their favorite taste tests? From what they learned this year, what gardening advice would they give others?

Preparation

1. Varied colored yarn
2. Scissors and hole punchers
3. Soil Scientists cover page

Vermicomposting is one of my favorite units to teach at this age level. By the end of the year, the students will have developed strong relationships with garden decomposers, have better science skills, and a firm grasp of the soil ecosystem. In the previous year of study, students looked at the living and nonliving elements within the soil system. Now, they will investigate one common, yet fairly unexplored and fascinating member of that community.

The class will still participate in seasonal garden work, but will do everything within a conversation about worms. During mulching, compost creation, and ground cover activities,

FALL
Becoming a Worm Expert

1. Garden Exploration
2. Sowing Winter Seeds
3. Garden Work and Handling Tools
4. Harvesting the Surplus
5. Mulching for Worms and Winter
6. Meet an Earthworm
7. External Worm Anatomy
8. Internal Worm Anatomy

9|10. Build Your Own Worm

WINTER
Mini-Worm Bin Project

1. Generating Questions

2|3. Experiment Design and Construction

4. Beginning Inquiry Project
5. Experiment #1: Light
6. Experiment #2: Smell
7. Experiment #3: Life Cycle of a Worm
8. Experiment #4: Habitat
9. Formulating Conclusions
10. Planting for Spring

SPRING
Worms Help Grow a Healthier Garden

1. A Recipe for a Worm Bin
2. Cafeteria Waste Audit
3. Measuring Worm Bin Health
4. The Worm Bin Community
5. Identifying Worms in Our Soil

6|7. Worm Waste to Fertilizer

8. Planting Starts
9. Planting Seeds in the Garden
10. Celebration

the students will keep these nutrient-makers in mind. The class will journey from the garden into a thorough scientific exploration of a worm's anatomy, its unique characteristics, and make their own large-scale model of a worm.

In the winter, the students will build mini worm bins, and care for worms as they gently discover their fascinating attributes. From the mini worm bin exploration, they will apply their lessons in the garden with the goal of fostering even more productive plants and soil ecosystem. From brewing worm tea for fertilizer to helping care for a school worm bin, the students will be engaged in memorable scientific and gardening experiences.

The sunflower yellow bright with

color hanging above me, bees flying

around when I get close. You

make me happy and filled with joy.

— Anonymous Student Poem

Next Generation Science Standards

While the vermicomposting unit leads away from the energy and motion science themes of this year, the skills that students develop and utilize are the same. Through studying worms, the students will explore the internal and external structures that function to support worm survival, reproduction, growth, and behavior. They will do this through testing, observations, and analysis of data. Small student groups will design weekly experiments to form predictions, test hypotheses, collect data, and draw conclusions and arguments from their study, as well as generate questions for further exploration. The class will design solutions for gardening and environmental issues and carry out investigations in the classroom and in the garden.

Unit extensions from these lessons can include healthy ecosystems and their effect on environments and humans, the effect of humans on natural systems, and nature's response to pesticides, changing ecosystems, and climate change.

Permaculture Principles

- Observe and Interact
- Catch and Store Energy
- Use and Value Renewable Resources and Services
- Produce No Waste
- Integrate Rather than Segregate
- Use Small and Slow Solutions
- Use Edges and Value the Marginal

Fall: Becoming a Worm Expert

1 Garden Exploration

Time Frame: 45 minutes
Overview: Students will review the behavior expected in the garden and go on a garden scavenger hunt.
Objectives: To understand what it means to Care for Self, Care for Others, and Care for the Land while exploring garden changes.

Introduction (5–10 minutes)

1. Welcome the students to the garden by reviewing garden expectations. What are the behaviors that they enjoy seeing their classmates exhibit in the garden?
2. What lessons have students learned about how to treat plants, animals, and insects that live in the garden?

Activities (25–30 minutes)

1. Students will explore the garden alongside a classmate and become familiar with its changes while practicing their best garden behavior. Together, they will pursue a Garden Scavenger Hunt activity, focusing on teamwork and good garden behaviors to accomplish their task.
2. After about 20 minutes, or when students begin to finish their activity, they can harvest late summer foods, such as cucumbers, tomatoes, beans, and berries, to taste test as a class.
3. The taste testing activity can be as guided as the students need. Having pre-picked plants to demonstrate careful harvesting techniques also helps the students remember how to be gentle with the plants.

Assessment (5 minutes)

1. After the activity, the students can gather back together and enjoy their taste test as they share what they discovered in the activity.
2. What did students discover in their explorations? What surprised them? Were there any plants and fruits that students noticed thriving?
3. What activities would they like to engage with over the course of the year? What plants would they like to see again in the garden next year?

Preparation

1. Garden Scavenger Hunt worksheet

2 Sowing Winter Seeds

Time Frame: 45 minutes
Overview: Students will sow cold-hardy seeds in the garden and prepare the beds for cold season weather.
Objectives: To learn what winter seasonal plants can thrive in the garden and what varieties to plant in the fall to provide spring crops.

Vocabulary

1. Seasonal
2. Cold Hardy
3. Cover Crop

Introduction (5–10 minutes)

1. Brainstorm with students what they know about cover crops and winter seeds: What are cover crops? What types of plants grow well in cool and cold weather? How deeply should students plant seeds?

2. Do students have any experience with reading the back of seed packets? Introduce the class to the information that is provided there. What are the important details on the back of a seed packet, and how will they help guide the class in the activity?

Activities (30 minutes)

1. Students will break into pairs with a seed packet to share. The partners will read the information on the seed packet, determining the depth, space, and light requirements for their seeds. Then, they will find the appropriate spaces in the garden to plant their seeds.

2. Good work will be careful and conscientious planting, as well as teamwork between the partners. Their tools will be their hands, unless the soil is too compacted. If the students finish with one seed packet, they can plant another plant variety that complements what they just planted. For example, if they planted carrots, then they can sow radishes among them. Other combinations could be a deep-root crop with a shallow one for a quick harvest.

> I have found many resources for free, organic, and mostly heirloom seeds in my time as a Garden Educator. I learned that local nonprofits who specialize in garden education will often get donations of seed packets from companies who can't sell last year's product, but those seeds are still viable if they've been stored correctly. Local food bank chapters often get large donations of seed packets from businesses that can't sell last year's seeds. They offer these seed packets for free to people who need them. Seed savers exchanges are also another great source of good-quality seeds.

3. The students can use wooden sticks or available materials to label what variety of seeds they planted and where. They should also make sure to cover their seeds with soil and ensure seed-to-soil contact.

4. After their activity, the students can wash their hands, clean up or return the materials from their activity, and enjoy a taste test of seasonal foods.

Assessment (5 minutes)

1. Students should be able to name the winter seeds they planted and in what combinations.

2. How did the students navigate reading the seed packets? Were they successful in carefully planting the seeds?

Preparation

1. Seed packets/seeds

3 Garden Work and Handling Tools

Time Frame: 45 minutes

Overview: Students will prepare the garden for winter and learn how to safely work with garden tools.

Objective: To practice safe gardening practices and seasonal maintenance.

Introduction (5–10 minutes)

1. Brainstorm with the students how to handle shovels, rakes, and wheelbarrows in safe ways: What behavior do students want to see from their teammates during this activity? Students should be able to demonstrate tool handling practices and create an agreement to safely use them. Emphasize working slowly and carefully, the way gardeners do.

2. Tool use and safety also includes putting the tools back where they are stored. The students should be familiar with where these storage spaces are and what "putting tools away" looks like.

3. In the activity, the students will be weeding along the garden paths and in the garden beds. Show them examples of the weeds they will be removing and demonstrate the best way to remove the weeds with the roots included. Encourage the students to brainstorm what they know about plants that are considered weeds. How is a weed simply a plant in an undesirable location? Are there examples of weeds that are so prolific as to be considered invasive?

Activities (30 minutes)

1. In small groups, students will weed, lay bark chips on the path, and rake out chips to improve the walking paths. Student groups will rotate throughout the activity, especially if there aren't enough tools for the whole class to use at the same time.

2. The small groups will practice teamwork, collaboration, and taking turns for this activity, as well as doing careful and quality work. When all the groups have rotated through each task, they can clean up and return the garden tools back to their designated area.

3. At the end of the activity, the students can enjoy a taste test from the garden, while walking on the beautiful paths they helped restore.

Assessment (5 minutes)

1. Gather the class back together and have students individually explore the garden to make sure that all the weeds are in the compost or in separate bins, the tools are picked up, and the bark chips are spread out.

Preparation

1. Bark chips
2. Wheelbarrows/buckets
3. Large shovels
4. Rakes
5. Weed examples

4 Harvesting the Surplus

Time Frame: 45 minutes

Overview: Students will practice harvesting different garden vegetables.

Objectives: To learn how to determine what ripe fruits look like and carefully harvest produce from the garden.

Introduction (10 minutes)

1. Students will be harvesting food that is ready to pick from the garden to share at a community harvest event, in the cafeteria, a schoolwide taste test event, or for the class. Plants to pick could include pumpkins, cantaloupe, cucumbers, squash, tomatillos, tomatoes, beans, edible flowers, or peppers.

2. Depending on what plants are available for harvesting, pre-pick examples and demonstrate how to carefully harvest them. For example, if harvesting pumpkins or cucumbers (and other thick-stemmed plants), rather than give the students a knife to cut them off, I have the students practice gentle and slowly twisting them off the vine. I always recommend that students use two hands to harvest fruit.

3. Additionally, have the students determine the qualities of a ripe fruit and an unripe fruit. What does a ripe or unripe fruit look like or feel like? What are the qualities the students will look for in ready-to-pick fruit?

Activities (30 minutes)

1. Students will harvest the ripe fruits from the garden, taking their time to do good work and carefully harvest from the plants. The students can work individually or in pairs to harvest one type of vegetable at a time and then carefully transport their pickings back to a table or washing station to clean and store.

2. Students can gather the fruits into separate harvest baskets. Rather than working with a competitive mind-set, students should practice careful, considerate gardening and discernment during harvesting.

3. After they have harvested in the garden, they can taste test some of the diverse foods they picked. They can comparatively taste test an unripe and ripe vegetable (like a tomato) and explore the differences between the two. As they eat their taste tests, have the students reflect on their activity in the garden.

Assessment (5 minutes)

1. What were student experiences in the garden? What did they learn as they were harvesting the fruits?

2. Are there fruits in the garden that are not ripe? How could they tell? When do the students think these fruits will ripen? What happens to the fruits and seeds if a fruit does not ripen before the end of the growing season?

A layer of leaves and mulch protects the soil ecosystem from the frost.

Preparation

1. Harvest baskets
2. Fruits for harvesting demonstration
3. Washing station

5 Mulching for Worms and Winter

Time Frame: 45 minutes

Overview: Students will spread straw or leaves over garden beds to decompose for healthier soil.

Objective: To learn about how mulching provides habitat for decomposers, such as worms.

Introduction (10 minutes)

1. Have the class brainstorm what they know about mulching: What does it mean to mulch? What kinds of mulching have they done in the garden? How does straw help build worm habitat in the garden? How does mulching and worm activity benefit the soil?

> Leaves and straw are both lightweight, often available materials which retain moisture in the soil by adding various particle sizes that catch and absorb rainfall. They provide habitat for the worms, decomposers, and even amphibians to move more freely across the garden soil.

2. Use a raised bed to demonstrate the best way to lay down leaves or straw without wasting it. It should be like tucking the plants in, rather than scattering it loosely across the bed.

3. If this activity has been performed previous years in the garden, demonstrate to the students what the dried materials look like as it decomposes by digging down into a raised bed and giving them each a handful or soil. Can they see old mulching pieces in the soil? Are there any decomposers, insects, or other creatures living in the soil? How long do the students think it will take straw or leaves to fully decompose? Is the decomposing material dry or moist?

Activities (30 minutes)

1. As a class, students will carefully gather available leaves or straw and spread it over the garden beds, taking care not to cover still growing or living plants. Good work will look like careful attention to detail. I recommend that students focus on one raised bed to do their work.

2. A few student volunteers can rake the mulching materials that have fallen on the garden paths in order to utilize all of it. Then, the class can clean up and enjoy a taste test from the garden at the end of the activity.

Assessment (5 minutes)

1. Have the students go on a quality control journey through the garden by making sure all the mulch is in the garden beds (not on the pathways) and not covering living plants.

2. Did the students encounter any worms or other decomposers during their activity?

Preparation

1. One straw bale
2. A large pile or bag of leaves

6 Meet an Earthworm

Time Frame: 45 minutes

Overview: Students will be introduced to worms and generate questions about them for future study.

Objectives: To become familiar with earthworms and learn how to interact with them.

Introduction (10 minutes)

1. During the activity, students will be able to interact with worms by holding them, gently touching them, and observing worm behavior. Brainstorm students' behavior expectations for handling worms: As scientists, how can students care for the creatures they are studying? What actions, words, and volume level do students want the class to respect? How can students Care for Land, Others, and Self during this activity?

2. Have the students meet worms collected from the garden. Demonstrate how to handle them and brainstorm with the students what they know about worms.

> "[W]e must accept that there are intelligent beings on this planet besides the human species...let's at least start by acknowledging them and understanding their roles in our ecosystem"[1]

Activities (30 minutes)

1. Each student or table group will get a container with a sample of soil containing worms. As carefully as they can, the students will find the worms in their sample, make observations about them, and generate questions about worm anatomy and behavior on their Meet an Earthworm worksheet.

2. Students will use the worksheet to draw what they see and generate scientific questions about the worms. The students can use magnifying glasses to help them observe more closely and illustrate a larger-than-life-size picture of

An adventuring worm moves across rich garden soil.

the worm as they pay attention to details and interesting features.

3. When the students are done, the class will return the worms and soil samples safely to a bin to be transported to the garden or a worm bin. They can clean up their areas, return any borrowed tools, and gather together to share their findings.

Assessment (5 minutes)

1. What took the students by surprise during the activity? Which student had a long worm or a small worm? What colorings or markings did the worms have? What questions do students have about worms after meeting them?

2. Could the students tell which end of the worm was the anterior and posterior? How quickly did the worms move? How does a worm move its body? What other behaviors did the students observe?

Preparation

1. Worms and soil sample
2. Meet an Earthworm worksheets
3. Magnifying glasses
4. Wooden sticks
5. Observation tray
6. Soil sample bin

Adaptations

If you don't have a worm bin or garden to collect worms, you can reach out to local nurseries that sell worms directly. They will often provide up to half a pound of free worms for student studies.

7 External Worm Anatomy

Time Frame: 45 minutes

Overview: Students will complete an anatomy worksheet and learn new key terms.

Objective: To explore the parts of a worm and practice this knowledge by actively observing a worm's movement and behavior.

Vocabulary

1. Setae
2. Clitellum
3. Prostomium
4. Anterior
5. Mouth
6. Muscle Segments
7. Posterior
8. Anus
9. Anatomy

Introduction (10 minutes)

1. Lead the class in a brainstorm: What is anatomy? Why do scientists study it? What parts of worm anatomy do students already know?

2. Introduce students to new vocabulary by illustrating a diagram of a worm's external anatomy on a classroom poster and exploring the purpose of each part through group discussion. There are a few new vocabulary words for them to learn, so they can use what they know about their own bodies to discover the worm body parts and their functions.

3. After the class worm diagram has been

labeled, the students will receive a worm for observation. Before receiving their worms and soil samples, be sure to review the agreed-upon class expectations for worm handling (e.g., no poking worms, and worms must remain in their containers).

Activities (30 minutes)

1. Each student will receive an External Worm Anatomy worksheet and a container with a cup of soil and a worm or two. They will observe the worm and identify the anatomy they can see on their worksheets.

> I make worksheet labels different from the diagram because I want the students to think about these parts and their function, rather than copy them from the board. I do leave the class poster up for students to practice their spelling and remember all the parts they are learning. Some students benefit from having the words written up on the wall and broken down by syllable so they are easier to pronounce.

2. At the end of the activity, all students will carefully return the worms and soil samples outside or in a bin to be transferred to the garden. They will then gather together to share what they discovered.

3. Mary Appelhof's *Worms Eat Our Garbage* provides many fun worm-themed puzzles and games that are great for any students who finish their work quickly (see Resources).

Assessment (5 minutes)

1. Students should be able to pronounce most of the words correctly and explain their function. They should be able to speculate about the maturity of the worms they examined using correct terminology. The worm's clitellum indicates the sexual maturity of the worm. Can they find the clitellum of the worm? How old do they think the worm is and why?

3. What findings did students make during their observations? Were there parts of the worms that they couldn't identify? How can the students tell if a worm is an adult? What methods did they use to determine the posterior and anterior of the worm? What questions did they have about the worms? Did they notice behavior that was difficult to explain?

Preparation

1. External Worm Anatomy worksheets
2. Worms and soil sample
3. Magnifying glasses
4. Observation trays

Resources

- Mary Appelhof et al. *Worms Eat Our Garbage: Classroom Activities for a Better Environment*.
- Binet Payne. *The Worm Cafe: Mid-scale Vermicomposting of Lunchroom Wastes.*

NGSS and Activity Extensions

Further in-class studies can include developing an argument "that plants and animals have internal and external structures that function to support survival, growth, behavior, and reproduction" (4-LS1-1: From Molecules to Organisms).

8 Internal Worm Anatomy

Time Frame: 45 minutes

Overview: Students will explore the internal parts of a worm's anatomy and learn new vocabulary.

Objective: To develop a better understanding of how a worm moves, eats, and reproduces.

Vocabulary

1. Gizzard
2. Esophagus
3. Nerve Cord/Nervous System
4. Intestine
5. Ventral Vessel
6. Circular Muscles
7. Epidermis
8. Hearts
9. Setae

Introduction (10 minutes)

1. Have the class review the key terms from the External Worm Anatomy lesson by writing the vocabulary words on a classroom poster illustrated by a diagram of a worm.

2. Introduce students to the new anatomical terminology and definitions, explaining how a worm functions with each element. Use a poster to illustrate these descriptions.

3. What body parts are used when a worm is eating or being pursued by a predator? How are worms' bodies adapted and equipped to handle the stresses and needs of their life? How can students easily remember the names of these parts?

Activities (30 minutes)

1. Students will fill out an Internal Worm Anatomy worksheet, labelling the worm's internal parts and using the poster as a guide. They can also color this diagram and focus on correctly spelling each anatomy part. When they have finished labeling and illustrating, the students can clean up their materials and gather back together for a new class activity.

2. In two groups, students will play out the structure of a worm, with each student taking on a role in order to build a living worm's system. Guide the students in building this large-scale worm by assigning students to each role, such as mouth or gizzard, and determining an action for each piece of anatomy, as a class. When all the students are assembled, have them act out their section, animating the whole worm.

> I like to assign students to be food particles as well, so that the worm has to "eat" them by having the food crawl under the worm—or in this case, crawl under the students' legs.

Assessment (5 minutes)

1. Students should comfortably be able to pronounce most of the terminology and properly use it in their descriptions of the activity.

2. What questions do students have about worms after exploring their internal parts? How is worm anatomy similar to human anatomy?

Preparation

1. Poster of internal worm anatomy
2. Internal Worm Anatomy worksheets

Resources

- Mary Appelhof et al. *Worms Eat Our Garbage.*
- Binet Payne. *The Worm Cafe.*

NGSS and Activity Extensions

Greater classroom studies can include discussions on how creatures receive various forms of information through their unique senses and use this information to respond and survive (4-LS1-1: From Molecules to Organisms and 4-LS1-2: From Molecules to Organisms).

9|10 Build Your Own Worm

Time Frame: 1½ hours (2 × 45 minutes)
Overview: The students will make a model of the internal and external anatomy of an earthworm.
Objective: To foster a growing knowledge of a worm's internal and external parts and how they function as a whole.

Vocabulary

1. Anterior
2. Prostomium
3. Esophagus
4. Clitellum
5. Circular Muscles
6. Intestine
7. Ventral Vessel
8. Setae
9. Gizzard
10. Nervous system
11. Posterior

Introduction (10 minutes)

1. Have students recall what they know about worms from their previous lessons, such as Meet an Earthworm, External, and Internal Worm Anatomy. Are there any remaining questions about worms that students have been wondering?

2. Using the posters of worm anatomy as a reference, what are the different parts of a worm the students have studied and what are their functions?

3. Demonstrate the activity steps for the students and the materials they will need for the project. What does craftsmanship and quality work look like?

Activity (1 hour and 10 minutes)

1. Using what they know of external and internal worm anatomy, each student will create a large three-dimensional version of a worm using poster paper and will label it with the key vocabulary words. They should write these words inside of the worm's body and can color the model using realistic or fanciful colors. (I often let them do realistic colors for the internal side and fun, free-choice for the external.)

2. One side of the paper will be labeled with external worm anatomy labels, and the other side will be labeled with internal worm anatomy vocabulary. Students will need two days to do quality work—one session for each side.

3. Students will design and color both sides of their worms. Afterward, they will cut out the worm, leaving an inch of blank paper around the image so that they can staple the two pieces of paper together and stuff them with scrap paper for a three-dimensional effect.

Assessment (10 minutes)

1. Students should be able to use worm anatomy vocabulary in conversation accurately or during a final group reflection after the lessons. They should be able to describe each part and explain its function.

2. What choices or challenges did students engage with when constructing their worms?

Preparation

1. Poster paper
2. Colored pencils
3. Stapler
4. Scrap paper
5. Diagrams of worm anatomy

Resources

- Mary Appelhof et al. *Worms Eat Our Garbage.*
- Binet Payne. *The Worm Cafe.*

Winter: Mini Worm Bin Project

1 Generating Questions

Time Frame: 45 minutes

Overview: During the winter trimester, the students will conduct a worm inquiry project in class. Time should be committed to generating student questions about worms and what experiments they may want to design.

Objectives: To review what the students know about worms and to begin designing scientific questions and experiments to study them.

Introduction and Activity

1. Introduce students to the winter inquiry project. They will be constructing mini worm bins and conducting non-harmful studies throughout the next few weeks. Examples of such studies include a worm's ability to process taste, sight, smell, light.

2. Review past questions that students have generated on their worksheets as a way to inspire students to wonder about worms. What are things students have been wondering about worms? What has surprised them in their interactions with worms and worm anatomy?

3. As a class, students will need to gather the materials needed for the experiment (see Preparation section below). Some materials will be easier for students to get than others. This is a class-wide project, so if students want to have individual worm bins, then they will have to bring in enough materials for everyone. If there is a small amount of materials brought in, then the students will work in small groups or partners.

> This project was inspired by Katie Boehnlein, a talented educator, from whom I first heard about indoor vermicomposting when I was searching for a winter science project.

Preparation

1. One-quart clean plastic soda bottles (one for each student or pair)
2. Black and white newsprint paper
3. Rubber bands

2|3 Experiment Design and Construction

Time Frame: 1½ hours (2 × 45 minutes)

Overview: Students will construct their mini worm bins and gather the necessary materials to construct a habitat.

Objective: To build a worm bin by completing each step and carefully following directions.

Introduction (10 minutes)

1. Provide the students with a small tutorial on how to assemble their mini worm bins. Emphasize the importance of craftsmanship and careful work. If the students take their time in following the directions, they will provide the best habitat for their worms. A class poster with an illustration of what a mini worm bin looks like will help the students visualize the layers of materials they will use to create habitat.

2. Arrange the supplies for the students to easily access. Students should practice safety and teamwork during this activity. Brainstorm with the class: What does good work look like today? What can students do before they start assembling in order for each team member to participate?

3. How will students make sure they are following directions? How should they behave, carefully handle worms, and take care of them?

4. What types of food should students feed they worms? What should they not feed them? What does a damp habitat look or feel like? What are factors the students can look for in creating a healthy worm habitat?

Activities (1 hour and 10 minutes)

1. In groups, pairs, or individually, the students will collect their Mini Worm Bin Experiments worksheets and follow the instructions on the first page. The groups will then gather the materials they need to build their project.

- One one-quart empty soda bottle (quart Mason jars could also work)
- One or two sheets black/brown construction paper
- Scotch tape
- Scissors
- Coffee filters
- One rubber band per bottle or jar
- Markers or colored pencils

2. Keeping safety in mind, students will cut the top part of the soda bottle off, just before it curves toward the mouth. It may be necessary to cut part of the bottle for the students to get them started. Using a marker, the students should mark their bottles in advance with lines that demonstrate how much to fill their bottles. An illustrated poster with a worm bin model and the ratios of each needed habitat material will help the students visualize what their worm bin could look like.

3. The groups will then gather the habitat materials to fill up the worm bins. The students should pay attention to the measurements/ratios of each material.

These materials include:

- Compost
- Shredded newsprint paper
- Grit (eggshells)
- Food

4. The students should use one or two dark construction paper pieces and fit them around the worm bin so that this wrapper can easily slide on and off. They can secure the wrapper with tape and should be sure to write their names on it. They can decorate this worm bin cover if they have extra time. The cover will mimic the darkness of the soil that the worms prefer. When students remove this cover for later discoveries, often the worms will be easier to see along the outside of the container and not be hiding in the center of the container to avoid the light.

5. Gathering worms: There are many ways to gather the worms for this experiment. They could be donated from a local nursery, gathered from a school worm bin, or purchased online. If there is an active worm bin available, demonstrate to the students how to gather worms using Binet Payne's methods in *The Worm Cafe.*

I usually have the students collect worms from the garden and gather a diversity of worms for the experiment. It is an opportunity for the students to actively engage with worm habitat and brainstorm ways to gather them gently.

6. Students will gather ten or more worms and gently transplant them into the worm bins. The students should then collect a prescribed amount of food scraps for their worms and layer these in the worm bin as well.

7. Finally, they will shred newsprint paper finely and lay a small amount of over the compost and soil in their worm bins. This newspaper should be heavily misted with a spray bottle until it is fully damp. Then, they can assemble a cover for the bin using coffee filters and rubber bands.

8. Students who finish early can illustrate the first page of the Mini Worm Bin Experiments worksheet.

Assessment (10 minutes)

1. Students should present their mini worm bins to be examined before class is over. They should have filled the bottles with the required materials and constructed the cover. Now is the time for the students to make sure they have all the materials gathered for future experiments.

2. Have the students check over each other's work, especially focusing on the size and quantity of worm food, as well as the thickness and quality of the newspaper layer.

Preparation

1. Poster of worm bins with ratios of habitat material
2. One-quart empty soda bottles or quart Mason jars
3. Black/brown construction paper
4. Scotch tape
5. Scissors
6. Coffee filters
7. Rubber bands
8. Markers or colored pencils
9. Sample mini worm bin
10. Compost/Soil
11. Grit (pebbles, eggshells)
12. Shredded newsprint paper
13. Spray bottle for moisture
14. Compost and worm food
15. Mini Worm Bin Experiments worksheets

Resources

- "Build a Mini Worm Bin." Oregon Agriculture in the Classroom Foundation. (online—see Appendix B).
- Binet Payne. *The Worm Cafe.*

4 Beginning Inquiry Project

Time Frame: 45 minutes

Overview: Students will check on the health of their worms and learn what to feed them.

Objective: To prepare the students and their science subjects for the upcoming experiments.

Introduction (10 minutes)

1. Students will be engaging in a 20-minute activity about what foods worms eat and then will spend the rest of garden class time checking the health of their worms, as described on page 1–2 of their Mini Worm Bin Experiments worksheet.

2. Brainstorm with the students what they know about foods worms prefer to eat in a worm bin. Do worms like hard or soft foods? Can they eat junk food or vegetables? What kinds of food would a worm eat in the garden?

3. A worm has a small mouth flap called a *prostomium* that it uses to shovel soft, decomposed, rotten, and wet food. Worms don't have teeth and instead use a gizzard to break down food particles.

Activity (30 minutes)

1. Each student will receive a piece of paper that has a food item listed on it, as found on the What Worms Eat worksheet. They should keep the identity of those items to themselves. Quietly, the students will read the paper and then decide which column their item belongs in and why. On the classroom board, set up a large poster with three columns on it, such as Love It, Like It, or No Way. The students will determine if the food item on their piece of paper will be eaten or uneaten by a worm, or should not be put in a worm bin at all.

2. While they are brainstorming, each student will get a small piece of tape to fold onto the back of the paper, and then they will put their paper under the column where they think it belongs.

3. When the students have assembled all the food items under the columns, they can look at the poster and see if there are changes they would make to where certain food items are placed. Are there foods that students argue should be in another column? Why should this food item be moved? What questions do students have about foods worms can eat? What patterns do students notice about the food items within the columns?

4. At the end of the activity, the students should discover that raw fruits, soft grains, and vegetables are preferred by worms, as well as a variety of other organic materials. Oils, meats, and dairy products should be avoided, and worms eat only small amounts of high-acid and pungent foods.

5. With this knowledge in mind, the students will gather in partner groups and check on their worm bins. Students should work together in teams and go through each part of their checklist together (pages one and two of their Mini Worm Bin Experiments worksheet). Students will need to gather food for the worms and make sure they are choosing the right types of food and small sizes of food. They can gather this food from the compost bins or garden, using gloves and tools if necessary.

6. When the students have finished the checklist, they can complete the illustration of the worm bin on page one of their Mini Worm Bin Experiments worksheets.

Assessment (5 minutes)

1. At the end of class, the student should have all items on page two of the Mini Work Bin experiments worksheet checked off and gathered appropriately sized food for the worms. All the worm bins should be moist and put back in their classroom spot.

2. Pages one and two of their Mini Worm Bin Experiments worksheet will be completed.

Preparation

1. What Worms Eat worksheet
2. Mini Worm Bin Experiments worksheet

Resources

- Binet Payne. *The Worm Cafe.*
- "Soil Decomposers." National Wildlife Federation. (online—see Appendix B).

5 Experiment #1: Light

Time Frame: 45 minutes

Overview: Students will design a test to determine how worms respond to light.

Objective: For students to begin learning the scientific method by making predictions and formulating conclusions to the experiments that they design.

Vocabulary

Photoreceptor

Introduction (5 minutes)

1. What do students know about how worms respond to light? How can worms process light or darkness without eyes? What bodily responses do worms have when experiencing light or darkness?

2. In partner groups, the students will use the materials provided to create their own experiments. They will follow the worksheet as a guide to designing and focusing their work time. How can students design an experiment with a partner? How would a scientist approach this activity?

3. Students should carefully remove the worms from their bins and onto the provided place mats or trays, practicing careful handling and worm safety. What does worm safety look like when the students introduce something, such as light, that worms are not completely comfortable with?

Activities (30 minutes)

1. With their partners, students will design a test to determine how worms sense light and observe their behavior in response to darkness, partial darkness, and light. After designing their test together, the groups can gather their materials, build the test site, and begin using the Mini Worm Bin Experiments worksheet to guide them.

2. The students should use the worksheet throughout the process, record their data, and formulate a conclusion.

> It's common for students to get distracted during the activity and only focus on the experiment, rather than data collection, so having an adult guide the activity time works best to help students complete their work.

3. After cleaning up their experiments and returning the worms to their bins, students should focus on writing their conclusions to the experiment.

Assessment (10 minutes)

1. As a group, or in smaller groups, students will share their findings and present any questions they developed over the course of the activity. What observations did the groups make on how the worms responded to light and darkness? Were they surprised by what they discovered? What body language did the worms display?

2. What are *photoreceptors*? Where on a worm's body may these receptors be?

3. If necessary, the students will finish by feeding or misting their worm bins with a spray bottle.

Preparation

1. Flashlights
2. Sheets of dark paper
3. Stopwatches
4. Place mats or trays (for worms)
5. Spray bottles

6 Experiment #2: Smell

Time Frame: 45 minutes

Overview: Students will design an experiment that explores how worms detect smell and which foods worms prefer.

Objective: To develop a greater understanding of worm anatomy by actively observing worms using their senses.

Vocabulary

1. Prediction
2. Hypothesis
3. Chemoreceptor

Introduction (5 minutes)

1. Before the activity, brainstorm with the class how worms find the food they like to eat or avoid what they don't like: Do worms eat everything they come across or do they pick what they want to eat? How are worms able to recognize food without a nose? Which foods do they prefer to eat and why?

2. What is a *prediction* or a *hypothesis*? What hypotheses do the students have about how worms smell? Write these hypotheses on a board or chart to reference later.

2. Students should remember to practice worm and human safety during this activity by carefully handling the worms and using gloves while picking up any rotting compost items.

Activities (30 minutes)

1. In their worm bin pairs, the students will use provided materials to develop an experiment that test how worms smell and what foods they prefer. They will use Experiment #2 on the Mini Worm Bin Experiments worksheet to guide their research.

2. Students will need to gather three food substances with very different smells from the compost or the garden to complete their study. They will perform two tests with different worms and record their observations on the worksheet as they go.

3. At the end of their activity, the students will have most of the worksheet filled out and the worms safely returned to the worm bins.

Assessment (10 minutes)

1. Upon cleaning up, students will have a few minutes to finish filling out the reflection page of their worksheet.

2. As a group, the students will revisit the hypotheses generated at the beginning of class. What would they change about their original ideas? Were their predictions correct?

3. What new questions do students have about worms after this experiment? After observing the worms' behavior, were students able to find out how they use their body to find food even without a nose?

4. Introduce the class to a worm's *chemoreceptors* that help sense chemical stimuli. How are these different and similar to the photoreceptors the class learned about last week?

Preparation

1. Place mats or trays
2. Spray bottles
3. Compost/food samples
4. Magnifying glasses

Resources

- Darci J. Harland. "Can Worms Smell? Wormy Experiment." STEM Mom. (online—see Appendix B).

7 Experiment #3: Life Cycle of a Worm

Time Frame: 45 minutes

Overview: The students will study the life cycle of a red wiggler worm and create a survey of worms in their bins.

Objectives: For students to be able to identify the different stages of a worm's life cycle and understand how to measure the health of their bins.

Vocabulary

1. Juvenile
2. Immature
3. Mature

Introduction (10 minutes)

1. Brainstorm with the students how many developmental stages a worm goes through in its life: What are key characteristics for young or adult worms? What does a worm cocoon look like? How can students tell if a worm is a juvenile or an adult?

2. Using a classroom poster, guide the students through the life stages of a worm and add images, illustrations, and words to help

A student fosters his scientific skills by making observations and recording data in the garden.

the students remember key features of each stage. When all the stages have been filled in, student volunteers can tell the story of a worm's life through playacting and verbal descriptions.

3. During the activity, students will continue practicing worm handling safety. Discuss what the students' behavior will look like in the context of removing all of the contents of the worm bin.

Activities (25 minutes)

1. The students will carefully remove all the materials from their worm bins onto their place mats or plates. They should follow the directions on their worksheet to explore the bins and identify each stage of a worm's life that they can.

2. They can use magnifying glasses to find the worm life stages and record the data on their worksheet as they go (Experiment #3). Before the activity time is complete, students should carefully reassemble all their bins, feed the worms, and add moisture to their bins as needed.

Assessment (10 minutes)

1. Coming back together as a class, have the students share their data on a larger chart for the students to observe. What do the numbers of cocoons, juvenile, and adult worms tell the class about the worm population in the bins?

2. What does this data say about the health of the bins and the success of these mini ecosystems? Are there changes students should make to encourage the health of their worms?

Preparation

1. Magnifying glasses
2. Place mats or plates
3. Poster on life stages of a worm

8 Experiment #4: Habitat

Time Frame: 45 minutes
Overview: The students will design a test that explores what soils worms prefer as habitat.
Objectives: To develop knowledge on the scientific method and learn more about the habitat worms thrive in.

Vocabulary

Hypothesis

Introduction (10 minutes)

1. Begin with a group discussion and brainstorm on what kinds of conditions and habitats the students know worms prefer to live in: Where do they find worms outside? What soils have students seen them live in? Are there some soils worms prefer over others?

2. Students will be developing a hypothesis on their own during this activity. What is a hypothesis? How is it different from a prediction or a guess? How do scientists use a hypothesis in their experiments? What does a well-constructed hypothesis look like?

Scientists develop hypotheses by considering what they already know about a subject and asking questions to extend that knowledge. A hypothesis is not just a guess without any supporting evidence. They are used to explore an idea that a scientist wants to know more about, rather than stating what they already know. For scientists, it's more fun to ask questions than state facts!

Activities (25–30 minutes)

1. Using the provided materials and soil samples (or ones that students gathered), the worm bin student groups will design an experiment that answers the question: What soils do worms prefer? They will use observations of worm behavior to develop their conclusion.

2. Students will record their hypothesis and data on the Mini Worm Bins Experiments worksheet under Experiment #4.

Assessment (5–10 minutes)

1. After returning the worms to the bins, the students will gather to share their findings. Overall, which soils did the worms gravitate to? What behavior did the students notice to support this claim? Did their experiments support or refute their hypothesis?

Preparation

1. Four or five soil samples (large particles, organic matter, clay, rocky soils, sand)

9 Formulating Conclusions

Time Frame: 45 minutes

Overview: Students will synthesize the data from their experiments and create conclusions about worms.

Objective: For students to use the scientific method to articulate their own discoveries about worms and their survival needs.

Introduction (5 minutes)

1. Now is the time for the students to review their Mini Worm Bin Experiments worksheets, finish any reflection pieces they may have missed, and formulate conclusions about the worms they have studied.

2. What were some of the discoveries students made during their experiments? What took them by surprise or inspired more questions? What experiments would they like to perform again? What new experiments could they do in the future?

Activities (40 minutes)

1. The students will review their worksheets and finish any uncompleted work. They should focus especially on their reflection sections of the worksheets—these will help the students generate questions for future discussions.

2. After 20 minutes, students should come back together and be prepared to share their questions and discoveries. They should strive to finish the phrases, "What I never knew about worms..." and "What I want to know more about is..."

3. How do the students feel about their projects (proud, frustrated, inquiring)? As scientists and students, what would they do differently if they were to do this project again?

4. What key things should the class take away from these experiments about worms? What should students remember and use in later classes?

5. After the discussion, students will disassemble their worm bins, compost organic materials, recycle what materials they can, and gently return the worms to the garden or worm bin.

10 Planting for Spring

Time Frame: 45 minutes

Overview: Students will plant early-spring and cold-hardy plants in the garden.

Objectives: To begin food production in the garden and encourage beneficial worm habitat by mulching the soil.

Introduction (10 minutes)

1. As a class, the students will be planting a variety of cold-loving plants in the garden, such as kohlrabi, cabbage, arugula, kale, lettuce, and onions. What are the most successful ways to plant these starts and seeds? What growing factors should students consider?

The first spring flower appears in the garden.

2. Each plant thrives with different amounts of space and companionship that students should consider. For example, how big does a cabbage grow? Will the population growth of pests be higher if too many similar plants are planted together (like brassicas)?

3. In addition to planting vegetables for a food source, the class will gently lay down straw around the starts. What are some of the impacts this straw mulch can have on the garden soil and the health of the plants?

4. Good gardening work for this activity will look like conscientious planting and slow, careful work when laying down the straw. Hands should get dirty!

Activity (30 minutes)

1. Individually, students will get a variety of plant starts to transplant into the garden or into a planting container. I prefer to inform the students about the preferences of each variety and then let them decide where would be the best place to plant. Depending on the amount of plant starts, the students can establish many new starts during the work time.

2. After they have planted the starts, the class can return all the empty planting containers before gathering straw during the last 10–15 minutes of the activity. They should focus on not wasting the straw by tossing it onto the beds or taking too much at once. Rather, they can practice taking small handfuls and carefully tucking a light layer of straw around each plant start.

3. When students have finished laying down straw around their plants, they can enjoy a taste test from the garden or early spring plants.

Assessment (5 minutes)

1. Before class is over, have the class perform a quality control check over the work done today in the garden: Are the plant roots fully in the ground? Is the straw scattered over the beds or tucked around the plants?

Preparation

1. Straw bale
2. Cold hardy vegetable starts

Resources

• "All About Organic Mulch." Bonnie Plants Organics. (online—see Appendix B).

Spring: Worms Help Grow a Healthier Garden

1 A Recipe for a Worm Bin

Time Frame: 45 minutes
Overview: Students will work together to build a worm bin for their class.
Objective: To apply student knowledge by creating a practical worm habitat.

Introduction (10 minutes)

1. Brainstorm with the students what key information they have learned from their experiments about worms and what the class should remember when assembling the worm bins. What types of habitat can students build for the worms, according to the environments worms prefer? What are beneficial materials and foods to add?

2. Worm bins can be constructed out of many different materials and can be as large or small as wanted. Some last longer than others. When building a worm bin, the students need to think about a container, bedding, oxygen, the decomposition process, moisture, grit, and food. In assembling a worm bin, the class creates a decomposing micro-ecosystem. The worms will be the focal point of this ecosystem, but other creatures will be a part of it too. The students will need to choose a material that can best handle the decomposition process, the moisture, and potential heat and cold throughout the year.

3. As a class, the students will be assembling a worm bin to care for over the next few months. Each person will have the opportunity to add materials to the worm bin.

Activities (30 minutes)

1. Using their Recipe for a Worm Bin worksheet, students will fill out the necessary ingredients for a worm bin and some examples of materials they can use to create the right environment. They can fill in these examples step by step, as the class works through adding each ingredient into the worm bin.

2. Beginning with a container, the class will then add actual bedding materials, compost, grit, soil, food, water, and worms, discussing the purpose and benefit of each material as they go.

3. Students can take these recipes home for their own use in the community or add the recipe to their garden journals.

Assessment (5 minutes)

1. How will the students be able to tell if the worm bin environment is healthy for the worms and successful at decomposing organic material?

2. Why is it valuable to keep worms as decomposers for student or cafeteria food waste?

Preparation

1. Mulch
2. Soil
3. Compost with grit (eggshells)
4. Newspaper (no colored ink)
5. Red wiggler worms
6. Watering can
7. Recipe for a Worm Bin worksheet

Resources

• "Worm Composting." Oregon Metro, Yard and Garden. (online—see Appendix B).

2 Cafeteria Waste Audit

Time Frame: 45 minutes

Overview: Students will begin to measure and study food waste from the school cafeteria or their classroom.

Objectives: To learn about the food worms can eat from the cafeteria and discuss how to manage school waste productively.

Introduction (5–10 minutes)

1. Brainstorm with the students about what kind of food is left over from lunch or meal time: What items do students see being added to the compost or garbage? From the waste they've seen, what kinds of foods do students know worms like to eat?

2. Encourage students to recall the earlier activity and poster they created about what worms like to eat. They will also want to identify foods that are eaten in the cafeteria but shouldn't be given to worms.

> If you are not able to access food waste from the school cafeteria, you may be able to gather classroom snack compost or bring in home compost. Students should always use gloves when handling food waste.

Activities (35 minutes)

1. Each student will fill out the Cafeteria Waste Audit worksheet for week #1.

2. For the initial test, the class should gather school cafeteria waste, or classroom waste, from compost bins and weigh them. Each student should record this data and document what types of food they notice in the bin, comparing it to what foods worms can eat.

3. The class should also observe if there is anything in the compost bin that shouldn't be there. They can also analyze the cafeteria garbage and discuss how best to sort out compostable cafeteria waste. Is there food in the garbage that could be composted and fed to the worms? Is there a way to raise awareness on the better composting practices?

4. After recording these observations, the students can pick out the food worms eat and use them in the class worm bin. This is an ongoing project that should continue for a few weeks, needing only 10 minutes a week after the initial test and instruction, to document cafeteria food waste.

5. With extra time, the class can make posters or announcements that raise awareness about vermicomposting to share with the school or community.

Preparation

1. Cafeteria Waste Audit worksheet
2. Scale
3. Gloves for anyone handling food waste

3 Measuring Worm Bin Health

Time Frame: 45 minutes
Overview: Students will learn ways to measure the health of a worm bin and encourage better habitat in the current bins.
Objectives: To learn how to monitor and cultivate a healthy worm bin ecosystem.

Vocabulary

1. Anaerobic
2. Aerobic

Introduction (5–10 minutes)

1. Brainstorm with the students what healthy soil looks like, smells like, and feels like. How would students describe the sensory experience of garden soil? What have the students mulched in the garden to add to its health? What does soil that is not ideal for garden plant growth smell, feel, or look like? What does healthy garden soil need to thrive?

2. Introduce the students to anaerobic and aerobic soils:

- *Anaerobic* soils have little to no oxygen and tend to have limited soil creatures, smell dank or stinky, contain a lot of water, and have limited decomposition.
- *Aerobic* soils have more oxygen for soil organisms to thrive on, leading to more decomposition, a lighter, earthier smell, and are moist, rather than waterlogged.

Activities (30–35 minutes)

1. Using the Worm Bin Assessment worksheet, students will study the smell, moisture content, and state of decomposing materials within the worm bin. In small groups, the students can receive a soil sample on a tray to investigate more closely.

2. Through focused observation, the students should make recommendations on how to improve the health of the worm bin. Students should be encouraged to touch and smell the composting system, keeping in mind how to describe sensory experiences as a scientist would.

3. The students will encounter worms in the soil samples they receive. In addition to practicing worm safety, the students can include brief comments about the various worm life stages they come across and what these stages say about the health of the soil.

4. The students can also study other soil systems, such as hot compost, cold compost, or mulched garden beds and compare what they have learned about these systems.

Assessment (5 minutes)

1. Gathering back together, the class will share their recommendations for enhancing the health of the worm bin. What words did the students use to describe the smell, texture, and moisture level of the worm bin?

2. What worm life stages did the students encounter? What does this survey say about the health of the worms? Overall, what grade would students give the worm bin's health?

Preparation

1. Worm Bin Assessment worksheet
2. Worm bin soil sample
3. Additional soil samples
4. Trays
5. Magnifying glasses

4 The Worm Bin Community

Time Frame: 45 minutes
Overview: Students will look for different decomposers and consumers in soil samples from the worm bin.

Objective: To explore the worm bin ecosystem and measure insect populations as an indication of diversity and health.

Vocabulary

1. Ecosystem
2. Population
3. Diversity

Introduction (5 minutes)

1. As a class, discuss what other insects and creatures help decompose plant material in the garden. Some of the garden compost was used in establishing the worm bin soil, so the creatures that were in it have been feeding and thriving and reproducing just like the worms. What is an *ecosystem*?

2. What are important roles played by insects and other creatures in the worm bin ecosystem? Not all the insects are decomposers. What is the importance of predators in an ecosystem? How can the students support the balance of the whole worm bin ecosystem?

Activities (35 minutes)

1. In pairs, students will remove two cups of soil from the worm bins and carefully pour it onto their place mats or observation trays. Using a toothpick, they will explore the soil samples and record any insects and creatures that they see on their Worm Bin Community worksheets.

2. They will record their findings on the worksheet and draw magnified pictures of creatures they can't identify, generating questions and conclusions about the soil samples in a reflection piece at the bottom of the worksheet.

3. Students who finish early can gather another compost example from a different composting source and perform the same observations, comparing and contrasting the different creatures in each system.

4. If possible, this would be a good opportunity to add a microscope for student studies, with each group taking turns to observe a soil sample.

Assessment (5 minutes)

1. Gathering students back together, give them time to finish up their worksheets and have them to share some of their most exciting discoveries with a partner or as a class.

2. Did the students discover more populations of one type of insect than another? What different types of creatures did each group find? Did any of the students see the insects and creatures at work or eating? What types of insects did they find that were predators or decomposers?

3. For the students who compared the worm bin with another compost sample, were there different insects or creatures that they discovered?

Preparation

1. Worm Bin Community worksheets
2. Soil samples
3. Trays
4. Magnifying glasses
5. Microscope

5 Identifying Worms in Our Soil

Time Frame: 45 minutes

Overview: Students will learn about how deep in the soil worms live and the different types of worms that thrive at various levels.

Objective: To explore and identify the diversity of worms in garden soil.

Vocabulary

1. Epigeic
2. Endogeic
3. Anecic

Introduction (10 minutes)

1. Discuss with the class what kinds of worms they have observed in the garden and around the school, on sunny or rainy days. Other than large and small worms, have the students observed different colors? For students who like to dig, how deep have they discovered a worm before? Have any students seen worms in unexpected places, like trees or ocean beaches?

2. In their inquiry projects, students should have noticed different sizes in the worms if they borrowed them from the garden. There are a large variety of worms that live in garden soil and many prefer to live in different environments. Some like the topsoil, soil surface, and others prefer to live deep in the soil; these are also known as *epigeic*, *endogeic*, and *anecic* worms (see Resources).

3. Discuss the differences between these three types of worms with the students as well as the characteristics that will make identifying them easier. What clues and features can the class come up with to remember these three types? What important role does each of the three types have in a garden?

Activities (30 minutes)

1. In pairs, students will use the Worms in Our Soil worksheet to explore three different varieties of worms in the school garden. They can be allowed to dig in specified areas of the garden or use selected soil samples that illustrate each layer of soil and the different worms that live there.

2. The students will record the number and life stages of these different varieties, identifying them as they go by using descriptions given on the worksheet. They can also measure the length of the worms, compare size of each type, and look for any evidence of worm tunnels under the soil and the direction (horizontal or vertical) that they were built.

3. When the students have finished their activity, they should gently fill in the holes that they dug and return any borrowed tools. Students who finish quickly can find a new site of garden soil to explore and compare it with the first site they analyzed.

Assessment (5 minutes)

1. Gathering together as a class, have the students share their collected data. What were the biggest worms that each group measured? What were the smallest worms? Did any groups find worm tunnels stretching horizontally or vertically?

2. Did the groups find worms that should have been in one category but were in another layer of garden soil that was unexpected? How many of each variety of worms did the students count? Were some worms difficult to identify and why?

3. What recommendations do students have to improve worm habitat in the garden?

Preparation

1. Worms In Our Soil worksheets
2. Shovels

Resources

• "Niches Within Earthworms' Habitat." Science Learning Hub. (online—see Appendix B).

6|7 Worm Waste to Fertilizer

Time Frame: 1½ hours (2 × 45 minutes)
These lessons work best as back-to-back sessions in two days.

Overview: Students will create organic worm tea fertilizer to nourish plant starts.

Objectives: To explore the natural fertilizer qualities produced by worms while discussing the use of synthetic fertilizers and the benefits of organic farming.

Vocabulary

1. Synthetic
2. Fertilizer
3. Nitrogen
4. Potassium
5. Phosphorus
6. Oxygen
7. Worm tea

Introduction (5–10 minutes)

1. Brainstorm with the class about what they know about fertilizers, either *synthetic* or *organic*, and address the different feelings around fertilizers. Some students may have strong opinions about the use of fertilizers from previous experiences and others won't be familiar with them. What do students know about the use of fertilizers in the garden? What are their purposes and what are they made of?

2. Many gardeners and scientists would argue that soil can be considered a living thing and a prized resource. Organic fertilizers, such as worm or compost tea, feed valuable soil microorganisms in addition to the garden's plants; they build soil health and stability by allowing beneficial microbes to flourish. Synthetic fertilizers do not have this effect and often deplete overall soil quality.

3. *Worm tea* is a liquid concentration of nutrients that have been processed by worms through the food they have eaten, which is then distributed to plants. Like many organic and compost-based fertilizers, worm tea contains a spectrum of vital nutrients needed by the plants, including *nitrogen*, *phosphorus*, and *potassium*. What do students know about the role each of these items plays in the growth of a plant to the production of fruit?

> "[H]arsh chemical fertilizers can dissolve the organic ties that bind particles into nice loamy crumbs. With this soil 'glue' weakened, the earth tends to disintegrate."[2]

Activities (30–35 minutes)

1. In pairs, students will follow the directions to make their own worm tea for garden plants or starts. They can gather the available materials and work together as teams to carefully create their tea.

> There are tutorials online on how to make compost or worm tea, though I recommend using one from a collegiate source. I have had students put the compost in a cheesecloth tea bag and used a small fish tank pump to push air in to feed the microbes. I have also had students stir the mixture every hour to incorporate oxygen.

2. The essential ingredients for worm tea are fresh compost (without worms) and *oxygen* (to feed the aerobic-loving microbes). "Tea bags" can easily be constructed using string and unbleached coffee filters, old socks, or cheesecloth.

3. Students can follow the compost harvesting methods detailed in Binet Payne's *The Worm Cafe*.

4. For the first part of the activity, the students should assemble the worm tea ingredients together, adding oxygen regularly to promote the growth of beneficial microbes.

5. During the second part, within a day or a couple of days of the first activity, the students can remove the compost, return it to a compost bin, and gently use the tea to water the roots of recently planted garden plants. Worm tea is a valuable resource that the students have cultivated, and they should respect this labor by using the fertilizer wisely.

Assessment

1. What does the tea smell or look like? Were students surprised by how it turned out or the process it took to create it?

2. What are other natural or organic fertilizers that students know about, such as eggshells, bonemeal, or blood meal?

Resources

- "Here's the Scoop on Chemical and Organic Fertilizers." Oregon State University Extension Service. (online—see Appendix B).
- "A Simple Way to Make and Use Worm Tea." Uncle Jim's Worm Farm. (online—see Appendix B).
- Sam Angima et al. "Composting With Worms." Oregon State University Extension Service, October, 2011. (online—see Appendix B).
- Binet Payne. *The Worm Cafe.*

8 Planting Starts

Time Frame: 45 minutes
Overview: Students will participate in a seasonal garden activity of planting favorite summer and fall vegetables in the school garden.

Creating polyculture garden beds starts with diligent planting, like this student is accomplishing.

Objectives: The class will practice considerate gardening behaviors toward each other, plants, and garden creatures.

Vocabulary
Transplant

Introduction (5–10 minutes)

1. For the activity, the students will be planting a variety of plant starts in the garden. Brainstorm with the students what each plant requires to grow and how big it will get when mature. Which plants will grow well next to get other? What does it mean to *transplant*?

> I have always preferred the wild beauty of a garden designed and planted by children. My goal is to teach them how to think like gardeners first—to know why before how. I like to let the students find the best place to plant starts and seeds on their own, after a class conversation about the needs of each type of plant. And, sometimes, the "garden fairy" can come back when the children are gone and move a few things around to better spaces...

2. Considering the size and shape of each potted plant start, how deeply will the students dig holes for the new plants? What does careful, considerate gardening look like during this activity?

3. Have the students practice the motion of removing a plant start from a pot. Some plants could be root-bound, and others could still be fragile. Brainstorm with the class what the best ways of handling these plants should be and how to treat root-bound plants so that they can thrive.

4. What types of soil organisms could the students encounter during the activity? How should they treat them?

Activities (30 minutes)

1. In pairs or individually, the students will transplant starts from planting containers into the school garden. The class should focus on careful digging and planting, rather than competitive or quick work. Good work in the garden will be plant starts that are in a hole of the right depth, that are carefully removed from their pots, and gently filled in and around with soft, crumbled soil.

2. After the students have planted their start, they can make a name marker for the plants using a wooden stick and marker.

3. At the end of the activity, the students should search the garden for any extra planting containers or shovels that were left behind. Some students can also do a quality control check to see if all the plants' roots are tucked into the soil. Then, they can enjoy a taste test from the garden before gathering back together to reflect on their experiences.

A student works carefully to sow seeds around and under established plants.

Assessment (5 minutes)

1. Did the students encounter any worms or soil organisms during their digging and planting?

2. Did students have difficulty removing the plant starts from their containers? If so, what did they do?

Preparation

1. Plant starts (brassicas, tomatoes, cucumbers, beans, herbs, flowers)
2. Shovels
3. Gloves (optional)
4. Wooden sticks
5. Markers

9 Planting Seeds in the Garden

Time Frame: 45 minutes

Overview: Students will be planting another crop under the starts they put in during the last activity.

Objectives: To learn about layering edible plants in the garden, companion plants, and to practice sowing seeds.

Introduction (5–10 minutes)

1. Brainstorm with the students what kinds of garden vegetables love sunlight and others that like partial shade. What plants have the students enjoyed eating in the garden through every season? What are the climate preferences of these plants?

2. The class will be planting seeds of plants that do well under and around the plant starts they already put into the garden. These plants will thrive in the cool shade of these places, help keep weeds down around the taller plants, and provide more food for the students and the garden ecosystem.

3. What will be the best method for the students to plant seeds? How will they ensure that the seeds are surrounded by soil, rather than sitting on top of the surface?

Activity and Assessment (35 minutes)

1. Individually or in pairs, the students will receive seeds or seed packets of a variety of plants (peas, lettuces, calendula, nasturtiums, mâche, radishes, mustard greens, dill). While being careful gardeners, the students will return to where they planted their vegetable starts and sow the seeds around each one, making sure to give them enough space around the plant.

2. The students will make sure to cover the seeds with soil and can water them with a watering can afterward. Depending on the type of seeds they use, the students can even mix a few varieties of seeds to plant together.

3. After they have planted, the students will mark the area with plant markers and the names of the seeds.

4. Before the activity is over, the class should explore the garden and pick up any seed packets left behind. Then, they can enjoy a taste test from the garden and go on a tour of other successful places where plants have been layered or thrive in the shadow of other plants. Can students find examples of plants that don't like shade? What garden plants love the sun?

Preparation

1. Seeds/seed packets
2. Wooden sticks
3. Markers
4. Watering can or cups

Resources

- Paul Alfrey. "How Much Food Can You Grow in a Polyculture?" *Permaculture magazine.* (online—see Appendix B).
- "Designing Polycultures for a Garden Setting." Learn, Garden & Reflect with Cornell Garden-Based Learning, Cornell University. (online—see Appendix B).

10 Celebration

Time Frame: 45 minutes

Overview: The students will celebrate their year of study by creating garden journals and taste testing in the garden.

Objective: The class will review all that they have learned and take away a memento of their work.

Introduction (5–10 minutes)

1. Each student will have accomplished a tremendous amount of work by the end of the year: planting seeds, mulching straw, studying internal worm anatomy, feeding worms, performing experiments, creating hypotheses, and brewing worm tea. The students will be culminating these experiences together in a garden journal from their worksheets, drawings, and puzzle sheets. Demonstrate the different ways to create this journal and where the materials are for students to use.

> I have students make garden journals in various ways. My preferred method is to have the class weave soft yarn strings through holes with a simple book bind, from top to bottom to top, and then tie them together at the top. I find the yarn is gentlest on the papers. It helps to have a knot tied on the yarn at the end the student uses to push through the holes.

Activity (30 minutes)

1. First, the students can gather taste tests from the garden to enjoy as they work. Then, they can return to the workspace to assemble their journals.

2. Each student should receive a packet of their worksheets and any papers they have gathered in the course of gardening class. Not every child will have all the papers, so they should celebrate what they have accomplished instead of focusing on what they missed.

3. Using the provided materials, they can each create their garden journal and then decorate the cover page when they are finished. An adult volunteer can be very helpful in working with students who struggle to tie knots or need a hand. Students that are finished can also offer help to their classmates.

Assessment (5 minutes)

1. Gathering back together, brainstorm with the class some of the things they believed about worms at the beginning of the year.

2. What did they never expect to accomplish? What memory will they never forget about caring for worms?

Preparation

1. Student papers and worksheets
2. Vermicomposting cover page
3. Rubber bands
4. Straight sticks
5. Yarn
6. Hole punchers
7. Colored pencils
8. Scissors
9. Adult volunteer

FIFTH GRADE—GRADE FIVE

COMPOSTING

Over their years of gardening, students will have delved into the science and systems of worms, soils, pollinators, and seeds. Now, they will study composting, an important gardening technique. Well-managed compost creates a valuable niche in garden systems, filled with microbes, predators, fungi, and decomposers who all help produce healthy garden soil. Past studies have taught students how to value soil health, but now they will take their learning a step farther by creating healthy soil from school waste and garden organic matter, and

FALL
Composting Basics

1. Garden Discovery
2. Sowing Winter Seeds
3. Dry Seed Saving
4. Wet Seed Saving
5. Green Manure
6. Mulching
7. Composting Introduction

8|9. What Does Decomposition Looks Like?

10. Compost Stew

WINTER
Compost in a Jar Experiment

1. Decomposition Timeline
2. Compost in a Jar
3. Variables and Assembling Experiments
4. Hot Compost and Cold Compost
5. Aerobic and Anaerobic Compost
6. Compost Trenches and Pits
7. Sifting Compost for Spring
8. Creating Potting Soil
9. Compost in the Garden
10. Reflect and Share

SPRING
Healthy Soils and Garden Plants

1. Planting Spring Seeds
2. Creatures in Our Compost
3. Discovering Microorganisms
4. NPK and pH Testing
5. Amending Garden Soils
6. Planting Starts
7. Mulching Garden Pathways
8. Trellises and Structures
9. Sowing Seeds
10. Garden Celebration

encouraging this process to capture energy and waste.

Seasonal garden activities will also be more complex during this year and take in account the physical and intellectual growth of these students. They will harvest smaller seeds, use more tools, and practice becoming garden leaders. Some of their activities will be familiar to them from their studies in the third grade soils unit, but these activities will now be understood in greater complexity and through deeper analysis.

In the winter, the students will transition from composting basics to a Compost in a Jar experiment and analyze how different materials degrade and change over time. They will compare systems, such as hot and cold compost, observe decomposition in nature, and learn to recognize the smelly side effects of anaerobic environments.

They will spend the springtime exploring the garden compost ecosystem, from the microscopic creatures to larger system members. Then, they will apply their studies in the garden by encouraging a healthier compost system and soil for the upcoming growing season.

© imray / Adobe Stock

"Produce No Waste: This principle brings together traditional values of frugality... the mainstream concern about pollution, and the more radical perspective that sees wastes as resources and opportunities."[1]

Next Generation Science Standards

This year, students will continue enhancing their already well-developed observation skills by studying the compost ecosystem. They will explore the energy cycling of organic matter and the changes in structure, weight, and nutrient availability that occur in compost piles/bins and throughout the garden ecosystem. The students will design experiments with a guiding question and test their hypothesis about the decomposition of everyday materials, as well as plan investigations, gather and illustrate data, and develop a conclusion and argument from their findings.

Classroom extensions from these lessons, for reading, writing, math, and social sciences, can include food waste and food systems, comparative agricultural practices, bacteria and microbe studies, and various ways organic matter cycles through different ecosystems.

Permaculture Principles

- Observe and Interact
- Catch and Store Energy
- Apply Self-regulation and Accept Feedback
- Use and Value Renewable Resources and Services
- Produce No Waste
- Use Small and Slow Solutions,
- Use and Value Diversity
- Creatively Use and Respond to Change

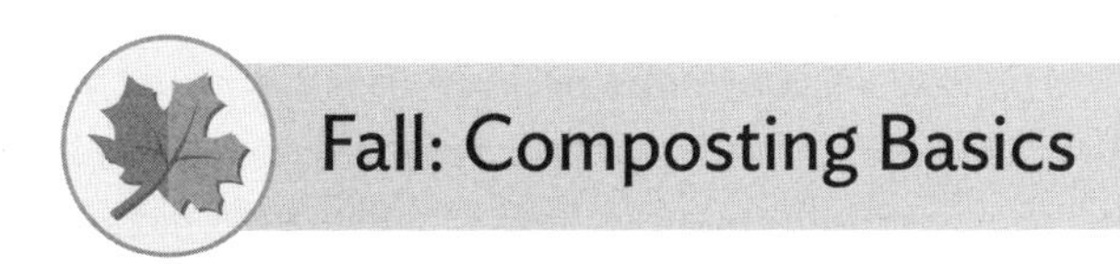

Fall: Composting Basics

1 Garden Discovery

Time Frame: 45 minutes
Overview: Students will get to know the garden, review garden behaviors, and explore seasonal changes through a scavenger hunt.
Objective: To foster positive interactions in the space and generate a discussion about seasonal changes and activities.

Introduction (5–10 minutes)

1. Welcome the students to the garden by reviewing garden expectations. What behaviors and actions do the students expect to occur in the garden? How should the class treat each other, themselves, the plants, and the animals?

2. Have students been into the garden since school started? What things have they noticed change or grow? Do they recognize anything they planted in the winter and spring of last year?

3. Did any students grow gardens this summer? What grew well?

Activities (25–30 minutes)

1. In partner groups, students will explore the garden and complete the Garden Discovery worksheets, focusing on teamwork and positive garden behaviors to accomplish their tasks.

2. When students have finished the scavenger hunt, or after 20 minutes, they can enjoy a taste test tour of the garden and share favorite garden foods with their partners. Students may need a reminder on how to harvest certain foods using a twist, snap, or two-handed method.

Assessment (5 minutes)

1. What were the students' favorite discoveries during their explorations?

2. What plants did they come across that they would like to see again in the garden next year?

Preparation

Garden Discovery worksheet

2 Sowing Winter Seeds

Time Frame: 45 minutes
Overview: Students will carefully plant cold-tolerant seeds in the garden.
Objective: To learn which edible plants prefer cold weather conditions.

Introduction (10 minutes)

1. Brainstorm with students about what kind of plants can be put into the garden in the early fall or late summer (depending on the regional climate): Does the class know of any that like cold weather? What plants have students eaten in the school garden during past winters?

2. Some plants (squash, pumpkins, and tomatoes) thrive in warm seasons, but others (kale, peas, and radishes) prefer cooler environments. Frost can kill some plants, but others will survive throughout winter. What characteristics do plants have to help them thrive in different environments?

3. The students will be planting a variety of seeds. How deeply should they plant them? What areas should they look for in the garden to plant their seeds?

Activity (30 minutes)

1. Students can be given a handful of seeds or a seed packet with a partner and should focus on finding the right conditions to plant their seeds. For example, cabbages need larger amounts of space than radishes, which can be planted close together.

2. Students should focus on planting their seeds in one area rather than scattered sporadically, so that they can make sure their seeds are planted at the right depth, covered, watered, and mulched well (if necessary).

3. When they have finished planting their seeds, the students can explore the garden for taste testing and, if possible, be offered tastes of mature plants they just seeded.

Assessment (5 minutes)

1. What challenges and successes did students have when planting their seeds? Did they find all the space their seed variety needed?

2. How was their experience planting seeds at different depths? Did they find any similarities between the seeds they planted?

3 Dry Seed Saving

Time Frame: 45 minutes

Overview: Students will collect dry seeds from the garden and label them for storage.

Objective: To practice an ancient and valuable seasonal activity by saving healthy seeds.

Vocabulary

1. Dry Seed
2. Wet Seed

Introduction (5–10 minutes)

1. Some plants produce fruits to hold the plant's seeds and others do not. Some plants have seeds in pods that are freely exposed to wind, rain, and animals. Can the students think of seed examples for either category? What are characteristics or descriptions of seeds that aren't stored in fruits? Show the students examples to help them identify key characteristics (dry, brown, brittle, rough, smooth). Chard or carrot seeds are good examples of seeds grown without a protective pod or fruit. Which of the examples could be considered a *dry* or *wet seed*?

2. Why is it important to gather some of these seeds in the fall? What would happen to these seeds if students did not collect them? What can the class accomplish through collecting the seeds?

3. Gardeners will often store the seeds they collect in paper bags and in a dark, cool, and dry area throughout the winter. Why is it important to use a paper bag versus a plastic bag? What are gardeners trying to accomplish or mimic with these storage conditions?

Activities (30 minutes)

1. In pairs, students will go into the garden to look for certain small seed varieties (flower seeds, brassicas, herb seeds) and collect them in their collection cups. They should only gather one variety at a time and not mix up them up. The students can work with their partner to identify the host plant and determine which seeds are ready and healthy to harvest.

2. When the cups are full or students are ready to collect a new seed variety, they can store the collected seeds in large envelopes or paper bags. They should work on labeling these envelopes with the right name of the plant type, spelled correctly, as well as the date the seeds were harvested.

This activity can be preceded by another where the students make seed packets to store their seeds. They can illustrate these packets with plants they will be harvesting. The packets can be sold at school sales or kept for use in the garden.

3. At the end of the activity, the students can taste test in the garden, especially from the plants they collected, if they are still edible.

Assessment (5 minutes)

1. As students nibble on their taste tests, have them discuss their experiences during the activity in the garden. What challenges did they have in identifying the seeds?

2. What did it feel like to break open the seedpods? Can they describe the feelings, sensations, and smells?

Preparation

1. Collecting cups
2. Seed and host plant examples
3. Large envelopes
4. Paper bags
5. Markers

4 Wet Seed Saving

Time Frame: 45 minutes (with optional additional time)

Overview: Students will dig into large fruits and gather healthy seeds to save for planting next year.

Objectives: To carefully identify, gather, and preserve healthy seeds for the spring.

Introduction (5–10 minutes)

1. Brainstorm with the students what they know about saving seeds from fruits: What is

Squash seeds dry for winter storage after being carefully harvested and washed.

Students work together to separate fruit pieces from squash seeds.

wet seed saving? How can students preserve wet seeds to plant them in the spring? How can students determine what a healthy fruit looks like in order to get the healthiest seeds?

2. Demonstrate the activity for the class, particularly the method of removing the seeds and drying them. Show the students where the materials and the tools they will be using are located.

2. Wet seed saving is a messy and fun project that encourages students to take their time and do good work. This activity will produce clean seeds and fully harvested fruits. The class goal should be to waste nothing.

Activities (30 minutes)

1. In two- or three-person groups, the students can gather their supplies for the activity (spoons, cups, coffee filters, masking tape). Then, they will carefully remove the seeds from a fruit like a squash or tomatillo, focusing on collaboration as they work to collect the seeds in a cup.

2. When they have gathered all the seeds, the groups can wash and scrub them in a sieve under a hose or sink. The students should try to remove all fruit membranes and particles as they scrub and sort what they have gathered. Fruit pieces can rot and ruin the viability of the seeds.

3. While some students in the group are scrubbing seeds, others should use the masking tape and a marker to label the type of seed they are saving and the date. They can also label the group name if the class plans to plant these seeds in the spring. Each label can be carefully attached to a coffee filter.

4. When the seeds have been washed, the groups will carefully lay out the seeds into a single layer on a coffee filter. It would be best if the coffee filter is in the designated drying space before the students lay out the seeds because they will quickly get wet. Putting a plate under the filter works well too. Afterward, the students can clean up and return their supplies before enjoying a taste test from the garden.

Assessment (5 minutes)

1. Students should be able to complete all the steps and correctly label their seed trays.

2. What was the experience like for the students? During their activity, what questions did they come up with about fruits and seeds? What observations did they make about the structure of large fruits? What parts of the fruit were they surprised to encounter?

3. Could students identify healthy seeds and unhealthy seeds within the same fruit?

Additional Activities

1. Now, what to do with all these hollowed-out fruit pieces? If the class has time, these would be an excellent opportunity to encourage healthy eating, nutrition, and cooking opportunities by making a favorite dish like curry, muffins, or mini-pies.

2. When the seeds have dried on the filters, the students can remove them and store the seeds in envelopes that are labeled for planting in the spring.

Preparation

1. Fruits (squash, tomatillos)
2. Containers/cups
3. Sieves
4. Coffee filters
5. Pie plates
6. Markers
7. Masking tape
8. Hose or sink

Resources

- Suzanne Ashworth. *Seed to Seed: Seed Saving and Growing Techniques*, 2nd ed.
- "How to Save Seeds." Seed Savers Exchange. (online—see Appendix B).

5 Green Manure

Time Frame: 45 minutes
Overview: The class will sow seeds of winter cover crops in certain garden beds.
Objective: To learn about how certain plants can add nutrients to the soil in the form of green manure.

Vocabulary

1. Cover Crop
2. Green Manure
3. Off-season

Introduction (5–10 minutes)

1. Brainstorm with the class what they know about manure: Why do gardeners add manure to garden beds? What nutrients does manure add to the garden soil and how can it create healthier soils?

2. Nitrogen is one element that is important in growing healthy annual garden plants. Summer plants feed on a lot of nitrogen throughout the growing season, and it is important for gardeners to replenish nitrogen in the soil for future planting. Applying manure is a way for gardeners to put nitrogen back into soil.

3. *Green manure* is not animal waste, but rather organic matter. Plants that are adept at capturing nitrogen from the atmosphere are especially valuable to gardeners and farmers. Their decaying bodies release stored nitrogen back into soil for the next growing season, as well as other nutrients and energy. These cover crop plants also benefit the soil by keeping weeds down and growing in the beds during the *off-season*.

4. Cold-tolerant seeds that capture nitrogen include: fava beans, clover, peas, vetch, and other legumes. Depending on the climate, oats and rye are other cover crop options and benefit the soil when mixed with legumes.

Activities (30 minutes)

1. Individually, students will receive handfuls of cover crops seeds to plant in the garden. They will plant them in prepared garden beds that have been cleared of top mulch so the students can press their seeds into the soil. When the activity is over, a little mulch can be raked back over the seeds.

2. Students should focus on carefully planting their seeds by making sure they are connected to the soil and are well spaced out. At the end of the activity, students can taste test available nitrogen-capturing plants in the garden (beans or peas).

3. Student volunteers can also water the seeds with a watering can during or after the activity.

Assessment (5 minutes)

1. Were there other seeds, plants, or organisms that students came in contact with during their planting?

2. How long do students think it will take the cover crop seeds to grow?

Preparation

1. Winter cover crop seeds/legumes
2. Examples of nitrogen-fixing plants

6 Mulching

Time Frame: 45 minutes

Overview: Students will learn how to prepare the garden for winter decomposition and for healthier soil in the spring.

Objectives: To learn about cold compost and what materials can be added in the garden to encourage the decomposition of vital, nutrient-rich plants.

Vocabulary

1. Cold Compost
2. Green and Brown Materials
3. Mulching

Introduction (10 minutes)

1. Brainstorm with the students about what they know about mulching: Did they do it before, and what materials did they use to mulch the soil? Why is mulching a vital process for garden soil?

My ear hears
the rabbits say time for fall. Time
for drizzles of rain. Time for
leaves to fall. Time for plants to
go to bed and rest their sleepy heads.
Time for bees to stop their busy
work. Time for rain showers.

— Anonymous Student Poem

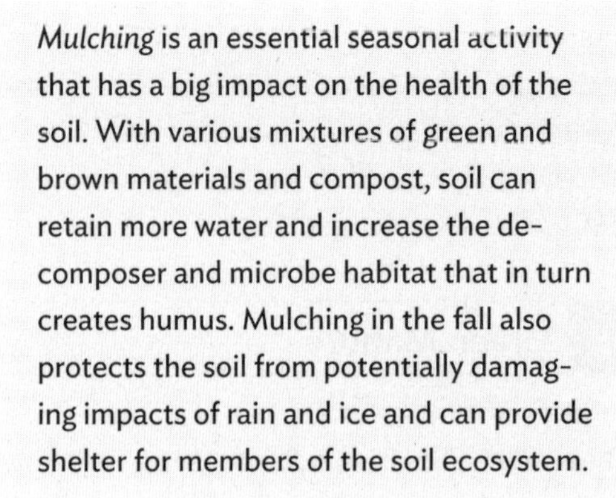

Mulching is an essential seasonal activity that has a big impact on the health of the soil. With various mixtures of green and brown materials and compost, soil can retain more water and increase the decomposer and microbe habitat that in turn creates humus. Mulching in the fall also protects the soil from potentially damaging impacts of rain and ice and can provide shelter for members of the soil ecosystem.

2. What are *green and brown materials*? What plants in the garden are examples of each? What does *cold compost* mean?

3. Demonstrate how to mulch a garden bed with dry or brown materials by carefully tucking the material around still-growing plants and not too thickly over seeds and sprouts. The students can also be introduced to any green materials that can be chopped down at the end of their life cycle and laid on the garden beds to decompose.

> "Nutrients in shed leaves are processed by soil organisms and converted to humus that can then feed the plant. This also has the effect of stimulating a very rich soil ecosystem."[2]

Activities (30 minutes)

1. As a class, the students will mulch the garden beds with brown and green materials. They should focus on doing quality work by mulching slowly and carefully rather than rushing to cover a lot of ground.

2. Students can also chop back green materials that are spent from the growing season, such as tomatoes or sunchokes, and break them down into small pieces using gardening tools. These pieces can be tucked into the garden beds and mixed with brown materials.

3. Other students can use a rake to gather any extra or loose materials that have dropped on the garden pathways. The class can accomplish this together or rotate through in groups. At the end of the activity, they can carefully put the tools away, clean up, and enjoy a late season taste test from the garden.

Assessment (5 minutes)

1. Have the students return to the garden and do a quality control check of the class's work, fixing over-mulched beds or plants covered by the brown material.

2. If there is time, have the students dig into a part of a garden bed to observe the many years of mulching materials in the soil. What is recognizable material and what isn't?

Preparation

1. Straw or leaves
2. Dead plant stalks
3. Clippers and loppers

7 Composting Introduction

Time Frame: 45 minutes

Overview: Students will be introduced to composting basics and learn key words.

Objectives: To begin understanding the composting process and what materials can decompose in the school compost.

Introduction (10 minutes)

1. Brainstorm with the students what they know about composting: Have students composted before? What important factors go into composting or are key players? Work with the students to develop their own definition of composting.

2. During the activity, the students will be separating various items into four categories: school compost, landfill, industrial compost, and recycling. Establish a basic understanding of each of these categories as a class before beginning the activity.

Activity and Assessment (35 minutes)

1. Students will work in pairs or individually to fill out their Where Does This Go? worksheets, by cutting out each section and identifying how each waste item will be disposed as school compost, landfill, industrial compost, or recycling. The students should follow the directions on their worksheet and not glue the pieces onto the columns until the end of the activity.

2. After students have cut their pieces and categorized them to the best of their ability, they will come back together as a class and discuss why they chose certain categories for the items on the Where Does This Go? worksheet. They can correct their answers as they go and glue the item names onto the correct spaces. During this time, the class can brainstorm why certain items should not be put in the school compost and clarify definitions of what each category means. For students who finish early, they can begin the Decomposition Timeline worksheet.

Preparation

1. Where Does This Go? and Decomposition Timeline worksheets
2. Scissors and glue

Resources

- "Trash Timeline: Exploring the Biodegradability of Trash." Alice Ferguson Foundation. (online—see Appendix B).
- "Approximate Time It Takes for Garbage to Decompose in the Environment." New Hampshire Department of Environmental Services. (online—see Appendix B).

8|9 What Does Decomposition Look Like?

Time Frame: 1½ hours (2 × 45 minutes)
Overview: To explore compost through a hands-on activity and identify its key properties.
Objectives: Students will recognize the features and stages of decomposition.

Vocabulary

1. Green Materials
2. Brown Materials
3. Humus

Introduction (15 minutes)

1. During the last lesson, students began brainstorming what goes into school compost and what doesn't. During this activity, students will look at school compost samples and identify the various items they find.

2. There are three key components to the garden compost system: green materials, brown materials, and microbes/decomposers. *Green materials* contain high amounts of nitrogen, are typically fresh and full of moisture. *Brown materials* provide carbon to the compost pile and are dry materials such as straw, leaves, and newspaper. Living creatures, particularly decomposers, are essential to the decomposition process and breaking down organic matter into *humus* (see Resources).

3. During the activity, students will look for items in the compost, work to identify them, and determine which of three categories the items belong to. What green, brown, and living things do students expect they will find?

4. Before students begin their activity, they should review the class expectations on how to treat and handle living creatures.

5. The goal for the first activity should be to finish Bin #1 of the worksheet. The second activity will be to finish Bin #2, a compost sample from a different compost bin, preferably one that is more developed than the first sample. The compost samples can be gathered in advance by an adult or student, or the student groups can gather their own samples from designated compost areas by digging six inches into the pile before harvesting their sample.

> "[O]rganic matter is the raw material you put into the compost heap, and humus is the rich, dark, crumbly finished product."[3]

Activities (2 × 30 minutes)

1. Each student will receive a worksheet and magnifying glass. Then, as individuals or in pairs, the class will gather a compost sample, explore it, and identify different items that they can recognize. They can categorize the items as green, brown, or living organisms on their worksheet.

2. Good work will look like a thorough analysis of the compost, labeling on the illustrations, and thoughtful reflection about the final questions.

3. After they have finished their worksheets, the students will clean up their stations, carefully return the compost samples, and gather together for reflection.

Assessment (15 minutes)

1. What living, green, and brown materials did the students discover in their samples? Which materials were most decomposed, the green ones or brown ones?

2. Were there items in the students' samples that didn't fit into either category? What kinds of living creatures and decomposers did they discover?

Gavin Blackwell

"Thistle Man and His Army"

A student poster created during one school's Worm Appreciation Week.

A Student's Portrait of Poppies From the Garden

Gavin Blackwell

"Flowers"

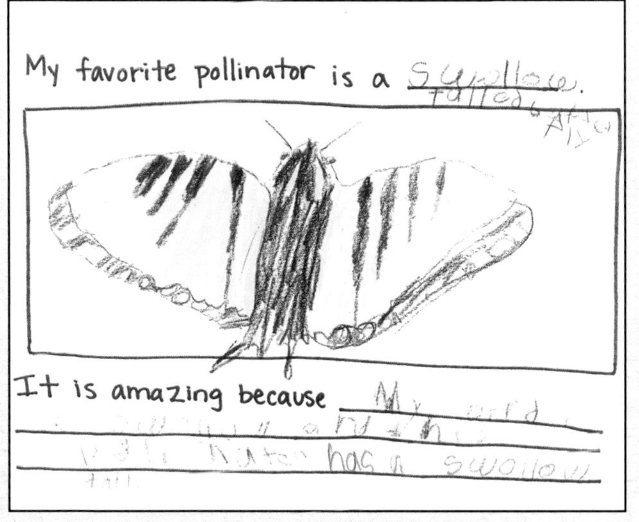

A young student illustrates the finer details of a swallowtail butterfly.

Observing the garden from a slug's perspective is an opportunity for students to foster an ecological mindset.

Teaching about life cycles can be as simple as letting an artichoke flower go to seed.

All predators are valued and celebrated in the learning garden, including this small garter snake.

Dirty hands are a sign of good gardening work.

Student-made garden signs identify specific edible plants.

The first frost comes to the garden and signals a natural shift in garden production and student activities.

While the end of the garden's life cycle is not the prettiest sight, many children have never seen flowers, such as this zinnia, during this life stage.

Ready to wear gardening gloves for all ages, from preschoolers through middle school students.

Students mix together the ingredients for their own potting soil.

Making seed balls is a fun, essential, messy, and community-building experience.

Integrated pest management is a goal and joy of the eco-minded gardener.

An adventurous child enjoys a whole pattypan squash from the school garden.

A polyculture planting with layered edible and beneficial plants for students, birds, and insects to all enjoy.

Indigenous flower species, like these beach daisies, provide nectar for pollinators and thrive throughout seasonal changes.

Winter in the school garden still offers great opportunities for exploration, learning and wonder.

Heat sensitive plants, such as brassicas, can thrive in the shade of the summer-sun-loving varieties with companion planting techniques.

By letting plants live out their life cycles, students have the opportunity to enjoy more edible parts of plants and diverse flavors, such as with these arugula flowers.

A Kindergartner plants bean starts around a student-made trellis in the learning garden.

Every pollinator, even this beetle, is celebrated for the role it plays in the garden ecosystem and for food production.

Students co-build non-harmful insects traps and learn more about surface level creatures at work in the garden ecosystem.

It's time to transplant student winter, indoor-grown experiments into the garden.

The calendula flower is a wonderful and vibrant learning plant by providing medicine, food, and nectar.

The story of the sunflower life cycle is a favorite for young students, and its hidden math is enjoyed greatly by older students.

A mason bee crawls across a student-made bee house in the learning garden.

This cold-hardy cauliflower will provide lunchroom food, but its leaves have also been taste-tested and enjoyed by inquisitive children.

A student plants flower starts to encourage beneficial insect activity around the garden foods.

Rhubarb is a fantastic perennial plant that is a favorite of children during taste testing opportunities.

A honeybee gathers pollen in its sacs and with its furry body.

Even a small insect can have a beneficial role in the garden. Here, ants eat away the peony's sugary coating and may help its flower bloom.

The first carrot of the spring season marks a special day in the garden.

A student advocates for worms by creating this poster for the school worm bin area.

3. What general comparisons were the student groups able to make about the two different compost piles? What materials took longest to decompose? Were there materials in the samples that would not decompose at all or for a much longer period of time? How did the students determine this?

4. What discoveries took the students by surprise during this activity?

Preparation

1. What Does Decomposition Look Like? worksheet
2. Magnifying glasses
3. Clipboards
4. Pencils
5. Compost samples

Resources

• US Environmental Protection Agency. "Composting At Home." (online—see Appendix B).

• How to Compost homepage. howtocompost.org, 2013. (online—see Appendix B).

10 Compost Stew

Time Frame: 45 minutes

Overview: Students will focus on what materials go into creating healthy compost.

Objectives: To better discern which composting materials are green and brown as well as what can or cannot be added to compost.

Introduction (10 minutes)

1. As a group, the class will brainstorm the green and brown materials students come into contact with at school, such as during snack, lunch, or in the cafeteria.

2. They will also brainstorm what materials shouldn't be added to the compost by using their observations during past compost explorations to guide them.

Activities (30 minutes)

1. In small groups, students will work on creating composting posters to put on or near the classroom, garden, or cafeteria compost bins. These posters will illustrate how each class can make their own compost stew and what brown and green materials can be added to the classroom compost bins.

2. Students can use available resources and books to enhance their posters and materials lists. They should work as a group toward the goal of creating clear, concise, and well-crafted posters to educate the school and community. All age groups should be able to clearly understand these posters.

Assessment (5 minutes)

1. After students have cleaned up their work areas, they can participate in a short tour of the class's work, observing how other students arranged their posters and what materials they added and used to illustrate their work.

2. As a class, what are the most important ingredients the students would recommend to add into the compost bins? What are common and valuable composting ingredients? What should not be added to the school compost?

Preparation

1. 8½ × 11-inch sturdy white paper for posters
2. Pencils, markers, colored pencils
3. Reference books

Resources

• Mary McKenna Siddals. *Compost Stew: An A to Z Recipe for the Earth.*

- David Squire. *The Compost Specialist: The Essential Guide to Creating and Using Garden Compost, and Using Potting and Seed Composts.*
- Michelle Eva Portman. *Compost, By Gosh!: An Adventure with Vermicomposting.*

NGSS and Activity Extensions
Further in-class studies can include having students "measure and graph quantities to provide evidence that regardless of the type of change that occurs when heating, cooling, or mixing substances, the total weight of matter is conserved" (5-PS1-2: Matter and Its Interactions).

Winter: Compost in a Jar Experiment

1 Decomposition Timeline

Time Frame: 45 minutes
Overview: Students will explore the different rates of decomposition for various materials and brainstorm about their winter science project.
Objectives: To begin understanding decomposition in the context of time and delve into a class conversation about landfills and waste management.

Introduction (10 minutes)

1. Students can be introduced to the focus of the new season: a composting experiment that they will design and conduct. At the end of class, the students should have two or three potential decomposing materials which they would like to study.
2. What are realistic materials that students could study? How would studying these items support a larger project goal of improving the school or community compost system?

Activities (30 minutes)

1. In pairs or individually, students will complete the Decomposition Timeline worksheet, using a pencil. When the students have finished, they will come back together as a class and review their answers.
2. As each material is announced with the time it approximately takes to decompose, facilitate student questions and discussions about why certain materials take so long and what effects this timing has on ecosystems, landfills, and the human waste stream. How do humans know how long it could take certain materials to decompose? What was most surprising to the students about the materials on the list? Did they match any times and materials correctly? How did they come to their conclusions?
3. During the last part of the activity, the students will break into partners for the first brainstorm about their Compost in a Jar experiment. Keeping in mind how long it takes some materials to decompose, the partner groups will come up with three items that they could want to observe decomposing over the next two months. The study of these items should benefit the community in some way, or answer a burning question that students or the class have about decomposition.

Assessment (5 minutes)

1. What kind of materials does each group want to study decomposing? What are key ingredients that will need to be added to aid decomposition?

2. How long do students expect it will take for their material to decompose? What will decomposition of their chosen materials look like?

Preparation

1. Decomposition Timeline worksheet

2 Compost in a Jar

Time Frame: 45 minutes

Overview: Students will begin their winter science experiment by creating a guiding question and a hypothesis.

Objective: Each group will develop a realistic and testable project to study for the next few weeks.

Vocabulary

Hypothesis

Introduction (10 minutes)

1. The goal of the class is for each group to have an experiment planned by the end of the work period, so that they can bring in their supplies over the week and assemble the project during the next class time. The students will set up the experiment by using two techniques that scientists practice before beginning an experiment: creating a guiding question and posing a hypothesis.

2. As a class, discuss what a guiding question is and how it helps a scientist prepare for an experiment. Brainstorm with the students about what is involved in creating a guiding question and what a strong question looks like. What is a *hypothesis*? Why do scientists practice this at the beginning of an activity? What is essential in composing a strong hypothesis? Record examples of guiding questions and hypothesize on a board or poster. The student groups can use this language and structure as a reference when developing their own sentences.

> During this activity, I typically have the groups generate a guiding question first, and then move on to a hypothesis afterward. Developing these ideas piece by piece is less overwhelming for the groups since this lesson can be the first time that many students have developed a strong hypothesis or guiding question.

Activity and Assessment (35 minutes)

1. In their small groups, the students will review the brainstorming sheet from the previous class, pick a project they want to pursue, and build a guiding question for their experiment

A student assembles the layers for the Compost in a Jar experiment.

on the Compost in a Jar worksheet. Finished questions can be reviewed for strength and clarity by a teacher or by another student group.

2. When the pairs have developed a guiding question, they will formulate a hypothesis about what they think will happen during the experiment. They will need to work together to make sure they have all the essential pieces of a hypothesis.

3. Students can share their group projects with the class and determine which partners will bring in the necessary materials for the next class.

Preparation

1. Compost in a Jar worksheets

3 Variables and Assembling Experiments

Time Frame: 45 minutes

Overview: Students determine the variables of the project and will build their experiments using the provided materials.

Objectives: To carefully construct the experiments and add the proper nitrogen/carbon ratio for decomposition.

Vocabulary

Variables: Independent, Dependent, and Controlled

Introduction (10 minutes)

1. Brainstorm with the class what variables will affect their experiments: What is a variable? As a class, define independent, dependent, and controlled *variables* and record examples of each for the class to reference as they record their own project variables on their Compost in a Jar worksheets.

The focus of this experiment is to observe the decomposition of certain materials in a carefully maintained compost environment. The emphasis on variables can be as detailed or vague as you would like. There are some deviations from a strict experimental process during the course of this experiment because I want the students to interact with it. Understanding variables, however, is important; students can learn and acknowledge the role variables play in an experiment and how they can best maintain the controlled variables throughout the decomposition experiment. I recommend giving the students time in the beginning of class to work as a group and record their own project variables on the worksheet before moving on to the second activity of assembling the experiments.

2. Brainstorm with the class what important ingredients will aid in decomposition: What kinds of materials can students use for good carbon/nitrogen or brown/green materials?

Many sources advise a 3:1 ratio of carbon to nitrogen, while other sources advise a higher ratio of carbon depending on the material. For this project, I typically have the main ingredient be mostly decomposed school compost that is already full of a healthy decomposer ecosystem.

Activities (30 minutes)

1. Each group will gather the materials they need and assemble the experiments by following directions on their Compost in a Jar worksheets. They will make sure to mix the materials together and have their composting item fully encompassed in the jar.

2. The groups should work together to make

sure that they have the right proportions of nitrogen and carbon materials, as well as compost. They can label their experiments and store them in a spot reserved for composting (I've often used a small shed or protected area outside just for composting and worm bins).

3. When the students have finished assembling their bins, they can clean up their stations and turn in their worksheets for review.

Assessment (5 minutes)

1. What was the experience for students assembling their compost bins? Was it difficult to fill the bins with the right ratios or mix up the materials?

2. What changes do they think their items will undergo over the next week? What will the first stages of decomposition look like?

Preparation

1. Plastic wrap for cover
2. Rubber bands
3. Extra containers
4. Tape
5. Markers
6. Compost
7. Straw/Shredded paper/Newspaper

Resources

- David Squire. *The Compost Specialist: The Essential Guide to Creating and Using Garden Compost, and Using Potting and Seed Composts.*

4 Hot Compost and Cold Compost

Time Frame: 45 minutes

Overview: Students will clean up the school garden by gathering organic materials and sorting them into piles for hot and cold composting.

Objective: To understand the differences between hot and cold composting and which garden materials do best in each system.

Vocabulary

1. Cold Compost
2. Hot Compost

Introduction (10 minutes)

1. Begin with a class brainstorm on different types of decomposition: What types of decomposition do students see happening in the garden? What methods are planned and what happen naturally? What are essential ingredients for healthy garden decomposition?

2. Introduce the students to the idea of cold compost and hot compost:

a. *Cold compost* is a slow, decomposing pile of organic materials (brown and green) that require very little maintenance or work. This method takes longer to break down and obviously, by its name, does not generate a lot of heat from a large organism base, as hot compost does. Cold compost usually doesn't kill weed seeds because it lacks heat. Depending on the climate, cold compost usually only requires rainwater to keep it active. Cold composting is a good method for gardeners who don't want a high-maintenance composting system, even if the compost won't be ready to harvest for many months or a year.

b. *Hot compost* is often contained in a closed system or barrel and requires manual watering and hands-on maintenance. By keeping a careful balance of brown and green materials, a hot compost system will heat substantially through the work of a thriving microbe and organism base. This heat can kill weed seeds and process food waste in a matter of weeks or months.

3. During the activity, the class will be encountering a variety of organic materials and sorting them into hot compost and cold compost piles. Demonstrate which plants and organic materials the students will be working with. How can students tell which materials should be in each pile? Should there be additional piles for items like diseased plants?

Activities (30 minutes)

1. As a class, students will be removing weeds, raking fallen leaves or branches, pruning shrubs, and gathering other organic materials from the garden to be used as compost in both hot and cold systems. They will gather these materials and then sort them into the designated piles.

2. The students should practice slow and careful gardening during this activity, making sure that they are correctly identifying plants and weeds. Other than learning about how to compost various materials, the class goal is to clean up the garden from the effects of winter and to prepare it for spring.

> In the past, I have had students collect branch clippings from the fruit trees I pruned, fallen leaves, dead vines, mulching straw that has fallen out or blown away from garden beds, and dead annual plants. Additionally, there may be persistent weeds that went to seed which could be removed.

3. After 20 minutes of gathering and piling materials, the class should come together to examine what was gathered and determine if all the materials should go in their designated area. Are there some materials that should not go into the hot or cold compost system at all? Are there items that would be better additions to other pile? After these discussions, have the students re-sort items into the right composting systems.

4. When the students have finished their activity, they will check on the decomposition progress of their experiments. Using the worksheet Measuring Decomposition, they will examine the changes of their decomposing materials and record this data on their worksheet. They should also be sure to check moisture level, smell, and decomposer activity during this time.

Assessment (5 minutes)

1. What changes did the students notice with their decomposition project? Were there changes in smell, heat, or color? Are there some project materials that are decomposing at a faster rate than others?

2. During the garden cleanup activity, were there some materials that the students encountered which they debated in finding a pile for?

Preparation

1. Plant and organic material examples
2. Clippers and shears

Resources

• David Squire. *The Compost Specialist: The Essential Guide to Creating and Using Garden Compost, and Using Potting and Seed Composts.*

5 Aerobic and Anaerobic Compost

Time Frame: 45 minutes

Overview: Students will learn about the differences between in aerobic and anaerobic decomposition and identify the processes in their compost experiments.

Objective: To learn about the process of aerating compost to encourage aerobic decomposition.

Vocabulary

1. Aerobic
2. Anaerobic

Introduction (10 minutes)

1. Brainstorm with the class what they know about the importance of oxygen in the decomposition process: What happens to organic matter and decomposers when there is little to no oxygen?

2. Introduce the students to the terms *aerobic* and *anaerobic*. Some students may be familiar with aerobic exercise, which encourages the body to move and absorb oxygen. Encourage the class to build a definition for each type of decomposition and write them on a board for students to reference later. The class can use the same classroom poster to share data at the end of the activity.

3. Where are examples of oxygen-low environments in nature? What do these ecosystems look or smell like? What type of decomposition does the class want to foster in the school garden and why? How can students support this type of decomposition?

Activities (30 minutes)

1. In pairs, students will work on the definitions of aerobic and anaerobic on their worksheets by reflecting on the class conversation. Then, they will visit four or five stations with a soil or compost sample and will make observations about the materials and the types of decomposition that are occurring. The groups will diagnose whether the samples are aerobic or anaerobic. These samples could include school compost, old wood mulch, food waste, two student volunteer samples, or even wetland soil.

Students observe examples of aerobic and anaerobic environments in the garden.

2. The groups can analyze and make observations on the smell, moisture level, visible materials, soil particle size, and texture of each sample. When the students have finished exploring the compost samples, they can record their findings on the class board and will conclude which samples are aerobic or anaerobic.

3. Then they will return to their compost experiments and use what they know about aerobic and anaerobic decomposition to determine the health of their compost so far.

4. The student groups will aerate their compost gently and record decomposition data on their worksheets. They may also need to amend their composting systems by using provided materials (brown and green material, water) and record these additions on their worksheets.

Assessment (5 minutes)

1. Coming back together as a class, the students will discuss their findings with the compost samples and their own experiments.

2. For those groups who discovered anaerobic decomposition in their group projects, what were the key indicators of that process? How did students discover aerobic compost? What characteristics did they look for before coming to their conclusions?

Preparation

1. Soil or compost samples
2. Aerobic and Anaerobic Worksheets

Resources

• Texas A&M Agrilife Extension. "Chapter 1, The Decomposition Process | Earth-Kind® Landscaping." (online—see Appendix B).

6 Compost Trenches and Pits

Time Frame: 45 minutes

Overview: The class will use roughly processed compost or food waste to create compost trenches or pits in garden beds.

Objective: The students will discover alternative ways to process food waste in the school garden.

Introduction (10 minutes)

1. One large problem that occurs when schools try to compost their waste is the large quantity of food scraps generated every day. Sometimes hot and cold composting systems cannot keep up with the constant flow fast enough, which may lead to unsafe and unproductive systems.

2. One solution is to dig compost pits or trenches in the school garden beds during the slow production season. By digging in compost, whether raw or roughly composted, at the root level of the summer and fall plants, school gardens can find a solution for the school's large food waste stream and provide nutrients and moisture for future plants. When the students are ready to plant seeds and starts in the garden beds, the compost will be significantly decomposed.

3. Brainstorm with the students about how deep the pits or trenches need to be for plant roots to reach future decomposed organic matter, at the same time avoid attracting pests. What is the best way to work with the plants still growing in the garden? What time of year would be best to dig a trench or a pit?

Activity (30 minutes)

1. In pairs or small groups, the students will dig trenches or pits about 1–1½ feet deep and fill them with 4–6 inches of rough compost or raw food waste. They should make sure that any fruit stickers, trash, or plastic are not buried in the garden beds. Students should use gloves when handling unprocessed or roughly processed compost.

2. The class can also consider what types of materials they are putting into the trenches. For example, some groups may choose not to bury all of the orange peels or citrus fruit in these trenches.

3. Working in designated garden beds, the students will need to be aware of any living plants in the garden. If they are working near cover crops, they can dig in trenches that incorporate that green manure into the mix. If working in beds with well-established plants, they should be considerate of the roots and dig pits, instead of trenches, in those beds.

4. When the students have finished digging the pits or trenches and have filled them with compost, then they can gently cover the holes with any garden soil they excavated.

5. As the activity comes to an end, provide

enough time for students to clean up their tools, gloves, and worksite.

6. Students should perform a quick check on their decomposition projects by measuring the rate and status of their materials and recording this data on their worksheets.

Assessment (5 minutes)

1. What composting materials did the students put into the garden beds? Were there more of one type of organic material than others? Did the groups discover evidence of other decomposing materials in their digging?

2. How long do students think it will take for these underground materials to decompose? What other benefits are there to trench/pit composting?

Preparation

1. Gloves
2. Shovels
3. Raw compost

7 Sifting Compost for Spring

Time Frame: 45 minutes

Overview: Students will sift processed compost into finer particles to use in the garden.

Objective: To learn how to harvest compost and determine when it is finished by recognizing key features.

Introduction (10 minutes)

1. Brainstorm with the class when they think the composting process is complete for garden use: Is the product different for hot or cold compost systems? Are there key features that gardeners use to recognize finished compost?

2. The goal of garden composting is to create rich humus whose biome and nutrients will feed the garden soil ecosystem and the plants. What does humus look, feel, or smell like?

3. Illustrate the differences between compost of various stages of decomposition by passing around samples of unfinished and finished compost. What smells, textures, colors, and features do the students notice about the finished compost? Encourage the students to use their senses in exploring the differences and to touch the samples for texture and moisture content.

> Finished compost should look like a rich dark or brown soil base, be moist to touch, and have an earthy smell. There should be very little recognizable organic material pieces in it, and it should have crumbly, small-medium soil particles.

Activities (30 minutes)

1. In pairs or small groups, students will harvest finished compost from a school compost pile. They will carefully gather compost by using a shovel and bucket and then transport it to a wheelbarrow or larger bin.

2. Each group will pour their compost onto a sifting tray and gently shake the tray back and forth between two or more students. Any larger organic matter pieces that don't make it through the tray will be brought back to the compost pile to decompose.

> My favorite sifting trays are a thick black plastic growing trays usually used for holding 4-inch plastic planters. The holes are about ¼-½ inch wide, and the tray is sturdy enough for students to use without the weight of compost making for messy work.

3. The students will repeat this process until the wheelbarrows are full. The groups should keep track of what materials are not completely decomposed. Clay or dirt clods can be broken up before being returned to the compost pile, and any insects or bugs should be carefully returned as well. The finished compost will be saved for activities over next two weeks.

4. After the activity, when students have cleaned up and returned their tools, they will check on the stage of decomposition that their project material is undergoing and record this data on their worksheets before gathering back together as a class.

Assessment (5 minutes)

1. What were the students' experiences in sifting compost? Was it difficult or easy to do? What less decomposed materials did the students come across during the activity? What does the school garden compost look, smell, and feel like?

2. Knowing what they do about finished compost, how close is each group's compost jar to being finished? How much longer would each group estimate their material to decompose? Is this time frame different than the group's original estimate?

Preparation

1. Screen or small-holed planting tray
2. Wheelbarrows
3. Shovels
4. Finished compost pile
5. Compost examples
6. Buckets
7. Gloves

Resources

- "Sifting Compost" in Patrick Lima. *The Natural Food Garden: Growing Vegetables and Fruits Chemical-free.*

NGSS and Activity Extensions

For further in-class extensions, the students can create a model that reflects the movement of matter among plants, animals, and ecosystems, such as a soil or compost system (5-LS2-1: Ecosystems).

8 Creating Potting Soil

Time Frame: 45 minutes

Overview: The class will be creating potting soil from the compost they sifted as well as other available organic materials.

Objective: Students will practice using compost to grow healthier plants and learn how to create a valuable seasonal resource for free.

Introduction (10 minutes)

1. Brainstorm with the class what they know about potting soil: When and why do gardeners use potting soil? What are the benefits of potting soil for seedlings and young plants? How much does it cost to buy potting soil?

2. Potting soil provides a soft, water-retaining, and nutritious soil base for new plants or potted garden plants. Rather than purchasing it, the class will be making their own valuable seasonal resource to start seedlings. They will use the finished compost they sifted as well as other available organic materials.

> It is helpful to have uniform potting soil that is consistent in water drainage and retention. Many gardeners have different mixes and recipes that they like. I prefer to use what I have on hand because I am used to gardening on a small or non-existent budget. I often recycle potting soil from spent pots, gather sand, and take whatever donations of organic materials I can use. Ground eggshells or organic bonemeal and blood meal are also good additions.

3. The goal in making potting soil is to create a growing medium that allows water to move slowly but consistently so that the plant roots can access it, but aren't drowning in moisture. Nor are plant roots getting too dry because the potting soil allows water to move too fast, or lets water sit on the surface because its particles are too small. Soil particle size is very important when creating potting soil, and the students will want a mix of different sizes, small to medium, to help retain and move water at a good pace. Particles which are too small (clay or silt) can lead to trapped water, and particles which are too large (dirt clods and bark chips) allows water to escape too quickly.

4. Demonstrate these experiences by having student volunteers pour the same amount of water into four-inch potting containers filled with small to large particles of organic material. What observations can students make about the way water drains through each pot?

Activities (30 minutes)

1. In pairs, students will create potting soil from the materials on hand, while following the recipe on the classroom poster. They will put their ingredients in three- to five-gallon buckets and make sure to mix all the ingredients carefully together with their hands or a shovel.

2. Student can test the water flow of their mixture by taking a sample of their potting soil and pouring an inch of water into it. They should observe how the water flows, sits, or drains through the container. They can also dig into the soil to see how deeply the water was absorbed. Have the groups troubleshoot any problems and brainstorm as a group how to adjust mixture and particle size to resolve any issues.

3. When students have finished creating their potting soil, they can store their mixtures in a safe space for a later activity. Then, they can clean up their work area and check on the decomposition process in their project, record their findings on their worksheets, and make scientific observations about any changes.

Assessment (5 minutes)

1. What problems did the groups solve with their potting soil mixtures during the activity? What solutions did they come up with?

2. What changes are the group project materials undergoing during decomposition? Are there significant changes in the project?

Preparation

1. Three- to five-gallon buckets
2. Hand shovels
3. Potting soil ingredients: sand, compost, eggshells, organic bonemeal or blood meal, recycled potting soil
4. Four-inch pots of different-sized soil particles
5. Watering can

9 Compost in the Garden

Time Frame: 45 minutes

Overview: The class will lay compost that they harvested on the school garden beds.

Objective: Students will utilize a valuable resource to grow healthier garden plants and cultivate a flourishing soil ecosystem.

Introduction (5–10 minutes)

1. Brainstorm with the class about what benefits the finished compost will have on the school garden: How will the compost help the garden plants and the soil ecosystem?

2. How can the class do careful and quality gardening work during this activity? What does respecting each other, themselves, and the garden look like?

Activities (30 minutes)

1. Individually or in pairs, the students will gather compost from the pile of harvested compost that they screened and spread it across designated garden beds. They will use buckets to transport the compost and will spread it evenly across the beds, one-inch deep, making sure to avoid pouring it over any established plants.

2. Student volunteers can also rake the compost across the beds as others empty the buckets of compost. Some compost should be reserved for later planting activities.

3. At the end of the activity, the students will check on the progress of their projects and record the final data from their decomposition study.

Assessment (5 minutes)

1. Have students return to the garden and do a quality control check of their work. Is the compost evenly distributed across the garden beds? Are living plants tucked in with the compost and not smothered? Are there any garden beds that still need compost?

2. Did the student groups notice any substantial changes in the decomposition of their materials?

Preparation

1. Harvested compost
2. Buckets
3. Rakes and shovels

Resources

- Neil Bell et al. "Improving Garden Soils with Organic Matter." Oregon State Extension Service, May 2003. (online—see Appendix B).

10 Reflect and Share

Time Frame: 45 minutes

Overview: The groups will make a final assessment of their decomposition projects and will share their findings with the class.

Objective: To draw scientific conclusions from student studies and make recommendations for the school compost system.

Introduction (5–10 minutes)

1. As a class, discuss how to develop a conclusion for the end of a science project. How should students generate engaging questions about the experiment?

2. The students should be prepared to either make recommendations for the school on how to best compost certain items or suggest what the school should keep in mind when it processes food waste.

Activities and Assessment (35 minutes)

1. In their project groups, the students will do one last assessment of their Compost in a Jar experiments, measure the decomposition of their item, and record their findings on the worksheets. The groups will also reflect on the experiment process and make recommendations for future experiments.

2. When the groups have finished their conclusions, gather the students back together for a class reflection and for the groups to share their findings. How much did each group's item decompose? Did the conclusion of their experiment match their hypothesis? What would they do differently if they were to do this again?

3. The students should focus on generating unanswered questions about this activity as well. What was successful about this experiment? What unanswered questions do they have? What recommendations do the groups have for the school compost system?

4. At the end of class, the students will clean up and dispose of their materials. If they have organic materials that can go into the school compost, they can dispose of it in there. If the students have meat or dairy, they can throw it away. The jars can be cleaned out in a sink or with a hose and stored for future use.

Preparation

1. Compost in a Jar worksheets
2. Compost bins
3. Hose/sink

Spring: Healthy Soils and Garden Plants

1 Planting Spring Seeds

Time Frame: 45 minutes

Overview: Students will plant summer and fall crops in pots using the potting soil that they created.

Objective: To practice seasonal garden skills and plant favorite foods for the garden.

Introduction (5–10 minute)

1. What are some of the foods that students like to eat in the garden during the summer and fall? In what season will these foods be ready to harvest?

2. Brainstorm with the students how deeply to plant seeds and how to fill up the planting containers with soft potting soil. What are the different steps that students should follow to plant their seeds? What is the best soil environment that students can offer their seeds, considering moisture level, softness, and soil level?

Activities (30 minutes)

1. Individually or in pairs, the students will plant a variety of seeds in growing containers. They will fill up the containers with the potting soil they created, making sure to keep it soft and uncompacted. The students can use the information on the back of seed packets to determine how deeply to plant the seeds, as well as how many will go into each pot.

2. When students have planted their seeds, they should label the pots with tape and permanent markers, or plant markers, and thoroughly moisten the pots with a watering can.

3. The students can repeat this activity with different seeds, making sure not to mix the varieties together. The class should focus on doing careful, considerate gardening during this activity.

4. At the end of the activity, after students have cleaned up and returned any extra seeds, they can enjoy a taste test in the garden.

Assessment (5 minutes)

1. What seeds did students plant and look forward to eating this year?

2. Did the students find it easy to plant in the potting soil they created?

Preparation

1. Seeds—summer and fall varieties
2. Potting soil
3. Four-inch planting containers
4. Planting trays
5. Watering can/hose

6. Masking tape
7. Markers

2 Creatures in Our Compost

Time Frame: 45 minutes
Overview: Students will explore a functional, healthy compost system and search for members of the soil ecosystem.
Objective: To see how a thriving soil ecosystem consists of decomposers and consumer invertebrates.

Small groups share in the excitement and wonder of finding diverse organisms in a compost sample.

Vocabulary
Invertebrate

Introduction (10 minutes)

1. What *invertebrates* do students know play an important function in the school compost bins? What creatures have students engaged with during their compost explorations? Brainstorm what students know about invertebrates living in the compost system.

2. Students may think that some creatures are pests, like fly larvae and spiders, but should understand that these serve a purpose in the ecosystem. How can students think like scientists when interacting with the compost ecosystem?

3. Introduce the students to the decomposers and the various levels of consumers that inhabit the compost system, help with the decomposition of organic waste, and build a healthy or balanced soil ecosystem. See Creatures in the Compost worksheet for examples of this soil ecosystem.

4. Create a food web map with the class on a classroom chart by having class volunteers label or name the invertebrates and roles played in the compost system. What do students predict they will find in two different samples of compost, one less decomposed and the other more? Will the creatures in the compost ecosystem be the same in each sample?

Activities (30 minutes)

1. In pairs, the students will harvest compost from school compost bins and explore two samples of compost, one from a less decomposed bin and another from a more decomposed sample. They will use the tools provided to find invertebrates and fungi at the smallest to largest level that they can. The groups will record these findings on their worksheet and can use the picture illustrations, as well as the class food web chart, to guide their exploration.

2. The students can pour a sample of the compost onto observation trays, making sure they practice safety and consideration when interacting with living creatures. If possible, this is an opportunity for students to work with microscopes and enhance their observations.

3. When the students have finished their observations, they can gather back together to share their discoveries and questions.

Assessment (5 minutes)

1. What observations did students make on the diversity and populations of invertebrates? What does the diversity and population tell the students about the balance within the compost ecosystem?

2. Did the groups observe any interaction between the different creatures? Did they discover any egg sacs? Were there any creatures that they couldn't identify? What were the key differences between the two samples?

Preparation

1. Observation trays
2. Magnifying glasses
3. Creatures in the Compost worksheets
4. Two different compost samples
5. Gloves

Resources

• Healthy Youth Institute. "Compost Bin Identification." Oregon State University, Linus Pauling Institute. (online—see Appendix B).

• "Do The Rot Thing: A Teacher's Guide to Compost Activities." Central Vermont Solid Waste Management District. (online—see Appendix B).

3 Discovering Microorganisms

Time Frame: 45 minutes

Overview: The class will explore a microscopic sample from a thriving compost system and illustrate the living things that they observe.

Objectives: To observe the smaller creatures in a decomposing system and use a microscope.

Vocabulary

Microorganism

> I recommend beginning this class by projecting a microscopic view so that the entire class can see a microorganism from a compost sample. This experience will fire up the class's enthusiasm for their own investigations. You can also guide the students in generating high-quality questions and describing their observations, which will benefit them later in the activity.

Introduction (10 minutes)

1. Introduce the class on how to use a microscope, if they are unfamiliar with the tool. How can the groups safely use this tool and respect the creatures they will be studying?

2. Brainstorm with the students on how to find microorganisms in a compost sample: What types of creatures might the students discover? How will they treat and interact with the living things in the soil?

Activity (30 minutes)

1. In small groups, the students will explore a small compost sample using a microscope. They will work to identify three different creatures and illustrate the features, movements, and characteristics of each living thing on their Discovering Microorganisms worksheet. The groups will collaborate and share their observations, practicing teamwork and cooperation.

2. The students can enhance their illustrations as a scientist would by

- labeling parts of the creatures that they recognize
- adding engaging scientific questions to their drawings
- using the margins of the worksheet to label and make observations

3. When the groups have finished their observations, they will gather back together to share their findings, or work with another group to compare their discoveries.

4. If possible, have reference and resource books available on creatures in the soil ecosystem for students to use for identification.

Assessment (5 minutes)

1. Were the groups able to identify any microorganisms in their compost sample? What were the most exciting creatures they discovered?

2. What unique features and behaviors did they observe? Did they notice microorganisms interacting with each other? What roles do these organisms play in the compost ecosystem?

Preparation

1. Resource and reference books
2. Microscopes
3. Discovering Microorganisms worksheets
4. Projector

Resources

- Nancy Trautmann and Elaina Olynciw. "Compost Microorganisms." Cornell Waste Management Institute, 1996. (online—see Appendix B).
- "The Science of Composting." University of Illinois Extension—Composting for the Homeowner. (online—see Appendix B).
- Healthy Youth Institute. "Compost Bin Identification." Oregon State University, Linus Pauling Institute. (online—see Appendix B).

NGSS and Extensions

Greater classroom studies can include having students develop a model that illustrates how matter is made of particles too small to be seen (5-PS1-1: Matter and Its Interactions).

4 NPK and pH Testing

Time Frame: 45 minutes

Overview: The class will survey pH and NPK levels of the school compost system and garden beds by using a testing kit.

Objectives: To build a class map of the growing conditions in the school soil system and determine what plants will grow best in each space, or make recommendations for how to alter soils to improve growing conditions.

Nitrogen (N) is vital to plant growth and is most visible in plants by the greenness and bushiness of plant leaves. As with most elements, plants need a certain amount of N to be accessible in the garden soil; too much or too little can be damaging depending on the preference of each plant.

Potassium (K) improves the health and strength of a plant, as well as fruit color and taste.

Phosphorus (P) is necessary for plants to grow their best seeds and to strengthen the genetic yields of the plant.

pH levels are important to plants because the acidity or alkalinity of soils impacts plant growth as well. All plants have a soil pH level that they prefer.

Introduction (10 minutes)

1. Brainstorm with the class what role nitrogen, potassium, phosphorus, and pH play in garden soil, in the environment, chemistry, and living bodies: Are the students familiar with these elements and the roles they play in the school garden?

2. Demonstrate how to use the pH and NPK testing kit. What different roles can each student play so that everyone has a job? What does good and scientific work look like during this activity?

> There are many types of pH sensors and kits available. Because of my minimal budget, I've often used simple soil test kits which can be found at garden centers.

Activities (30 minutes)

1. In small groups, the class will use a simple NPK and pH testing kit to measure the growing conditions of certain garden beds and the school compost bin. For example, the class can be split into teams, one for nitrogen, and another for potassium and so on. These groups will share responsibilities in gathering a soil sample from the appropriate soil depth.

2. The students will measure and record data on a class chart or garden map for later reflection, making sure to identify the place they took the sample. They can perform the experiment twice, if there is time, and compare their results.

3. When the experiments are done, have the students clean up, gather a taste test, and come back together to share their discoveries.

Assessment (5 minutes)

1. What do the levels and measurements say about the growing conditions of the garden beds? Some plants prefer different levels than others, but are there garden soils that are depleted of NPK or that have an acidic or base soil?

2. What can the students do as gardeners and scientists to amend the garden soils and create more ideal growing conditions for annual plants?

Preparation

1. NPK and pH testing kits
2. Class poster or map
3. Shovels
4. Watering can

Resources

- "Natural Fertilizer" in Patrick Lima. *The Natural Food Garden: Growing Vegetables and Fruits Chemical-free.*

5 Amending Garden Soils

Time Frame: 45 minutes

Overview: Student groups will add organic materials to certain garden beds and adjust the pH or NPK levels for better growing conditions for annual vegetables.

Objective: To use organic ingredients to encourage healthier soils and garden plants.

Introduction (10 minutes)

1. Review with the class the discoveries and measurements the class groups took during the last activity. What does the NPK and pH levels in the garden beds say about the growing potential for plants? Introduce the students to some plants that like rich and nutritious garden soils and others that don't.

2. There are simple, natural materials such as coffee grounds, banana peels, compost, and organic blood meal and bonemeal that can be added to garden soils to adjust NPK and pH levels of soil. These materials are comparatively easy to procure and for students to handle.

> With the exception of a few crops, many annual plants like the same range of soil pH and NPK, though they also consume varying amounts of nutrients during their life cycles. Keeping this in mind, I don't perform this activity in all the garden beds, since some plants prefer less rich soils and other soils need an additional nutrient boost for future plants.

Discuss the nutrient value of each ingredient with the class. How should the students mix these materials into the soil?

Activity (30 minutes)

1. Looking at the class chart/map of the NPK and pH levels in the selected garden beds, have the student groups determine which garden beds they should add their group material to in order to create more balanced soil. For example, the nitrogen group can look at the map/chart and add a nitrogen-rich material to the garden beds that are deficient or lacking nitrogen.

2. Working in teams, the groups will amend the garden soils with organic materials, taking time to do considerate and conscientious gardening. The students should be careful when amending the soil around living or established plants.

3. When all the garden beds have been amended or the materials used up, the students can clean up and then enjoy a spring taste test from the garden.

Assessment (5 minutes)

1. Were the groups successful in working as a team and amending all the garden soils which were deficient? How long do students think it will take for these materials to decompose and for nutrients to become available to established or new plants?

Preparation

1. Coffee grounds
2. Banana peels
3. Compost
4. Blood meal/bonemeal
5. Shovels

6 Planting Starts

Time Frame: 45 minutes

Overview: The students will transplant the garden starts that they began in homemade potting soil.

Objective: To plant healthy annual, biennial, and perennial plants in the school garden.

Vocabulary

1. Annual
2. Biennial
3. Perennial

Introduction (5–10 minutes)

1. Brainstorm with the class about what seeds they chose to plant in their potting soil weeks before: What types of plants will the class transplant into the school garden? What does it mean for a plant to be *annual*, *biennial*, or *perennial*?

2. Discuss the growing conditions needed for each plant, particularly space and nutrient requirements. The students should keep these conditions in mind when they plant their starts. As they plant, the students can discuss with their neighbors where would be the best place to grow their plants.

3. How deeply should students dig their holes and plant their starts? What other activities can the students do to promote the health of their plants?

Activity (35 minutes)

1. Individually or in pairs, the students will transplant plant starts they began weeks before into the garden beds that they have amended. They will make sure to dig holes deep enough for the plants' roots and gently fill in the holes with soft soil. When the plants are installed, then the students can water them with a watering can.

2. When the students have planted their starts and watered them, they can use markers and wooden sticks to make name markers for the plants.

3. At the end of the activity, the class should explore the garden and pick up any remaining planting pots and tools. Then, the students can enjoy a taste test from the garden.

Preparation

1. Shovels
2. Potted plants
3. Markers
4. Wooden sticks/markers
5. Watering can

7 Mulching Garden Pathways

Time Frame: 45 minutes

Overview: The class will work together to mulch the garden pathways with wood chips and beautify the school garden.

Objectives: To perform a seasonal activity and use garden tools safely.

Introduction (5–10 minutes)

1. During the activity, the students will be using wood chips to mulch the garden pathways and amend the paths from the effects of winter and spring weather. What other mulches have students used in the garden? What is the role of mulches in organic gardening?

2. Review tool safety and handling with the class before passing out the equipment. How do students practice shovel and rake safety around their classmates?

Activity (30–35 minutes)

1. Either as a class or in small groups, assign students to the different roles needed to accomplish the activity, such as students to shovel the wood chips, carry buckets and wheelbarrows, and rake the chips. The groups can rotate through these jobs throughout the activity time, as directed by an adult.

2. The goal of this activity is to do quality gardening work, so the students should focus on being safe and doing careful work in every role. The students who rake can direct the students who carry wood chips to areas where the mulch is most needed.

3. At the end of the activity time, the class can clean up and return all of the tools before gathering a taste test from the garden.

Assessment (5 minutes)

1. Have the students return to the garden and reflect upon the work they accomplished as a class. What differences does this activity make to the garden, before and after?

2. The wood chips will decompose over time. Knowing what they do about composting, how long do students estimate the chips will take to break down? What creatures and invertebrates will help in the decomposition?

Preparation

1. Shovels and rakes
2. Wheelbarrows
3. Buckets
4. Wood chip mulch

8 Trellises and Structures

Time Frame: 45 minutes

Overview: Student groups will create trellises for garden plants and fix garden structures for the growing season.

Objective: To make garden improvements for a safer and sturdier garden.

Introduction (10 minutes)

1. Brainstorm with the class about the role of trellises in the school garden. What containment and guiding structures have students made or seen before?

2. Using a classroom chart, illustrate different trellising methods to the class, explaining how to construct each one. Brainstorm with the students the best ways to build a trellis: How can the students work as a team to build a sturdy structure?

Activity (30 minutes)

1. In small groups, students will construct a simple trellis for climbing garden plants such as peas or beans, using the materials provided. They can use the classroom chart illustrations as a guide and choose one that they would like to make. After the trellis is constructed, it will need to pass a "stability" test from a staff member or adult before being established in an open space in the garden.

Students weave new garden beds and make the garden look its best for spring planting.

2. Students who finish early can fix and repair structures in the garden, such as raised beds, fences, and other trellises. If the students use tools, then they should practice safety and slow, careful work.

3. At the end of the activity, the class can clean up their tools and materials before going into the garden for a taste test.

Assessment (5 minutes)

1. Were there structures and trellises that were easier to build than others? What were the groups' experiences in constructing the frames?

Preparation

1. Willow, bamboo, or hazelnut suckers/branches
2. String
3. Scissors
4. Hammers
5. Nails or staple gun

9 Sowing Seeds

Time Frame: 45 minutes

Overview: The class will sow climbing seed varieties under the trellises that they built.

Objective: To fill the school garden with edible foods and pollinator forage.

Introduction (5–10 minutes)

1. During the activity, the students will seed climbing plants in the garden using presoaked seeds. Discuss the benefits of using presoaked seeds over dried seeds.

2. With seed packets as a reference or through class discussion, have student pairs study the width and depth requirements of the seeds they will be planting. What are the needs of bean seeds or peas?

Activity (30 minutes)

1. In their small groups from the previous activity, the students will plant climbing garden plant seeds around the trellises that they built. They can pay attention to the depth and width of seed plantings, making sure to cover seeds with enough soil to prevent birds or other foragers from finding them.

2. The groups should work as a team and do careful gardening work during the activity. They can use a watering can to thoroughly soak the seeds after they are planted. They can also label the planting area and plant variety with wooden sticks or markers.

3. When all the seeds have been planted, the class can explore the garden for a taste test of spring foods.

Assessment (5 minutes)

1. Have the students return to the garden and check the seeds that they planted. Are the seeds completely covered by the soil?

2. Did the water penetrate to the seed level or just on the soil surface?

Preparation

1. Presoaked bean or pea seeds
2. Watering cans
3. Seed packets

10 Garden Celebration

Time Frame: 45 minutes

Overview: The class will construct their garden journals and enjoy a large taste test from the school garden.

Objective: To celebrate and reflect on student work over the year.

Introduction (10 minutes)

1. What are favorite activities or memories that the students have from gardening this year? What was the most memorable lesson that the students learned? What will they never forget about composting?

2. What are seasonal foods that the students would like to taste test from the garden?

Activity and Assessment (35 minutes)

1. Before they construct their journals, the students can go on a five-minute taste test in the garden. They can gather and wash certain spring foods that they can eat as they make their journals.

2. When the students return, they will construct their garden journals by collecting the worksheets they filled in during experiments and other garden activities. They can use the provided cover sheet and available materials to construct the journals.

3. When the students have assembled their garden journals, they can decorate the cover page using colored pencils and provided materials.

4. As the students work, discuss what plants they would like to plant in the following year, what activities they would like to do again, and what lessons they learned that they could pass to younger students.

Preparation

1. Hole punchers
2. Yarn
3. Scissors
4. Colored pencils
5. Composting cover page

SIXTH GRADE—GRADE SIX

RAIN GARDENS

This year's exploration into Rain Gardens is an exciting and creative time for the students. At first, most have no idea what a rain garden could look like. But after exploring the beautiful designs, water-capturing techniques, habitat opportunities, and unique benefits of rain gardens in community spaces and schools, the rest of the year's activities are propelled by student enthusiasm.

Rain gardens can be as big or small as a community needs, and the lessons can be applied to creating a new space or updating an existing garden. The opportunities for student learning are incredible, no matter the scope of the project. Students will observe the way water moves, learn about passive and active catchment systems, find permeable surfaces, and develop their spatial awareness through

FALL
The Way Water Moves

1. Garden Exploration
2. Dry Seed Saving
3. Wet Seed Saving
4. Planting Winter Starts
5. Preparing the Garden for Winter

6|7. The Way Water Moves

8. The Water Catchment Race

9|10. Active and Passive Water Catchment

WINTER
Rain Garden Design

1|2. Mapping the Schoolyard: Topography and Surfaces

3. Water Flow
4. Land Use and Traffic
5. Designing a Rain Garden

6–10. Landscaping and Building the Rain Garden

SPRING
Building a Rain Garden

1. Rain Garden Plants
2. Preparing the Soil

3|4. Planting in the Rain Garden

5. Seeds and Starts
6. Building Habitat
7. Creating Signs
8. Student Docents
9. Making Journals
10. Garden Celebration

mapmaking. They will use garden maps to enhance their understanding of interactions among organisms and elements in an environment, as well as analyze how humans can interact in a space and still provide habitat for insects, amphibians, and birds. And the students will learn to identify local plants and the environments those plants prefer.

The class will design a dream rain garden and then use their knowledge of seasonal water flow to develop a realistic model for the school or a community space. In late winter, they will see their plans become reality by working together to landscape a rain garden. They will enhance the space by caring for the structures, features, soils, and plants that make this garden a special place.

At the end of the year, the students will have learned how to capture a valuable resource, slow its movement across the landscape, and harvest water for increased habitat and water retention. They will have looked at how humans impact landscapes, alter or change natural systems, and they will develop a solution to larger issues that can connect to modern concerns, such as climate change and wasted natural resources. This knowledge will extend into the vegetable garden as the students use what they have learned to increase soil water-holding capabilities that promote better summer growth.

Why rain gardens and not butterfly gardens or hedgerows? Rain gardens are valuable for multiple reasons, the first being to slow down and capture rainwater runoff to sink into the ground. Plants in this design can have many functions, including edible, pollinator-friendly, and food-producing plants. Additionally, students of this age group often study water cycles, climatology, and landscape changes in other areas of the curriculum. When enhanced by rain garden studies, these units can be connected to the effects of global climate change and the extreme effects of water across the planet (droughts, flooding, water scarcity in cities, wildfires).

Next Generation Science Standards

With rain gardens as a focus, students will be designing and engineering solutions to larger environmental issues. These lessons provide students the opportunity to engineer and design a response to rainwater runoff, on either a large or small scale. By taking measurements, making observations of a physical space and seasonal water flow, mapping human interaction in the space, as well as sunlight and soil type, the students will build substantial skills and valuable knowledge. This information will better inform their garden designs and provide opportunities for students to use evidence in making arguments for different design elements.

Further in-class studies that enhance these lessons include the causes and impacts of water erosion, the effects of climate change on global water movements, the impact of human activity on Earth's systems, the cultural impacts of water in climate change, and ecosystem response and resilience.

Permaculture Principles

- Observe and Interact
- Catch and Store Energy
- Use and Value Renewable Resources and Services
- Design from Patterns to Details
- Integrate Rather than Segregate
- Creatively Use and Respond to Change

Fall: The Way Water Moves

1 Garden Exploration

Time Frame: 45 minutes
Overview: Students will discuss behavior and expectations before embarking on a focused exploration in the garden.
Objectives: To make observations about seasonal changes, practice positive behaviors, and explore healthy garden foods.

Introduction (5–10 minutes)

1. After welcoming the students back into the garden, review key garden behavior expectations: What do students expect from themselves and each other in the garden while interacting with the food, creatures, and other people? What does Care for Self, Others, and the Land look like from students?

2. What changes do they anticipate will have happened in the garden over the summer? What new garden elements have they already noticed?

Activities (30 minutes)

1. Students will complete their Welcome to The Garden worksheets with a partner, working as a team and exploring garden changes as they go.

2. After they have finished, students can participate in a taste test of garden foods such as tomatoes, cucumbers, and nasturtiums while gathering back together to reflect on their activity and exploration. The class should be encouraged to gather the foods they enjoy eating, but challenge themselves by trying at least one new food for taste testing.

Assessment (5 minutes)

1. With their worksheet as a guide, what did the students discover in the garden? What evidence of changing seasons did they observe? Were the students surprised by any plant growth or discoveries?

2. What foods from the garden do students enjoy the most? What new food did they try and what was their experience eating it?

Preparation

1. Examples of foods for students to taste test
2. Welcome to The Garden worksheet

2 Dry Seed Saving

Time Frame: 45 minutes
Overview: Students will practice how to save dry seeds from healthy plants.
Objective: To identify and preserve healthy seeds for planting in the spring and summer.

Garden seeds are carefully stored in paper envelopes until the spring.

Introduction (10 minutes)

1. Brainstorm with the class what they know about saving seeds: What types of plants produce *dry seeds*? What do ready-to-harvest and healthy seeds look like? What do seeds look like that gardeners don't want to save? What seeds have students saved in the past? What lessons do they remember from those experiences?

2. Provide examples of seeds for students to collect and demonstrate how to identify and harvest the seeds correctly. For older students, dry seeds could include small ones like dill, hollyhocks, marigolds, carrots, herbs, and brassicas. The students will be providing the school a service by carefully collecting, sorting, and labeling different types of seeds from the garden. They should harvest the seeds one variety at a time so the seeds aren't mixed up when being stored. The goal of this activity is to do careful, considerate gardening.

Activities (30 minutes)

1. Working individually or in pairs, the students will explore the garden and use collection cups to gather each seed variety, one at a time. They will need to correctly identify the plants and determine if there are healthy dry seeds to harvest. When their cup is halfway-to-full, the students will deposit the seeds in the small envelopes provided, labeling each packet as they go with the plant name and date of harvest.

2. Before the end of the activity, the students can return their supplies, clean up, and enjoy a taste test from the garden. If there are edible samples of the plants they just saved seeds from, then the students can be encouraged to try them.

Assessment (5 minutes)

1. What textures and smells did students discover during the activity? Were some seeds easier to harvest than others? Were there seeds easier to identify than others?

2. What seeds did the students enjoy harvesting? Can the class recall names of the seeds they collected?

Preparation

1. Markers
2. Small envelopes
3. Cups for collecting
4. Seed examples

Resources

• Suzanne Ashworth. *Seed to Seed: Seed Saving and Growing Techniques*, 2nd ed.

3 Wet Seed Saving

Time Frame: 45 minutes + 15 minute follow-up activity

Overview: Students will be saving wet seeds for next year's planting.

Objectives: To learn about fermentation, natural yeasts, and how to save small wet seeds.

Introduction (10 minutes)

1. Brainstorm with the students how to save seeds from juicy fruits like tomatoes and cucumbers that hold seeds inside their fruit bodies and protective membranes: What types of *wet seeds* have students harvested before? How can students harvest and save the seeds from wet fruits without rotting? How will they manage to remove the gelatin and sugars that surround the seeds?

2. Have the class brainstorm other common food items that humans ferment, such as bread, kimchi, sauerkraut, hot sauces, and chocolate.

Introduce the students to the fermentation process, where natural yeasts (and sometimes molds) in the air feed on the sugars surrounding the seed and remove the casing. Why do gardeners save wet seeds through fermentation, rather than let them rot naturally?

3. After a brief introduction to wet seeds, students can be shown the differences between seeds that are easily removed and dried and those that need to be fermented before they are dried for winter.

4. All students should be sure they have washed their hands before this activity, since they are handling produce they can later eat.

Activities (30 minutes)

1. Individually or in pairs, the students will carefully remove the seeds from a piece of fruit such as a tomato or cucumber. They can collect the seeds into a cup, being careful to gather all the seed material that they can. No fruit pieces, other than the seeds, should be gathered in the cup—these will rot. The fruit waste can be collected on clean plates. As they work, the students should pay attention to the texture of the sugar membrane around the seeds.

2. When they have collected all the seeds, the students will fill their cups halfway with water and place them in a designated area for fermentation. Ideally, the students will come every day to check on their cups, agitate them with stirring sticks to disturb any mold growth, and observe changes occurring in their cup.

3. After the students have cleaned up their stations, they can work together to make salsa or another recipe from the remaining fruit pieces, using the teacher or volunteers to help. The class can adjust the flavors of this small meal and work together to create a snack that they can all enjoy.

Assessment (5 minutes)

1. As students enjoy their taste testing, have them reflect on the experience of removing the seeds from the fruit. What observations did they make about the protective covering over the seeds? Why would a plant develop a covering over its seed?

2. What words would students use to describe the texture and flavor of the taste test?

Follow-up Activity (15 minutes)

1. Two or three days after the students have put their seeds in water to ferment and have stirred them to add oxygen to the fermentation process (an anaerobic environment can lead to the seeds rotting), the class will wash the fermented seeds and lay them out to dry.

2. With a sieve and a water source, each student will carefully pour out the contents of their cup with all the seeds and wash them thoroughly under water. When the seeds have been washed, the students will lay them out on coffee filters and store them in a safe place to dry.

3. Using tape and markers, the students can construct a small label recording both plant variety and the student's name. When the seeds and filters are dry, the seeds will be stored in a cool, dark space for the winter. Have the students compare the appearance of the dried wet seeds with their appearance when they were covered in the sugar casing. For example, tomato seeds look hairy when dry, while cucumber seeds look sharp and slick.

Preparation

1. Tomatoes or cucumbers (enough for each student to have half a fruit)
2. Knife
3. Seed collecting cups
4. Spoons to dig out seeds

5. Water/watering can
6. Salsa-making tools and additional ingredients
7. Paper plates or place mats
8. Coffee filters
9. Sieve
10. Masking tape
11. Permanent markers

Resources

- Chuck Burr. "How to Save Tomato Seeds." Permaculture Research Institute. (online—see Appendix B).

4 Planting Winter Starts

Time Frame: 45 minutes

Overview: Students will plant starts to provide sources for winter food.

Objectives: To learn about seasonal produce and varieties of plants which thrive in the winter.

Vocabulary

Biennial

Introduction (10 minutes)

1. Brainstorm with the students what edible garden plants grow well in the winter: What are the benefits of winter produce? What kinds of plants can withstand cold temperatures? What plants cannot handle the cold? Many winter plants are *biennial.* How does this affect the school garden?

2. How should students handle and plant vegetable starts? What are the planting requirements for the available plants? How deeply should the starts be planted? Demonstrate how to carefully remove the starts from their trays and plant them in the garden. This activity can also be done in containers.

Activities (30 minutes)

1. In pairs, students will pick one variety of produce to plant. They will keep in mind the planting requirement of their variety and find appropriate space for it in the garden.

2. Good work during this activity will include careful removal of the plants from the trays and quality planting. There should be no exposed plant roots, and the soil should not be too compacted after planting.

3. The groups can deeply water the plants afterward and label the variety with markers or wooden sticks.

4. Before the activity time is over, the groups can return their tools, collect any remaining trays or planting containers, clean up, and enjoy a taste test from the garden.

Assessment (5 minutes)

1. Have the students do a quality control check of their work in the garden. This is a time for craftsmanship and attention to detail.

2. Are all the tools and planting containers returned? Are the plants all thoroughly watered? Which plants do the students predict will thrive the best over the winter?

Preparation

1. Winter vegetable starts (lettuce, chard, onions, kale, cabbage, lettuce)
2. Wooden sticks or plant markers
3. Pens
4. Watering can

Activity Extension

Larger conversation topics for in-class extensions could include the importance of all-season food production and the environmental impacts of imported and distantly grown food.

5 Preparing the Garden for Winter

Time Frame: 45 minutes
Overview: Students will lay down burlap and wood chips on the garden paths or lay straw on garden beds.
Objective: To prepare the garden for the winter weather by protecting the paths and beds from erosion and seasonal wear and tear.

Introduction (5–10 minutes)

1. Discuss with the class what they know about mulching and what kinds of materials they have used as mulch in the past. What are the effects of seasonal weather on the garden? Mulching is when gardeners lay organic materials down in the garden in order to provide additional nutrients for the soil, suppress weeds, protect the soil from wind and rain erosion, and slow rainfall for the plant roots. How can mulching and laying down organic materials help protect the garden soil and pathways during the winter? Laying down burlap and bark chips on garden pathways prevents mud and hard-packed soil during the winter season.

2. Before laying down bark chips on the pathways, the students should review safety expectations with using tools like shovels and rakes. Additionally, the students should be shown how to lay out the burlap (if being used) by overlapping the pieces a few inches so weeds don't rise up through the cracks. This is a good project for detail-oriented students.

Activities (30 minutes)

1. Carefully focusing on slow gardening work, students will lay out mulch on the garden pathways. They can work in teams to gather the bark chips, or rotate through jobs such as shoveling, using wheelbarrows, carrying buckets, and raking out the bark chips.

The students use tools to lay bark chips on the pathways, reducing mud and soil compaction.

2. Additional jobs may involve students raking up the fallen materials so as to conserve and use them in the garden, as well as pruning and chopping spent plants such as tomatoes or sunchokes. Students can also remove sturdy and perennial weeds that grow on the pathways, which usually thrive and strengthen during the garden's off-season.

Assessment (5 minutes)

1. Gather the students together and go around the garden for a quality control check of the class's work.

2. Is there enough straw or bark chips in the necessary spaces? What spots need more attention?

Preparation

1. Straw, leaves, burlap, wood chips
2. Rakes
3. Shovels
4. Wheelbarrows
5. Trowels
6. Clippers
7. Adult volunteers

6|7 The Way Water Moves

Time Frame: 1½ hours (2 × 45 minutes)
Overview: Students will map spaces in the schoolyard where water collection and runoff pose a problem.
Objectives: To begin thinking of potential rain garden sites and how poor drainage or oversaturation can lead to human and water issues.

Introduction (10 minutes)

1. In preparation for planning out and building a rain garden area, students will first identify potential sites for a rain garden. They will explore the schoolyard and examine spaces that pose problems for the community when it rains or where there is obvious seasonal water activity.

2. As a class, have students brainstorm how water moves across the local landscape or school ground: What areas have students noticed to be flooded or soggy during the winter or rainy times of year? What evidence have they seen (standing water, puddles, a drain overflowing), and where did that problem originate?

> "Water movement, within ecosystems is the result of interplay between water, soil, gravity, living organisms, temperature, and air. We need to understand the nature of these relationships if we are to manage water effectively."[1]

Activities (30 minutes)

1. Students will use The Way Water Moves worksheet to guide their exploration of three problem areas in the schoolyard. They will work in pairs to accomplish this task and make detailed observations about water movement across the land.

2. This activity time will be repeated in order to give students the time to make quality observations about water movement and to identify the places they want to record. Alternatively, the class can go on a tour of water activity across the campus or community space together and make their observations as they go.

Assessment (5 minutes)

1. What were some of the sites students noticed were problem areas? What elements and clues indicated heavy water activity?

2. What potential effects could this water movement have on the schoolyard? What were similarities and differences between the sites?

Preparation

1. The Way Water Moves worksheets
2. Clipboards and pencils

Resources

- "Stormwater in the Desert: Middle School Activity Book." (online—see Appendix B).
- "Curriculum" in *Rainwater Harvesting for Drylands and Beyond* by Brad Lancaster. (online—see Appendix B).

NGSS and Activity Extensions

Greater in-class studies can include having students develop models that illustrate the cycling of natural materials and the flow energy through ecosystems or building a model that describes the water cycle driven by the sun's energy and gravity (MS-ESS2-1: Earth's Systems and MS-ESS2-4: Earth Systems).

8 The Water Catchment Race

Time Frame: 45 minutes
Overview: The class will explore key terms for

talking about water and how it interacts with different surfaces.

Objectives: To interact with impervious, pervious, and saturated surfaces and begin understanding active and passive methods of collecting water.

Vocabulary

1. Impervious
2. Pervious
3. Saturation

Introduction (10 minutes)

1. Each student will receive The Water Catchment Race worksheet. Have students fill in the key terms and their definitions on the worksheet during an initial class brainstorm. What do students know about these key terms and what definitions would they give them?

- Impervious: a surface that does not allow water to pass through it
- Pervious: a surface that allows water to pass through it, permeable
- Saturation: the state that occurs when something is filled to capacity

2. During the activity, the students will need to identify five different surfaces around the schoolyard, local landscape, or community. Some of these surfaces can be organic or naturally made while others can be constructed (asphalt or cement). Do students know where these different surfaces are located?

Activities (30 minutes)

1. In pairs or small groups, the class will gather the testing materials: a stopwatch, a pipette, and a cup to fill with water. Each group will test the way water interacts with five common surfaces by pouring the same amount of water on them.

2. From the moment that the first water droplets hit the surface, the students will time how long it takes for the water to pass into the surface and will record this time on their worksheets. From this test, and their observations on the movement of water, the students will determine whether the surfaces are pervious, impervious, or saturated.

3. When the groups have finished their tests, they can return their supplies and discuss and record their conclusions on the worksheet.

Assessment (5–10 minutes)

1. Did a current weather pattern affect student test results? How would students describe the surfaces on which water tended to splash or soak up water? Were these surfaces impervious or pervious? How can students tell if a surface is impervious or simply saturated?

2. Did any groups notice water runoff during their tests? What does this activity tell students of how water moves across different surfaces?

Preparation

1. Water Catchment Race worksheets
2. Stopwatches
3. Pipettes
4. Cups
5. Pencils
6. Clipboards

Resources

- "Stormwater in the Desert: Middle School Activity Book." (online—see Appendix B).

9|10 Active and Passive Water Catchment

Time Frame: 1½ hours (2 × 45 minutes)
Field trips encouraged

Overview: Students will explore different ways to actively and passively capture rainwater and snowmelt.

Objectives: To learn about the weight of water and the various methods and tools for harvesting and storing water.

Vocabulary

1. Active Water Catchment
2. Passive Water Catchment

Introduction & Activity (40 minutes)

1. Brainstorm with students what they know about the water cycle in their community—locally, regionally, and internationally: What ways is water moved every day, through rivers, streams, pumps, faucets, toilets, or drinking water?

2. Introduce the idea of active and passive water catchment systems and generate examples of each method, as well as potential pros and cons depending on climate and time of year. Review the key terms (pervious, impervious, saturation) from the Water Catchment Race with the students. What models have students seen to catch rainwater or runoff?

- Active water catchment: collecting, filtering, storing, and pumping harvested water from stormwater/rainwater and snowmelt. Examples are cisterns, water barrels, and tanks above and below ground.
- Passive water catchment: shaping or altering the landscape in order to slow water down and allow it to soak into the ground.

3. Are there examples of even more active ways that humans capture water? What is the weight of water? What are ways that water is wasted? Where does this water go?

4. Through field trips, in-class visits from local organizations, or a simple presentation, demonstrate the features and differences of active and passive water catchment systems. Refer to the websites and examples in Resources for in-class presentation images and information. Have the students record their observations about each system and their benefits and accessibility in the community. Later, they can share their observations as a class or in small groups.

5. Discuss the physical differences between catchment systems. What elements stand out and how are they being utilized? How do the different rain gardens and catchment systems interact with people and larger ecosystems? How do passive catchment systems interact with and affect ecosystems? This activity can also be done using a student reflection sheet completed during or after a field trip to observe these systems.

> I highly recommend planning field trips for this activity, if possible, to observe active and passive rainwater and snowmelt collection methods in the area. Local businesses are building passive water catchment areas into their designs more often, and there may be local companies that sell active water catchment systems (barrels, cisterns). Renovated government buildings and conservation groups are a good place to reach out to as well as Green/LEED-certified buildings, colleges and other schools, local permaculture groups, or local gardeners. These groups can offer examples and tours of their work for school groups. If field trips are not an option, invite these businesses and organizations into the classroom to discuss water catchment models.

Assessment

1. Students should be able to correctly use water catchment terminology to describe what they see and to argue the benefits of one system over another.

2. Where could passive water catchment systems and rain gardens benefit the school grounds or the community?

Activity Extensions

Some educators may want to extend these lessons on water catchment into themes of water access, privilege, conservation, storytelling, cultural history, and use. Weekly lessons are good for hands-on engagement with water catchment systems, and I challenge educators to give space for respectful engagement with this valuable theme and its cultural significance.

Resources

- "Rain Gardens in Grass Lawn Park." City of Redmond. (online—see Appendix B).
- Cado Daily and Cyndi Wilkins. "Passive Water Harvesting: Rainwater Collection." University of Arizona College of Agriculture and Life Sciences—Cooperative Extension. (online—see Appendix B).
- "Water Harvesting." Permaculture Research Institute website. (online—see Appendix C).
- Terry Babb. *Harvest Rain: The Movie.* (online—see Appendix B).
- Seattle Public Utilities. *RainWise Program.* (online—see Appendix B).
- King Conservation District. "Rain Garden Care: A Guide for Residents and Community Organizations." 12,000 Rain Gardens. (online—see Appendix B).

Additional Activity—Making Maps

Time Frame: 3 × 45 minutes

Overview: Students will make maps of the school grounds or a community space from different perspectives.

Objectives: To build student spatial awareness and mapmaking skills in preparation of creating a rain garden map.

Introduction (5 minutes)

1. What kinds of maps have students made or used before? What are key features on a map that help users understand it? What different kinds of maps are there, and when do people use them?

2. The students will use the provided tools to build maps of the school garden, grounds, or rain garden site. They can focus on an assigned perspective to act as a guide for their mapmaking practice, such as bird's-eye, landscape, or its key features.

Activities (35 minutes)

1. Students will draw different maps of the same space by using various scales and perspectives. They should keep in mind the importance of keys and symbols to provide clarity to their map.

2. Students can make a movement map of the places they will visit throughout the day. They can go a step farther by color-coding how often they visit each space or by marking the permaculture zones (see Resources).

3. Students can design a dream rain garden of their own creation. They should keep in mind the difference between creative and silly pieces. All maps should include a title, legend/key, and the writing should be legible.

Assessment (5 minutes)

1. Reflection is an important step in the mapping process. Have students display their maps and survey the class's work, taking time to look at each other's maps and thinking about the successful components of each. Give students time to share their thoughts, admirations, and successes. After seeing their classmates' work, what elements did students feel were most successful about the work? What would students do differently or continue to do on their own map?

2. If possible, have the students do a second map of the same space to practice their drawing, attention to detail, and include of all necessary elements.

Resources

- David Walbert. "Building Map-Reading Skills and Higher-Order Thinking." Learn NC. University of North Carolina School of Education, 2010. (online—see Appendix B).
- "Water Harvesting." Permaculture Research Institute. (online—see Appendix B).
- "4. Zones and Sectors—Efficient Energy Planning." Deep Green Permaculture. (online—see Appendix B).

Winter: Rain Garden Design

Picking a rain garden site: How do educators allow student input to be balanced with the practicality of garden spaces? Having a rain garden in a community or school space often requires permits, permission, and consideration of larger systems like septic and underground piping. These realistic elements can be contrary to student dreams and expectations. After weeks of discussing and exploring rain gardens, the students will already have many ideas about where to install one. Student ownership of the project will be the fuel to engage them in their future lessons and activities. I usually set aside time before this mapmaking activity for students to suggest site locations they have in mind and consider points about the values of each. Class discussions around potential rain garden sites will encourage the students to use what they know about the land, soil, water drainage properties, and human and animal interactions. I have allowed students to vote on their preferred site, as long as they know all votes will be taken into consideration by adults who ultimately decide on the site.

1|2 Mapping the Schoolyard: Topography and Surfaces

Time Frame: 1½ hours (2 × 45 minutes)

Overview: The class will make a foundational map of the new or current rain garden site.

Objective: Students use the skills they have been developing to create a clear and concise map.

Introduction (10 minutes)

1. Before going out to the rain garden site, introduce the students to the project they will be engaging with over the next few weeks. The class will be developing bird's-eye view maps of the rain garden site. The goal is to map the important elements of the rain garden in order to best determine how people and animals will be interacting with it and where the plants and landscape features should go.

2. For students, what are some of the key lessons to keep in mind during this activity? What are essential or basic elements they may find in the rain garden space?

3. When building this initial map, students

will need to stay focused on their individual work. They will be expected to make maps that are clear, yet provide enough detail to be interesting and understandable. Students should notice when they are getting too bogged down in details and be encouraged to take a step back to look at the bigger picture; the details will come in time.

> "The site design tools of zone, sector, and slope allow us to organize information about the site into useful patterns and provide a starting point for an overall concept plan."[2]

Activities (1 hour)

1. Introduce the students to physical space of the garden, setting down clear boundaries. On graph or blank paper, students should make a quick sketch of the garden space, looking at:

- Structures and pathways
- Already existing plants and features
- Clear boundaries for the garden site
- Physical shape

They will make a second draft of this initial sketch in a later class period, so the sketch does not have to be their best and final work. However, they should pay attention to key details and shapes. This should take approximately 20 minutes.

2. Using their sketch as a guide, each student will create an initial big picture map of the rain garden site. They should include elements such as a compass rose and a legend or key. The students can draw their map on graph or blank paper according to their needs and preferences.

Assessment (10 minutes)

1. Students should make sure they have all the basic elements in place before both activities have ended.

2. Additional time can be given for students to return to the space and double-check their measurements and the placement of established features on their maps.

Preparation

1. A sample map

Resources

- Phil Nast. "Teaching with Maps: Lessons, Activities, Map-Making Resources, and More." (online—see Appendix B).

3 Water Flow

Time Frame: 45 minutes

Overview: Students will look for evidence of water flow and outlets in the rain garden space.

Objectives: To identify key zones of water movement and determine what influences they could have on the rain garden shape and choice of plants.

Introduction (5–10 minutes)

1. As a class, brainstorm with the students what are the key elements to look for in the rain garden site that will help them identify water movement, such as divots, slopes, current water sources, drains, or downspouts.

2. How can students tell where water has moved or drained? What clues or evidence helps them determine where water has rested or traveled? Are there past examples or memories of water activity in the area that the students can recall?

Activities (30 minutes)

1. Students will return to the rain garden site and make a quick sketch of key water flow areas. Where does water enter the rain garden? Where does it rest and exit? The students can draw or write notes to help them remember the important places.

2. The students will transfer this information to the maps they drafted in the previous activity. The class should determine what symbols will be used in order to illustrate the movement of water across the map.

Assessment (5 minutes)

1. Students will make sure they have the corresponding symbols from their maps on their legend/key.

2. Students can share their maps with a partner to receive constructive feedback and questions on their work.

Preparation

1. Sample map and symbols

4 Land Use and Traffic

Time Frame: 45 minutes

Overview: Students will map out footpaths and other human activities that might affect the garden.

Objective: To identify ways a human community can interact with a rain garden.

Introduction (5 minutes)

1. What are some of the ways the class knows that the community interacts and moves through the rain garden site? What are clues and evidence of human activity they can observe?

> "Pattern recognition... is the necessary precursor to the process of design."[3]

Activities (35 minutes)

1. The students will journey into the rain garden site and make notes or sketches of human pathways and other activity in the space. Then, they will transfer their notes onto their final draft maps.

2. The class should determine how to best represent data in ways that don't confuse or distract another reader. What symbols can be used to represent key areas on student maps? The distinction between effective and ineffective symbols is in how recognizable they are to someone else reading the map. Symbols should be common and simple shapes (such as triangles, distinct circles, squares, raindrops, flower shapes, x's). When the symbols are repeated, they should be consistently the same.

Assessment (5 minutes)

1. Looking at their maps, what stories can students tell about the ways the school community interacts with the rain garden?

2. How do the pathways and waterways interact with each other? How will these observations influence the way the class designs and builds the garden?

Additional Activities

Sun & Shade

Students can also measure the amount of sunlight and shade in different areas of the rain garden and transfer this data onto their maps. The information can influence which plant varieties are chosen, according to their site conditions.

Soil Sampling

If I plan on having the students develop a rain garden in a space with soil variations, then they can perform a NPK and pH test using a simple testing kit. They should do multiple tests in different plots to develop a sense of soil variation range. The class can also dig a test plot in the garden site and make observations about soil color, moisture, and texture.

3-D Models

I have had students build models of the rain garden before, out of paper, which were very successful and often better scaled than their maps!

5 Designing a Rain Garden

Time Frame: One hour

Overview: Students will share their ideas and form conclusions on what design elements should be included in the garden.

Objective: Each class member will have their voices heard and generate a list of class concerns and agreements on design elements and garden shape.

Introduction (5 minutes)

1. The students should gather the different maps they have worked on over the past few months. This includes their dream garden maps and realistic rain garden map. They will use these to illustrate their ideas and generate new ones.

2. The class will be participating in a group brainstorming activity that will require them to be active listeners and to ask questions of their partners. They will need to converse at a working voice level and be aware of the teacher's voice when it's time to transition into the next activity.

Activity (50 minutes)

1. 10 minutes: In pairs, students will share their rain garden ideas from their dream and realistic maps. They will take turns discussing their concerns for the garden and what design elements they would like to see.

2. 10 minutes: In pairs, the students will review and fill out the checklist on the Eight Rainwater Harvesting Principles worksheet, to focus their ideas into a grounded and organized argument.

3. 20 minutes: Each group will present their ideas, while the teacher records notes on a single poster. Students should summarize the key ideas and design elements their group agreed upon or discussed.

4. 10 minutes: Individually, the students will write a reflection for the teacher to review later. They should seek to answer what they think are essential design elements for the garden and how they want people to interact with the garden and the plants.

Assessment (5 minutes)

1. After finishing their reflections, students can use the last minutes of class to take a vote on their favorite ideas.

2. Using the brainstorm poster, students can choose to either write down one more idea they didn't share or to put a tally mark next to their favorite idea.

Preparation

1. Poster paper
2. Student maps
3. Eight Rainwater Harvesting Principles worksheet (adapted from Brad Lancaster—see Resources below)

Resources

- Heide Hermary. "Working with Water: Wetlands, Bog Gardens and Rain Gardens" in *Working with Nature: Shifting Paradigms: The Science and Practice of Organic Horticulture.*
- Brad Lancaster. *Rainwater Harvesting for Drylands and Beyond.*
- Sustainability Ambassadors. "Low Impact Development Manual for Schools." (online—see Appendix B).

6|10 Landscaping and Building the Rain Garden

Overview: Students will begin building the physical space of the rain garden.
Objectives: Students develop the garden space, while practicing safety, communication, and teamwork to accomplish the goal.

At the beginning of each class, I spend 5–10 minutes brainstorming with the students about the changes they see in the garden, what natural shapes they notice emerging, and what purposes these design elements have. I also give the students the chance to take ownership of their time and effort by setting the schedule and goal of the day's work. I facilitate discussion on what design elements need to be finished, how many people should work on each project, and what tools should be used.

As students build the basic shape of the garden, new landscape features emerge unexpectedly. It can help to be flexible and work with these new elements and opportunities. In one rain garden, my students were excavating "the creek" and ended up piling some of the soil and rock debris into impromptu "islands." They took advantage of these shapes and developed islands as a new feature of the garden.

During the first landscaping class, I remind the students about the conclusions reached at the end of their previous design time. I take student ideas and consult with the other adults overseeing the project, including professional landscapers. Using student ideas, everyone can help make final decisions about what to include in the garden.

6 Breaking Ground

Time Frame: 45 minutes

Activities

1. Students will be introduced to the general design of the garden space as determined by the class and the community input. The garden development will start with the most foundational pieces and then working toward the finer details.

2. The students will work on establishing the depth and width of the rain garden, using shovels and rakes to excavate, and carting extra material in wheelbarrows to a predetermined space.

3. Work groups and jobs can be traded throughout the activity period. If the class is working or restoring an older garden site, this would be the opportunity to focus on the larger

As a class, students delegate tasks and build the rain garden landscape together.

structural features: moving plants, digging out shallow spaces, or addressing features that are no longer useful or functional.

Assessment

1. At the end of the work time, have the students reflect on what they accomplished. Have the students achieved the goals of the day? What are the next steps in building the garden?

2. The students should gather and store away any tools they used and construct a barrier around the construction site.

Preparation

1. Shovels
2. Rakes
3. Wheelbarrows
4. Caution tape
5. Hammer
6. Stakes

7 Finishing the Foundation

Time Frame: 45 minutes

Activities

1. Students will finish the shape of the garden and add important structural elements, such as islands, berms, swales, or terraces (either free-form or with landscape edging).

2. The students should move carefully throughout the garden at this phase and work in pairs or take turns at different activities.

Assessment

Take time for the students to check in with each other as a class on the development of the rain garden. What features need more work in order to be stable? How is the class working together to accomplish their goals?

8 Establishing Permanent Features

Time Frame: 45 minutes

Activities

1. The class will enhance the secondary design features of the garden, such as terraces, berms, and swales, by reinforcing these spaces with boulders, logs, or rock features.

2. Student groups can also begin laying down bark chips, gravel, pebbles, soil, or mulch.

9 Adding Aesthetic Elements

Time Frame: 45 minutes

Activities

1. Students will continue adding rock or log design elements to the garden, such as stepping-stones or log rounds, while finishing key features and spaces.

2. They can also begin to fill any swale spaces with rock or gravel and add mulch to garden beds.

10 Completing the Garden

Time Frame: 45 minutes

Activities

1. The class will finish landscaping the garden by laying down soil, mulch, bark chips, gravel, and accent rocks and logs as needed.

2. Now is the time to fix any remaining issues or concerns with the structure of the space.

Assessment

1. What were student experiences in building this rain garden space? What surprised them, and how does the space fit into the initial vision of the garden? What changes were made and unexpected obstacles or concerns arise?

2. If the students were to build another rain garden again, what would they plan or do differently? What advice do they have for community members who want to build a rain garden?

Spring: Building a Rain Garden

1 Rain Garden Plants

Time Frame: 45 minutes
Overview: Students will conduct research on rain garden plants to determine the best zones to plant them in the garden.
Objective: To learn about how plants can hold soil structure, slow down water, retain nutrients, and filter pollutants.

Introduction (10 minutes)

1. Lead the class in a brainstorm of what they know about how plants filter, slow down, and retain water and soil: What types of plants thrive in wetlands or seasonally wet places? What are the benefits of planting with native plant species? What are beneficial, non-native species for the rain garden?

2. As a class, identify the different zones in the rain garden space. What are the different spaces in the rain garden according to slope? Where will the wettest and driest spaces be? If the students were to create three or more zones in the rain garden with different characteristics, what would the boundaries of these zones be? How would the class define the water flow, sunlight access, and soil quality of these separate zones?

3. Zones are site-specific, but one way to define them could be:

- Zone 1: In areas with well-draining soil, the highest points of the garden require plants whose roots can reach moisture farther below the surface. Plants in this zone may not receive any water during the summer.
- Zone 2: The mid-level section of the garden has well-draining soil and can support plants with shorter roots that prefer more water.
- Zone 3: The lowest level of the garden that remains underwater or saturated throughout the seasons. This soil is often not well-draining and consisting of clay or sand. Plants in this zone love water throughout the year, but can survive surface dryness during the summer.

> "Beyond the design of simple earthworks to allow biological processes of water purification to work, the ecological succession of streamside vegetation can effectively reinforce these water-purifying functions."[4]

Activities (25 minutes)

1. In pairs, students will be assigned a native plant that will go into the rain garden. Using available resources, the groups will determine the water and sunlight preferences of the plant assigned to them. They should record their findings, as well as the scientific name, benefits and uses, sun and water preference, and ecosystem range. Later, this information will be included in a sign for the garden.

2. Students will then use their maps to decide where in the rain garden this plant should be established. The groups will need to determine multiple locations and write them down on their maps.

3. The students may also need to update their maps with current landscape features they have built.

Assessment (10 minutes)

1. When the students have completed their research, they can gather together with another group from the same zone, compare the potential planting spots they have mapped, and resolve any conflicting ideas.

2. Then, the students can find larger groups in their zones and continue comparing the planting spots, brainstorming how they will landscape each group's plant throughout the zone.

Preparation

1. Native plant identification resources (books, internet)
2. Student maps

Resources

- Scott Kloos. *Pacific Northwest Medicinal Plants: Identify, Harvest, and Use 120 Wild Herbs for Health and Wellness.*
- Jim Pojar and Andy MacKinnon. *Plants of the Pacific Northwest Coast: Washington, Oregon, British Columbia & Alaska.*
- Robert Emanuel et al. *The Oregon Rain Garden Guide: Landscaping for Clean Water and Healthy Streams.* Oregon Sea Grant, Oregon State University, 2010. (online—see Appendix B).

NGSS and Activity Extensions

Further classroom studies can include researching and presenting a "scientific explanation based on evidence for how environmental and genetic factors influence the growth of an organism" (MS-LS1-5: From Molecules to Organisms).

2 Preparing the Soil

Time Frame: 45 minutes

Overview: The class will add mulch and soil to the rain garden and prepare the garden for planting.

Objectives: To provide a quality growing space for the garden plants and finish any landscaping projects.

A nearly finished rain garden, with mulch brightening the garden terraces.

Introduction (10 minutes)

1. Brainstorm with the students what kind of landscaping elements should be added to the garden. This is an opportunity to change any defining features in the garden. What areas can be addressed and how will the students change these features?

2. The students will be applying compost/mulch/soil to the garden in order to create better drainage and improve the quality of the soil for the plants. How can students carefully handle the available tools and practice safety and communication with their classmates?

Activities (30 minutes)

1. As a class, the students will work to add quality soil to the garden, predig holes for their plants, and finish landscaping the garden by laying down any additional elements (pathways, rocks, stepping-stones, logs, signs, bird feeders, and birdbaths).

2. The students should continue practicing teamwork and careful gardening during this activity by respecting their labor and the labor of others. After they have finished, the class can clean and return any tools that they used before gathering back together for a reflection.

Assessment (5 minutes)

1. Is there any landscaping work left for the students to accomplish?

2. Is the garden prepared for planting or are there additional projects that need to be finished?

Preparation

1. Soil, compost, mulch
2. Landscaping supplies (pathways, rocks, stepping-stones, logs)
3. Wheelbarrows
4. Shovels/trowels

3|4 Planting in the Rain Garden

Time Frame: 1½ hours (2 × 45 minutes)

Overview: Students will plant in the zones they determined were the best habitat for their plant species.

Objectives: To follow instructions for quality and successful transplanting of rain garden plants.

Introduction (15 minutes)

1. Brainstorm with the students how to gently remove plants from their containers and how deep to dig the holes: What are some key lessons to remember about removing these plants and transplanting them from their pots? How deep should the holes be for the plants? How should students handle the plant roots, stems,

The class diligently plants native plant species in their preferred conditions.

and plant markers? What does tool use and safety look like during this activity?

2. Before the students work in the garden, they should review their maps and their information about the plants they were assigned. They should reevaluate which spaces are best for their plants and be prepared to have multiple spaces to choose from, in the case that some spots are better for other plants. The students can also meet with other teams in their zone to remember their agreements on where is best to plant the starts.

3. The students should be encouraged to use kind words, creative problem-solving, and teamwork when deciding where to plant. They will encounter other students who want to plant in the same spaces, so they should be prepared to work collaboratively with their classmates while also keeping in mind the growth preferences of their plant species.

Activities (1 hour)

1. In their small groups, the students will gather their plant starts and establish them in the rain garden. They will have to work carefully alongside their classmates to determine the best places to dig holes, according to their plant's growing preferences.

2. Students can use provided compost to incorporate into the local soil for increased soil health. Quality work will look like well-dug holes, use of hands to break up soil clumps, and plants buried fully into the ground and covered up gently with soil.

3. The student groups can deeply water the transplanted starts when they have finished planting.

Assessment (15 minutes)

1. The students should clean up all the planting pots and tools. They can go back over the whole rain garden to do a quality control check of the plantings and to make sure all the plants have been placed according to their zones.

2. What do students imagine these plants will look like in a few months, a year, or five years? How big will some plants get? What changes can be expected over a long period of time?

Preparation

1. Plant starts
2. Trowels
3. Compost
4. Watering can or hose

5 Seeds and Starts

Time Frame: 45 minutes

Overview: The students will plant seeds and/or additional starts in the rain garden.

Objectives: To practice essential gardening skills and fill open soil spaces in the rain garden.

Introduction (10 minutes)

1. Review with the students what types of native flowers and plants they will install in the rain garden, such as poppy, flax, meadow foam, pearly everlasting, columbine, asters. How will they spread these seeds when many of them are small? Demonstrate how to mix small seeds with sand/soil and broadcast them carefully over the open soil spaces.

2. What supportive roles will these seeds play in the rain garden for birds, insects, weeds, soil health, and humans?

Activities (30 minutes)

1. Individually or in pairs, students will mix a variety of native flower species with a set amount of sand and soil mixture. They will carefully distribute this mixture over the open soil spaces in the rain garden. If there are additional plant starts, then the students can plant them first before spreading seeds under and around them.

2. Then, the students can gently press their hands into the soil to ensure seed-to-soil contact. The students can water the rain garden plants and the seeds they planted.

Assessment (5 minutes)

1. What pollinator, insect, bird, and mammal habitat will the rain garden provide?

2. Are there additional changes or features that could be added to the rain garden?

A student-built amphibian home constructed from found materials

Preparation

1. Flower seeds
2. Sand/soil mixture
3. Watering can or hose

6 Building Habitat

Time Frame: 45 minutes

Overview: The class will work together to build amphibian habitat from recycled and found materials.

Objective: The students will learn how to provide habitat for valuable rain garden species.

Introduction (10 minutes)

1. Brainstorm with the students what types of mammals, amphibians, and insects they expect might live or migrate through the rain garden: What types of creatures have they already seen on the schoolyard and what roles do they have in the garden? What features and materials could the class provide to create habitat for them?

2. The rain garden will most likely attract many amphibians to live and reproduce in it. Introduce the students to a variety of frog and salamander homes they can create themselves, focusing on the key characteristics of different spaces (shade, holes, crevices, covered spaces).

Activities (30 minutes)

1. Individually or in pairs, students will gather natural or found materials to build into amphibian habitat spaces. They will need to provide the essential characteristics that they identified as a class in the initial brainstorm. They can be encouraged to use dead organic materials, found objects, and provided materials to build amphibian habitat in the rain garden.

2. Students are expected to use teamwork, craftsmanship, and creative problem-solving when building their habitats.

Assessment (5 minutes)

1. After they have cleaned up the work areas and returned their supplies, the class will tour the habitat designs, taking time to ask questions about them. Why did the student choose a particular style of habitat over another? What materials did they use to accomplish their goals?

2. Students will have time at the end to adjust and provide final touches to their habitats.

Preparation

1. Bark pieces
2. Old pots
3. Sticks
4. Assorted rocks
5. Moss and natural materials

Resources

- Russell Link. *Landscaping for Wildlife in the Pacific Northwest.*
- National Wildlife Federation. "Attracting Amphibians." (online—see Appendix B).
- National Wildlife Federation. "Schoolyard Habitats." (online—see Appendix B).

7 Creating Signs

Time Frame: 45 minutes

Overview: The class will create signs for the native plants, amphibian habitat, and key features in the rain garden.

Objective: To welcome the community into the rain garden by creating accessible signs that engage various age groups.

Introduction (10 minutes)

1. Brainstorm with the class about what makes an engaging sign: How can the students create signs that will be clear to every age group in the community? What languages should be used in the signs?

2. As a class, create a list of the plants, rain garden facts, and additional information that are important for the school community to know about. When the class has decided on what signs should be in the garden, have the students volunteer or be assigned to create a sign for one of these items, in addition to a sign about the plant species which they researched weeks before.

3. Discuss how the students practice craftsmanship and teamwork during this activity.

Activities (30 minutes)

1. In pairs, the students will create clear and engaging signs for the rain garden. The groups can sketch out what their sign will say in pencil, before painting or drawing their work. The goal is to create crafted pieces that will educate the school or community on the plants, features, and behaviors expected in the rain garden.

2. Before painting or drawing, the students will hammer a wooden stake onto their signboard, practicing safety when handling the tools.

3. When the groups have finished painting, they can leave their signs to dry in a safe place, clean up, and then explore the school garden for seasonal taste tests.

Follow-up Activity (5 minutes)

1. When the signs are dry, the students can install them in the rain garden with their hands or with a hammer.

Preparation

1. Wood boards approximately 5 × 8 inches each
2. Wooden stakes
3. Paintbrushes
4. Outdoor paint (or acrylic with sealant)
5. Water
6. Cups
7. Hammer
8. Trowel
9. Pencils
10. Nails

8 Student Docents

Time Frame: 45 minutes
Overview: The students will give tours of the rain garden to the school community.
Objective: To share the class vision about the rain garden's growth, its special features, and the hard work of the students.

Introduction (10 minutes)

1. Brainstorm with the class what key information the community should know about the rain garden. The class has worked hard to create this space, but now many children and families will be interacting with it.

2. What messages can the class give to the community about how to treat the rain garden plants and species? What important key features should they be aware of? How could students describe how to develop a rain garden to someone who may want to build one? Are there students who feel passionately about sharing certain messages?

3. Have the students reflect back on their first impressions of what a rain garden contained. Many students and community members won't know what rain gardens are, so how can the class explain the purpose of the garden clearly?

Activities (30 minutes)

Individually or in pairs, the students will lead tours of the rain garden for community members and answer questions, explaining the features and plants in the garden, and the importance of catching and slowing down water.

> The goal of this activity is for each student to share a piece of the rain garden that matters to them most and to pass on this project and legacy to younger students, as well as express what they know through engagement and presentation. There are many ways for students to share their knowledge of the rain garden and the behavior expectations for the school. Each school community will have a method of information sharing that works best for them. Some might want to have small groups go into each classroom and do a presentation, or have classes come out to the garden one at a time so the whole class can share their work. The class can also give tours during a larger community event for students and their families.

Assessment (5 minutes)

1. When the students have finished their tours and presentations, encourage them to reflect on how they feel about sharing their messages. What would they do differently? Were there messages that they did not get across?

2. What were community and student reactions to the garden? What kinds of questions did community members ask?

9 Making Journals

Time Frame: 45 minutes
Overview: The students will create garden journals and compile their year of worksheets and maps on the rain garden.
Objective: To celebrate the large amount of learning and work the class accomplished in building the rain garden.

Introduction (5–10 minutes)

1. Encourage the students to reflect on their experiences over the past school year, from watching the way water moves, making maps, and designing a rain garden. They have accomplished so much and should feel proud of their achievements.

2. What does quality work and craftsmanship look like during this activity? How have students made garden journals in the past? Are there innovations that they would like to try when assembling this year's journals?

Activities (30 minutes)

1. The students will gather a cover page and their worksheets, maps, and sketches from the past year and organize them into a garden journal. They can hole punch the papers and weave them with yarn to create a book bind. Students can work with a partner to accomplish this task (pulling yarn through the holes can be difficult).

2. Before the students color their cover pages, they can gather a taste test that they harvested from the garden to enjoy as they work.

Assessment (5–10 minutes)

1. Looking back on their work, how do students feel about the breadth of their learning? What do they notice about their work at the beginning of the year? How did it change over time?

3. What are their favorite memories and activities from the school year and in the rain garden?

Preparation

1. Colored yarn
2. Scissors
3. Hole punchers
4. Colored pencils
5. Rain Garden cover page

10 Celebration

Time Frame: 45 minutes
Overview: The class will harvest seasonal garden foods to create meals and share them with the class.
Objective: For students to explore new foods and flavors while celebrating their work during the school year.

Introduction (5–10 minutes)

1. What are the students' favorite spring and early summer plants in the garden? How do

Strawberry

The flowers and

food are starting to

grow on this lovely

plant today. All the

rain showers from the

beautiful sky are working

their way to this

lovely day.

— Anonymous
Student Poem

they like to eat them? What are other ways of eating these plants or incorporating them in recipes that students know?

2. Introduce the students to different spring recipes they can make from the food in the garden. How can the class create these recipes while practicing food safety in handling, washing, and preparing the foods?

Activities (30 minutes)

1. Small groups of students will follow a simple spring recipe, gather the plants from the garden, and prepare food for the class. Additional spring plants not grown in the garden can be added too.

> A class's ability to cook depends on each school's access to a kitchen and its rules on student food handling. I suggest some sample recipes in Resources that can be made using spring foods, for educators to use as they will. Another variation could be to have students bring in recipes to share.
>
> Recipe #1: Pea Green Pesto
> Recipe #2: Roasted Radishes
> Recipe #3: Beet and Ricotta Hummus
> Recipe #4: Strawberries and Cream

2. Each group will gather the ingredients and materials that their recipe calls for and then prepare their meal. They will use provided serving cups to prepare enough food for every student in class. They will work as a team and safely handle the food and the tools.

3. They can also gather garnishes for their meals from the garden, after they have washed the food. Finished groups can clean up their supplies and station.

4. When all the groups have assembled their meals, each student can receive a serving of each recipe, gather them onto a plate, and enjoy a taste test of prepared garden foods. Additionally, they can pick raw taste tests of their favorite foods from the garden.

Assessment (5 minutes)

1. As the students enjoy their taste tests and meal, have them describe the flavors and textures of the spring food. How does a radish taste raw versus roasted? How does cooking change flavors and textures?

2. What are student experiences with the different foods? Do they prefer them raw or cooked? What are the students looking forward to eating in the fall? What foods will be seasonally available?

Preparation

1. Recipe ingredients
2. Cutting boards
3. Washing station
4. Bowls, cups, plates
5. Utensils
6. Mixing bowls
7. Recipe cards (see Resources)
8. Adult volunteers
9. Blender (2)
10. Oven

Resources

- Love and Lemons. "Pea Tendril & Pistachio Pesto." (online—see Appendix B).
- Tasha de Serio. "Roasted Radishes with Brown Butter, Lemon, and Radish Tops." Epicurious, *Bon Appétit.* (online—see Appendix B).
- Claire Saffitz. "Beet and Ricotta Hummus." *Bon Appétit.* (online—see Appendix B).
- Adriana Martin. "Strawberries and Cream a Classic Mexican Treat." Adriana's Best Recipes. (online—see Appendix B).

Introduction

The needs of middle school students differ from those of other age groups due to rapid student growth, shifting social dynamics, and rigorous studies. Many students begin seeking their independence and gaining more awareness of the social, economic, and political conditions surrounding them. I have found that students of this age succeed with adaptable programming that complements their concerns and passions while providing training in practical skills. A balance of structure, larger concepts, and opportunities for creativity help these young people feel empowered to make the changes they want to see in their communities.

The lessons for the seventh and eighth grade students differ from the previous lessons in three major ways. First, these lessons are project-based and encourage physical engagement through tool use, building, and labor. This focus provides healthy physical activity for students who are increasingly sitting down for longer study periods. It also respects the students' general garden knowledge. When students have been gardening for years, many need additional challenges to be engaged, independent, and creative in their work.

Second, the seventh and eighth grade lessons offer a wider scope of gardening concepts. Previous lessons for younger grades can be accomplished in small spaces, but seventh and eighth grade students are able to manage large-scale projects with community support, like building garden beds, benches, and trellising structures. Middle school students can be integral to cultivating radical change in their school and community, while also engaging with larger social and ecological principles, such as the Permaculture Design Principles *Use Small and Slow Solutions* or *Use Edges and Value the Marginal* (see Holmgren's *Permaculture Pathways*[1]). After years of engaging with the language of Care for Self, Others, and the Land, students can now experience the direct origin of these principles and apply them beyond the garden.

The third difference in these lessons is the opportunity to encourage community engagement through extensions beyond the learning garden. The goal of this community-building focus is to help middle school students make positive changes in their communities by providing outlets for student concerns and skill-building opportunities. The seventh and eighth grade lessons I provide are a framework for potentially larger and more integrated student work in community organizations (food banks, nursing homes, community gardens) and with environmental and political issues (global warming, land use zoning). I believe it is invaluable for older students to have opportunities to practice personal responsibility in their communities with the goal of becoming the next generation of changemakers.

I have tried various methods of structuring gardening lessons, and my current plan has been the most successful. I use this basic framework with both the seventh and eighth grade classes, but the themes and opportunities for extended lessons vary at each grade level. This model provides some structure for educators to guide larger conceptual work, while being flexible to student needs, interests, and unit studies. For each grade, I balance garden lessons with three goals, or tiers, including:

Tier One	Tier Two	Tier Three
Practical Garden and Life Skills	Self-directed Projects	Garden Leadership

The relationships between the three tiers are flexible according to student interests, leadership opportunities, and seasonal garden work. The goals of the middle school gardening experience should emphasize:

- building community
- living a healthy lifestyle
- creative problem-solving
- craftsmanship
- engagement in nature through science
- personal responsibility
- stewardship

"Personal responsibility implies full awareness of the structure of our individual dependence on, and effect on, the local and the global environment, and local and global communities."[2]

Tier One: Practical Garden and Life Skills

Each class will engage in approximately four organized gardening lessons each season in order to build student skills and knowledge of organic gardening practices and permaculture principles. This format provides students with an opportunity to develop new interests, discuss relevant environmental and social issues, foster practical skills, and be physically active outdoors. For example, by having all the students participate in building raised beds, they can develop their engineering skills and their comfort and confidence in handling tools. This translates into high-quality, self-directed projects in the future and potential job skill training.

The activities for the seventh grade students are guided by permaculture principles that can extend to greater conversations on social, political, and environmental issues. I use the principle of *Produce No Waste* to engage with garden waste management, but part of student studies can include the human waste stream, social inequalities that intersect with polluted spaces, and human health issues. Eighth grade lessons encourage students to apply permaculture design practices and ecological mind-sets through creative problem-solving and innovative solutions. The scope of student impact and learning is determined by the passions and needs of each community or educator that uses these resources. The lessons I provide begin in the garden and use the space as a classroom and laboratory, but the concepts students learn there can extend far beyond the garden's grounds. These garden lessons are intended to inspire innovation.

Tier Two: Self-directed Projects

Many middle school students want to deeply pursue their own interests in the garden. Some students want to learn how to garden intensively while others want to build habitat boxes, raised beds, or construct signs. Others enjoy fixing broken things or want to exercise their bodies by hauling materials with wheelbarrows or using shovels. I plan for around four Self-directed Projects each season.

Having a constant list of possible seasonal projects for the students to pick from can help direct student work, but the students can also be encouraged to advocate for their interests and design their own "special project." This is a great opportunity for students to work with Permaculture Design Principles that would contribute more value to the garden.

For example, students may have an innovative idea for that messy, neglected, or frustrating section of the garden that seems to always encourage ill-placed plants. How could students *Use Edges and Value the Marginal* by applying a new design? How could they create a positive interaction with that space which acknowledges its value? Self-directed projects can be most effective when it arises from the passions and interests of each student.

These projects require a level of training and trust in student tool handling and focus. There is a privilege to small group and individual work, rather than worksheets and lessons. Many students are capable of being the agents of their own learning, time, and work. I prefer to block the four Self-directed Project times together, one class a week for four weeks. This schedule gives the students a chance to design their projects gather materials, build them, and troubleshoot any issues.

Self-directed Projects are the most flexible tier in the middle school garden lessons and are open to the needs and opportunities of any community or school. One community may have enough space for students to do their Self-directed Projects in an onsite school garden. Other communities may not have the space for many students to work outside at once. These Self-directed Projects can be accomplished in the local community, as service work for community centers, homes for the elderly, nonprofits, hospitals, community gardens, or food banks. The potentials for student work, innovation, and creativity are endless and should be directed by the students' ideas and interests.

Student Gardeners

Some students want their own garden beds or containers to nurture. They may have cared for the same spaces for years and want to continue working in them. These students must understand that a school garden is a community garden, so there will be other students interacting with "their" garden space.

The agreement I create between myself and these *student gardeners* is that they can make garden experiments to their own standards, but their produce will feed the students and

families in the community, just as the whole school garden does. Some middle school students choose to harvest the produce they raise for school taste testing or offer that harvest to the school lunchroom. If they do self-directed work in their gardens, the students inform me of their plan of action before each self-directed gardening class.

Tier Three: Garden Leadership

When students have learned gardening from Kindergarten through seventh and eighth grade, most will have received between five and eight years of garden education. They will have planted thousands of seeds, designed and built innovative garden structures, and understand the importance of seasonal garden chores. While there are still lessons and life skills these students can learn from the garden, they handle themselves with the confidence and grace of well-seasoned gardeners.

This is the time when a student's leadership skills can be encouraged to grow. Before they graduate, middle school students can develop leadership skills by teaching a new generation of younger students. Older students actively apply the knowledge they have gained and offer it back to the community, manifesting their knowledge in tangible ways, relearning lessons from the perspective of a teacher, and giving back to the community. I set aside approximately two gardening classes each season for leadership development. Seasonal garden work—seed saving, planting, mulching, and weeding—are all opportunities for older students to share the community's garden culture with younger students. Older students pass on their values and knowledge by exhibiting positive behavior actively to their partners. Together, younger and older students enrich the garden culture that is so valuable to the school and community.

Providing Structure

Ideally, I envisioned the middle school students creating actual lesson plans and activities from their own knowledge and experience. However, considering the limited time students usually have in gardening class per week, structured lessons create space for the older students' teaching and social skills to flourish.

When middle school students are partnered with younger students rather than with children their age, I have observed enthusiasm and excitement in the older students that they don't typically show with their peers. With the younger students, they can learn how to keep the students' attention, engage and interact with them, and build relationships. All the students focus with diligence on the work with their partners. It is community-building in action.

SEVENTH GRADE—GRADE SEVEN

RENEWABLE RESOURCES

This year in the garden, the seventh grade students will embark on a yearlong study of the cultural and ecological value of renewable and nonrenewable resources, such as seeds and soils. They will reengage with seasonal gardening activities, save seeds and screen compost, but with new perspectives and rigor in their studies. In the fall, the class will have the opportunity to learn about the permaculture principle *Catch and Store Energy*, discuss carbon sequestration in soils and the role that photosynthesis plays in processing energy

FALL
Catching and Storing Energy

1. Garden Skills: Planting Seeds for Winter
2. Garden Skills: Catch and Store Energy
3. Leadership: Save the Best, Eat the Rest
4. Garden Skills: Sowing Cover Crops
5. Leadership: Planting Bulbs
6. Garden Skills: Transplanting Cold-hardy Varieties

7–10. Self-directed Projects

WINTER
Valuing Renewable and Natural Resources

1–5. Research Project

6. Leadership: Screening Compost
7. Garden Skills: Cultivating Soil for Container Plants
8. Leadership: Planting Heirloom Seeds
9. Garden Skills: Mulching Green Manure
10. Garden Skills: Separating Perennial Plants

SPRING
Using Small and Slow Solutions

1–2. Garden Skills: Building Planting Containers

3. Leadership: Transplanting Heirlooms
4. Garden Skills: Garden-based Fertilizers

5–8. Self-directed Projects

9. Garden Skills: Designing Trellises
10. Leadership: Building Trellises

through an ecosystem. The students will engage with the cultural and historical legacy of plant genetics, as well as the science behind it.

In the winter, the students will thoroughly research heirloom seeds and discuss them in the relation to hybrid seeds, modern agrobusinesses, and global seed production. They will learn about the permaculture principle *Use and Value Renewable Resources* and practice techniques to accomplish this in the garden. The class will create profiles of valuable, culturally relevant heirloom seeds, delve into botanical classifications and illustrations, and create an informational sign and seed packet that can be used by the school and community for many years. They will build potting soil for container planting and become Garden Leaders through their work with younger grades. In the spring, the class will design and build planting containers and explore how to create *Small and Slow Solutions* in the garden and their community.

Over the year, the seventh graders will have many opportunities to practice personal responsibility, hone their leadership skills, and engage with craftsmanship. The gardening activities they will explore provide a framework for even greater in-class conversations and studies that expand into relevant global issues and concerns. Engaging with these conversations will only make the students even more effective when they continue gardening in the eighth grade.

© Sergey Novikov / Adobe Stock

Next Generation Science Standards

Throughout this year of gardening lessons, the students will enhance their research skills and have opportunities to practice engineering, design, and application of their design within the garden. They will have to use creative problem-solving to build and troubleshoot their projects, as well as engage in rigorous and thorough research. The overall goal is for the students to become knowledgeable and confident in their subject matter.

In-class themes and extensions could include how organisms pass on traits from one generation to the next, forming explanations about how genetic and environmental factors affect the growth of organisms, genetic factors and local conditions that affect plant growth and their offspring, exploring the role of photosynthesis in the flow of energy within ecosystems, and selective breeding of plants.

Additionally, the students can explore the role of soils and forests in carbon sequestration, the importance of plant diversity, agroecology practices, the origins of early agriculture, transgenic and GMO crops, the relationship of modern agriculture with nonrenewable and renewable resources, botany, open and self-pollinating plants, and the process of pollination.

Permaculture Principles and Perspectives

- Share the Surplus
- Catch and Store Energy
- Use and Value Renewable Resources
- Use Small and Slow Solutions

Fall: Catching and Storing Energy

1 Garden Skills: Planting Seeds for Winter

Time Frame: 45 minutes

Overview: During their first gardening activity, the students will celebrate their return by planting food for the winter.

Objective: For students to increase all-season food production and explore the garden's seasonal changes.

Introduction (10 minutes)

1. Welcome the students back into the garden. Discuss the changes they have noticed in the space, plants that have thrived or failed, and how the work that they accomplished in the spring has organically altered.

2. What are the students most excited for in the upcoming year? What does being a Garden Leader look like, and what behaviors do the students expect from themselves and their classmates? In what new ways will they continue to care for themselves, each other, and the land?

3. Why is it important to try and grow food all year long? What kinds of food-producing plants prefer growing in the cooler seasons? What impact can growing all-season food have on the community?

Activity (30 minutes)

1. In pairs, the students will receive seeds for a cold-season, fast-growing plant variety that can grow first in a small pot and then be transplanted. These could include chard, collards, kale, certain varieties of onions, and lettuces. The groups will work together to gather their planting containers, fill them with screened soil, read thc planting requirements on the seed packet, if available, and carefully sow the seeds.

2. When they have planted their seeds, the students can make a plant marker or label for their variety, water them deeply with a watering can, and continue planting more seeds.

3. Afterward, the students can enjoy a taste test of seasonal foods from the garden. They should remember to harvest with care and only take as much as they can eat.

Assessment (5 minutes)

1. Gathering back together, have the class discuss the changes in the garden that they noticed during their activity. What amazed and excited the students about the garden's growth? What plants did they taste test that they would like to save seeds from and grow again next year?

2. What were student experiences in planting their seeds? What plants do students expect to germinate and grow before others?

Preparation

1. Seed packets
2. Watering can
3. Screened soil
4. Plant markers
5. Four-inch planting containers

Resources

- Eliot Coleman. *Four-Season Harvest: Organic Vegetables from Your Home Garden All Year Long*, 2nd ed.
- Linda Gilkeson. *Backyard Bounty: The Complete Guide to Year-Round Organic Gardening in the Pacific Northwest*, revised and expanded 2nd ed.

2 Garden Skills: Catch and Store Energy

Time Frame: 45 minutes

Overview: The students will save seeds from heirloom plants while beginning to discuss the differences between heirloom and hybrids seeds, a conversation that they will continue in the winter.

Objective: Students will participate in a seasonal and cultural garden activity while delving into new regional and global conversations and histories.

Vocabulary

1. Thresh
2. Winnow
3. Heirloom
4. Hybrid

Introduction (10 minutes)

1. Brainstorm what the class knows about seeds and their genetics: What is an *heirloom*? What is a *hybrid*? Why do students save seeds? What is the cultural, economic, and ecological value of saving seeds? The class should understand that seed saving is fundamental for survival and can occur onsite in the garden. Saving seeds is also a celebration of unique, cultural values and features of a place and a way to preserve a seasonal surplus. Ecologically, it is also a way to preserve the genetic diversity of seeds.

2. Introduce the students to the principle *Catch and Store Energy*, as described by David Holmgren in *Permaculture Perspectives*. What kinds of energy have students captured and stored into the garden soil and ecosystem? How do seeds relate to catching and storing energy? How are seeds an onsite renewable resource?

3. Even the smallest seeds hold incredible potential for food, habitat, nourishment, and medicine. Cultivating and cherishing a seed line is an important way to catch natural energy and store it away. It is a practice in self-reliance, a cultural celebration of place, offers stability for yearly food sources, and also promotes autonomy and independence over reliance on industrial seed production systems.

> "By catching and storing the energy in seeds, growers maintained a genetic and cultural lineage from ancestors to descendants."[1]

4. Review with the class what heirloom plants they will be saving seeds from and what healthy plants look like. What are qualities in a plant, like a tomato, that gardeners wish to save? What grew very well in the garden, according to productivity, taste, and popularity with students?

Activity (30 minutes)

1. Individually or in pairs, the students will harvest seeds from selected heirloom plants in the garden. They can gather these from plants that are marked for seed saving or in a garden bed with plants saved for that purpose. They can use collection cups or bowls and clippers or scissors to gather the seeds. Dry seeds (like lettuce, radishes, members of the brassica family, and carrots) are good options for this seed saving activity because the students will be encouraged to *thresh* and *winnow* them.

2. Before they harvest the seeds, the students should assess the health and condition of the whole plant for evidence of disease, poor growth, taste, fruit or vegetable shape, and color.

3. When the students have gathered the seeds, they can rub and roll the seed heads between their palms to *thresh* the seeds from their pods and seed heads. Then, they can use a small screen to *winnow* the seeds by gently tossing the seeds in the air to remove them from the chaff, or by swirling them in a bowl to separate and picking out the debris. The screens can be made before this activity by students or adults (see "A Handful of Seeds" in Resources).

5. After the seeds have been separated from other organic matter, the students will store the seeds in a paper bag or glass jar that is labeled with the plant variety and date of harvest. Then, they can enjoy a taste test from the fall garden.

Assessment (5 minute)

1. What were student experiences when gathering the seeds? What challenges did they face and overcome when identifying healthy plants? How many seeds (surplus) did they gather? Was it more or less than they expected?

2. Considering the size of the seeds they harvested and the size of the plant they harvested the seeds from, what conclusions can students reach about the energy contained within a seed? How are they valuable renewable resources?

3. How did students enjoy threshing and winnowing seeds? What skills and techniques did the students learn when doing these activities?

Preparation

1. Paper bags or glass jars with lids
2. Markers
3. Examples of seeds to harvest
3. Winnowing supplies

NGSS and Activity Extensions

In-class extensions from this activity and its themes include: plant genetics and heredity (the inheritance and variation on traits), the environmental and genetic factors that affect plants and seeds, local conditions that affect plant growth, the structure of a flower, plant reproduction, the energy input of commercial and industrial seed production, and an introduction to agroecology.

Resources

- Suzanne Ashworth. *Seed to Seed: Seed Saving and Growing Techniques*, 2nd. ed.
- David Holmgren. *Permaculture: Principles & Pathways Beyond Sustainability.*
- Tina M. Poles. "A Handful of Seeds: Seed Study and Seed Saving for Educators." Occidental Arts and Ecology Center. (online—see Appendix B.)
- William Woys Weaver. *Heirloom Vegetable Gardening: A Master Gardener's Guide to Planting, Growing, Seed Saving, and Cultural History*, rev. ed.

3 Leadership: Save the Best, Eat the Rest

Time Frame: 45 minutes for older students, 30 minutes for a younger class

Overview: The older students will lead their younger partners in a wet seed saving activity and a taste test experience.

Objective: Older students will learn by teaching others about the cultural and ecological value of saving seeds from heirloom plants.

Introduction (10 minutes)

1. Before the younger students come out to the garden, gather the older students together to review the expected behaviors and goal of the activity. In what ways can the seventh graders be Garden Leaders? What does leadership look like? What does *Save the Best, Eat the Rest* mean in the garden?

2. The goal of the activity is for the older students to teach their young partners how to identify healthy heirloom plants, how to harvest their fruits, and how to collect the seeds. During the activity, the older students will keep their partners engaged and teach them why seed saving is an important cultural activity (self-reliance, heritage) and ecological activity (preserving plant diversity).

3. Review with the older students what they have learned about how to identify and harvest healthy fruits, as well as the steps and materials needed for the activity.

4. When the young students arrive, they can be partnered with an older student and should review the expected behaviors in the garden and during the activity.

> "Even with increased use of perennial crops, annual and biennial vegetables and field crops remain essential to sustenance and culture."[2]

Activity (30 minutes)

1. Before beginning the activity, introduce the students to the food handling expectations of their community. Since they will later be preparing salsa or a dipping sauce from the fruits they gather, they may need to wash their produce according to community or district standards.

2. In their partnerships, the older and younger students will harvest fruit that contains seeds protected within it, such as cucumbers or tomatoes. Peppers are another fruit that has seeds inside, but pepper seeds are not protected in a gelatinous membrane and don't need additional fermentation time to separate membranes from the seed. For this activity, communities can choose whether they want students to work with gelatinously covered seeds and the fermentation process or not.

3. The older students should instruct their younger partner on how to identify healthy plants and fruits, how to harvest them, how to carefully remove the seeds from the fruit, and why this activity is important culturally and ecologically.

4. When they have harvested their fruit, the groups will carefully cut them open, remove the seeds, and gather them in collection cups. They should try to remove any fruit pulp and pieces from their cups and just leave the seeds behind.

5. If the seeds have a gelatinous layer protecting them, the groups can fill the cups with water and set them in a safe space to gather yeasts and molds to eat away at this layer (around three days to a week). Seeds that don't have a gelatinous layer can be laid out on a coffee filter or a sheet of paper to dry before being stored (see Resources).

6. All the groups should carefully label their collection cups or drying area with the plant name and the date, using tape and markers.

7. When the groups have saved their seeds, they will pour the remaining fruit pulp into a blender, add additional ingredients as they would like, and create a simple salsa or dipping sauce to enjoy with the class. The salsa can be made by students or by adults while the groups gather additional taste tests from the garden.

8. Finally, the two classes can share in a taste test from the fruit they harvested, with the older students helping the younger ones find accurate words to talk about the texture and flavor of foods.

Assessment (5 minutes)

1. The older students can help clean up the workstations and then can gather back together to share their experiences.

2. What struggles or accomplishments did the older students have when being leaders in the garden? What techniques did they use to inspire their partner's interest and sustained attention? How effective do the students think they were in communicating the messages of the activity?

Preparation

1. Collection cups
2. Sieve
3. Hose or sink
4. Coffee filters
5. Tape and markers
6. Cutting boards
7. Knives
8. Blender(s)
9. Prepared additional salsa ingredients (lime, onion, cilantro)
10. Spoons and bowls
11. Chips
12. Adult volunteer

Activity Extension

Now is a great time for the students to engage with the community-building aspect of gardening by reading and discussing Paul Fleischman's *Seedfolks.*

Resources

- Suzanne Ashworth. *Seed to Seed.*
- Paul Fleischman. *Seedfolks.*

4 Garden Skills: Sowing Cover Crops

Time Frame: 45 minutes

Overview: The students will sow cover crop seeds in designated garden beds or containers.

Objective: The class participates in a seasonal garden activity that is vital to energy cycling in the garden and that supports the agroecology initiative of the garden community.

Vocabulary

Agroecology/agroecosystem

Introduction (10 minutes)

1. Brainstorm with the class about what they know of the garden energy cycle: What renewable natural resources process through the garden? What are nonrenewable resources within the garden system? Through what methods do these resources cycle? As a class, begin drawing a map of the garden energy cycle on poster paper.

2. What role do living plants and nonliving organic matter, or biotic and abiotic factors, play in this cycle? What relationships do they have with each other? Introduce the students to the term *agroecology* and *agroecosystem* (see Resources). While the term may be new to students, they should be very familiar with the concept of a garden or food production system being perceived as a whole or ecosystem. This idea has been ingrained in all their past gardening studies on soils, seeds, decomposers, compost, and mimicking natural systems. They should be able to discuss the roles that humans play in this ecosystem, through relationships, as consumers, and within the process of energy cycling.

3. What activities have the students, as gardeners, engaged with that support the processing of energy within the garden system? How does cover cropping and generating green manure relate to agroecology? What part of the garden's energy cycle does cover cropping affect, and how is cover cropping a response to human food consumption?

4. Other than adding organic matter and nutrients into the soil, what other benefits are there to covering soil with a winter crop? These benefits include fixing nitrogen, suppressing weeds, controlling erosion, managing insect pests, and attracting beneficial insects.

5. What cover crops seeds have students planted or sown before? What methods did they use to plant them and ensure seed-to-soil contact? Demonstrate the planting activity for the day and the spaces where students will be sowing seeds.

I recommend checking with local organic farmers and gardeners in your community to see what they plant as a cover crop in your region. Good places to plant cover crops include empty beds, containers, or raised bed edges. I also recommend planting multiple varieties of cover crops so that students can observe the different growth rates and successes of each variety in relation to environmental and local conditions.

Activity (30 minutes)

1. In pairs or small groups, the students will sow cover crop seeds in designated garden beds or containers. They should work slowly and carefully to press the seeds into the soil firmly, and make sure the seeds have seed-to-soil contact, without compacting the soil.

2. In a larger garden, the students can prepare the soil by raking it, spreading the seeds, and then gently pressing the seeds in using the rake. Afterward, the students can use a hose with a sprinkler setting to gently water the seeds.

3. When the class has finished, they can clean up and gather any late-season foods they would like to taste test.

Assessment (5 minutes)

1. What observations did the students make about the seeds they planted? How difficult or easy was it to ensure seed-to-soil contact?

2. In what ways will the cover crops contribute to the garden energy cycle, and what changes can the students expect to see over the course of the year? How does cover cropping support the principle of catching and storing energy?

Preparation

1. Poster paper
2. Markers
3. Cover crop seeds (clover, grains, legumes)
4. Hose
5. Rake

NGSS and Activity Extensions

In-class conversations and studies can include research into practices that promote and support agroecology, the process of photosynthesis, and the flow of energy within ecosystems (MS-LS1-6). The students can also research the additional benefits of using cover crops for soil management, reducing the effects of erosion, and supporting a valuable ecosystem.

Resources

- Stephen R. Gliessman. *Agroecology: The Ecology of Sustainable Food Systems, Vol 1*, 3rd ed.
- Fred Magdoff and Harold van Es. *Building Soils for Better Crops*, 3rd. ed.
- Barbara Pleasant. "Use Cover Crops to Improve Soil." *Mother Earth News*, October–November, 2009. (online—see Appendix B).

5 Leadership: Planting Bulbs

Time Frame: 45 minutes with older students, 30 minutes with younger class

Introduction (10 minutes)

1. Begin with a ten-minute introduction with the middle school students about the goals of the activity and how to be Garden Leaders. The goal for the older students is to teach how deeply to plant a seed or bulb and how plants catch and store energy. The goal for the young student is to follow directions and to get their hands dirty.

2. Brainstorm with the students how a tulip or daffodil plant captures and stores energy throughout the growing season: What role does the bulb play in this process? What does it mean that the bulb is dormant? What will happen to the bulb over the winter?

3. The middle school students will stay with their young partner throughout the allotted time and encourage them to participate in all the tasks. What questions and concerns do the seventh graders have about the overall activity? What tools will they need to use, and how can they prepare the garden site for planting? How deep should they teach the young students to plant seeds?

I tell my students to use a rule of doubles when planting seeds; this means that they measure the seed (with their fingers or mentally) and then double the size of the seed or bulb. This should be the depth that the seed is planted. Some sources recommend tripling this depth for flower bulbs.

4. When the young students arrive, each can be assigned to a seventh grade student partner, be introduced to the day's activity and behavior expectations, and then go into the garden site to plant a certain amount of tulip or daffodil bulbs.

Activity (30 minutes)

1. Each partnership with gather their bulbs and tools before beginning to plant in the designated garden plot or container. During the planting, all the students will stay in their partner groups and work together to do high-quality gardening work. They should dig appropriately deep and wide holes for their bulbs and use their hands to fill in the holes with soft, non-compacted soil.

2. The older students should be able to describe the energy storage capacity of a bulb, the process by which its plant captured this energy, and why the students won't need to water these plants.

3. After planting, the older students can show their younger partners how and where to put away garden tools and then take a taste test tour in the garden together.

4. When the younger grade returns to class, the middle school students can double-check the quality of the planting and pick up any remaining bulbs and tools.

Assessment (5 minutes)

1. Gather the seventh graders together and briefly reflect on their experiences with their partners. Who remembered the name of their partner? How many hands got dirty (participated) in the planting? How many older students feel they communicated the lesson of the day?

2. What challenges did partners face? What ideas does the class have to make the next group activity more successful? What strategies and successes did some groups develop with their partners?

6 Garden Skills: Transplanting Cold-hardy Varieties

Time Frame: 45 minutes

Overview: The class will transplant the vegetable starts that they began on the first day of class.

Objective: To cultivate more all-season foods and increase the garden's ability to capture solar energy.

Introduction (10 minutes)

1. Brainstorm with the class what they know about *photosynthesis*: How do plants convert sunlight into energy? What value does this process have within an ecosystem? How does this energy transfer into other parts of the garden ecosystem?

2. Beyond the science of photosynthesis, what social and cultural benefits are there to growing cold-hardy plant varieties and increasing access to all-season food energy within the community?

> "[T]he basic lesson of biological sciences: that all life is directly or indirectly dependent on the solar energy captured by green plants."[3]

3. Review the plant varieties that the class seeded in their first fall activity. What varieties do the students remember planting? What is unique and valuable about each plant? What

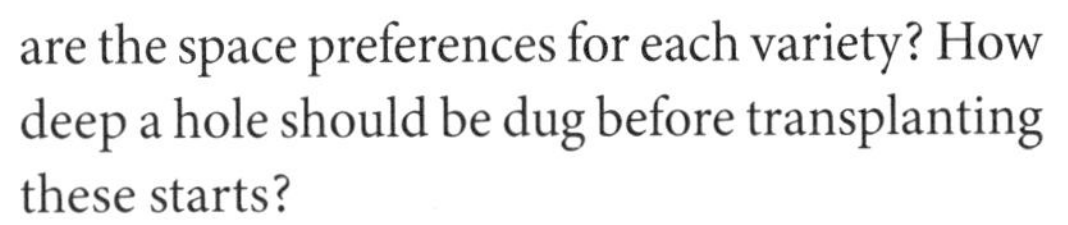

are the space preferences for each variety? How deep a hole should be dug before transplanting these starts?

4. Introduce the students to the tools they will be using and spaces that they will be gardening in during the activity. What language and behavior can students use to work as a class, or team, and to help each other in sharing space during planting?

Sun

I watch the sun rise and set.

I am sure the sun is tired, I bet.

I see the sun's bright colors,

I see the sun is paler than a

four leaf clover. I see the birds

fly in the shadow of the sun.

If I could pick a person who has

a bright personality, the sun would

be the one.

—Addie

Activity (30 minutes)

1. Individually or in pairs, the students will transplant their vegetable starts into the school garden beds or containers. Using what they know about the growth and space preferences of each variety, the students will identify the best place to plant and will work alongside their classmates to organize the plantings together.

2. Then, they will dig a hole of the best depth for transplanting, carefully remove their start from the container, and plant it in the garden. They can water it deeply afterward and tuck it in with any mulch in the beds. The students will continue this activity until the time is up, all the plants are established, or the garden spaces are full.

3. Surplus winter plant starts can be distributed in the community to increase food access to its members.

4. After they have cleaned up, returned their tools, and collected all the planting containers, then the students can enjoy a taste test from the garden.

Assessment (5 minutes)

1. Did the students find it easy or difficult to remove the starts from the containers? What challenges did students face when arranging their plantings?

2. How did they communicate with their classmates and practice teamwork? What seasonal conditions could affect the growth of these plants?

Preparation

1. Trowels
2. Watering can
3. Plant starts

NGSS and Activity Extensions

Continuing conversations and in-class studies can include the process of photosynthesis, energy laws and the flow of energy within ecosystems, as well as diverse forms of energy, such as solar, wind, biomass, and water runoff.

Resources

- Eliot Coleman. *Four-Season Harvest.*
- Linda Gilkeson. *Backyard Bounty.*

7-10 Self-directed Projects

Time Frame: 45 minutes for each class period
Overview: The students will design and implement self-directed projects to benefit and enhance the school garden and its community.
Objective: For students to advocate for their interests and learning by applying the principle of *Catch and Store Energy* into community and garden building projects.

Introduction (10 minutes)

1. During the next four activity periods, the students will work on self-directed projects of their own design. These projects should enhance the community to *Catch and Store Energy* and be focused in the garden. The students can develop projects from activities they participated in over the past few weeks, or they can generate a new project that inspires them and arises from a special interest.

2. What projects are students inspired by that will benefit the garden and the community? In what ways do these projects fulfill the principle of *Catch and Store Energy*? What does a realistic project look like?

3. How can students stay focused, behave like Garden Leaders, and respect the trust and freedom that comes with being an older, experienced gardener?

Activity (30 minutes)

1. Individually or in small groups, the students will design a manageable project or series of projects to fulfill the goals of the self-directed work. They should communicate their goals before each class period and have a clear work plan. Some students may want to work within specific garden beds while others may design a project for the whole community.

2. For students that need more direction, additional seasonal activities could include: mulching brown and green materials, seed saving, revitalizing the garden compost system, planting additional winter food varieties or seeds, and sowing cover crops.

Assessment (5 minutes)

1. How have these projects benefited the school and community, while also Catching and Storing the garden's Energy? What challenges and successes did the students face during their projects?

2. What self-directed work did students design that they enjoyed or wouldn't do again? What factors will they consider during future opportunities for self-directed work?

Winter: Valuing Renewable and Natural Resources

1-5 Research Project

Time Frame: 45 minutes for each class period

Overview: In place of self-directed winter work, the class will embark on a research project about heirloom seeds.

Objective: For students to foster ownership of the community's seed saving initiative and to value an important renewable resource.

Introduction (5–10 minutes)

1. During the next five garden classes, each student will engage in a research project around an heirloom plant that can grow in the garden. They will be conducting research to build a thorough profile of the plant and eventually use this information to build a sign for the garden and to design a seed packet that will be used by the community. By the end of the research period, the class will have collectively acquired profiles on dozens of heirloom plants.

2. Brainstorm with the class about the behavior expectations during the research and group work sections of the project: How will students stay focused, perform thorough research, and practice craftsmanship?

The students will be planting their variety at the end of the research project, so I recommend starting a list of plants whose seeds are already being saved from the garden. If seeds need to be purchased, I recommend acquiring open-pollinated seeds, requesting donations from local seed banks, or having students attend seed exchanges in their community.

Activity (30 minutes)

In pairs, the students will complete three major tasks:

1. Class Period #1—Research: Each student pair will gather information about a specific garden heirloom variety and create a profile of the plant, including the plant's botanical/Latin name, genus and species, the origins of the variety, and its wild and domesticated ancestors. The groups should create a map of how the plant has been dispersed across the globe, and draft planting directions, a flavor profile, directions on the best ways to seed save from the plant, and what specific time of year is recommended to harvest it. They can also include favorite cultural recipes and meals that feature this heirloom plant.

2. Class Periods #2 and #3—Informational Sign: Students will design and build a sign that features this information and that will be established in the garden in the spring. The groups should practice craftsmanship in creating this sign and generate work that will last for years.

3. Class Periods #4 and #5—Seed Packet: The groups will design and create a seed packet for their variety which provides a botanical illustration of the whole plant, planting directions, sunlight and space preference, and description of plant flavor, size, and coloring. They should also include the botanical name. They can use commercial seed packets as an outline for this project.

Assessment (5 minutes)

1. What interesting histories and stories did the students learn about their heirloom plants?

What makes them valuable and unique to the school garden?

2. How can these plants continue to be cherished and shared in the community? What should be known and remembered about them? How are heirloom seeds a renewable resource?

NGSS and Activity Extensions

Further in-class studies can include the work and life of Nikolai Vavilov or Vandana Shiva, studies on genetic diversity, the origins of early agriculture, transgenic and GMO crops, and the relationship of modern agriculture with nonrenewable and renewable resources. The class can also study botany, open- and self-pollinating plants, the process of pollination, industrial food and seed production, agribusiness, and agrobiodiversity.

Resources

- Slow Food Foundation for Biodiversity. "Ark of Taste." (online—see Appendix B).
- Suzanne Ashworth. *Seed to Seed.*
- "Community Seed Donations." *Seed Savers Exchange.* (online—see Appendix B).
- Cindy Conner. *Seed Libraries: And Other Means of Keeping Seeds in the Hands of the People.*
- Carol Deppe. *The Resilient Gardener: Food Production and Self-Reliance in Uncertain Times.*
- Stephen Facciola. *Cornucopia II: A Source Book of Edible Plants*, 2nd. ed.
- "Food Sovereignty." *Navdanya.* (online—see Appendix B).
- Baker Creek Heirloom Seeds. "Growing Guide." (online—see Resources).
- Botanical Artists of Canada. "History of Botanical Illustration." (online—see Resources).
- Tina M. Poles. "A Handful of Seeds." (online—see Resources).
- Community Seed Network. "Resources & Information for Seed Saving, Sharing & Networking." (online—see Resources).
- Seed Savers Exchange. "Seed Saving Chart." (online—see Resources).
- Seeds of Diversity. "The Ecological Seed Finder." (online—see Resources).
- Benjamin A. Watson. *Taylor's Guide to Heirloom Vegetables: A Complete Guide to the Best Historic and Ethnic Varieties.*
- William Woys Weaver. *Heirloom Vegetable Gardening.*
- Katie Willis and Katie Scott. *Botanicum: Welcome to the Museum.*

6 Leadership: Screening Compost

Time Frame: 45 minutes

Overview: The class will work together to harvest and process finished compost from the garden.

Objective: Students will practice gathering a valuable, nutrient-rich resource for future planting.

Vocabulary

Carbon sequestration

Introduction (10 minutes)

1. Brainstorm with the class what they know about compost, from its composition and ecosystem, to harvesting it. What does the compost cycle look like? Who are key participants in this cycle? Why is this process so valuable to ecologically minded gardeners?

2. Discuss with the class the relationship between soils and *carbon storage*, or *sequestration.* How is carbon captured and stored by plants? How does photosynthesis engage with this process? How is carbon converted into a soil base through plant decomposition? What role does garden composting play in this process?

> "Soil is the most important storage for nutrients in temperate climates. Humus... increases the capacity of soils to store mineral nutrients (as well as water and carbon)."[4]

3. Demonstrate the activity for the day and the tools the students will use. How can they work as a team to do careful, good gardening? How can they respect themselves, others, the garden space, tools, and ecosystem?

Activity (30 minutes)

1. In pairs or small groups, the students will gather, screen, and harvest finished compost from the school garden's hot or cold compost system. Using shovels, a wheelbarrow or bucket, and a sifting tray or screen, the students will work together to dig below the compost pile, shovel the soil onto the sifting tray, and gently sway the tray or screen back and forth between two people, letting the finer particles fall through and gathering the larger, less decomposed pieces to return to the compost pile.

2. The groups will continue this activity until their wheelbarrow or bucket is full. As they harvest the compost, they can observe the members of the soil ecosystem that they encounter and can identify. Using what they know about compost system organisms, the students should keep note of these creatures to share with the class at the end of the activity.

3. When the compost is harvested, the students can put away their tools, clean up, and enjoy a seasonal taste test from the garden.

4. The students can also perform a pH and NPK test on a compost sample and determine if there are any surpluses or deficiencies in nutrient levels.

Assessment (5 minutes)

1. Gathering back together, have the class share their experiences interacting with the soil ecosystem. What creatures did the students discover as they sifted the compost? Where there living things that they could identify or ones that they could not? How many of each species did they find?

2. What is the texture, smell, and particle size of the harvested compost? What kinds of organic materials did the groups find that were partially decomposed?

NGSS and Activity Extensions

Further in-class studies can focus on the importance of soils and forests in carbon sequestration, agroecology practices, and the globalization and importing of nonrenewable resources.

Preparation

1. Shovels
2. Wheelbarrow
3. Screen/sifting tray
4. Gloves
5. pH and NPK test
6. Watering can

Resources

- "Carbon Sequestration in Soils." Ecological Society of America. (online—see Appendix B).
- Pam Dawling. "Screen Compost Now to Make Your Own Seed Compost for Spring." *Mother Earth News*. (online—see Appendix B).

- Jeff Lowenfels and Wayne Lewis. *Teaming with Microbes*.
- Fred Magdoff and Harold van Es. *Building Soils for Better Crops*.
- Sage Publications. "Compost Can Turn Agricultural Soils into a Carbon Sink, Thus Protecting Against Climate Change." ScienceDaily. (online—see Appendix B).

7 Garden Skills: Cultivating Soil for Container Plants

Time Frame: 45 minutes

Overview: The students will create foundational potting soil for their future planting projects.

Objective: Students continue to cultivate another valuable natural resource in the garden.

Introduction (10 minutes)

1. Gathering the class together, have the students brainstorm how soil is a life-giving natural resource: What are the key traits and ecosystem members that gardeners look for in healthy soil? What materials are there in the garden and community to help students nurture soils?

> Students who have engaged with *The School Garden Curriculum* in third and fifth grade will have created variations of potting soil before, both focusing on the water-retaining capabilities of these homemade potting soils. The seventh grade students need to keep in mind how water will flow or be retained in their mixtures, but should focus on and participate in conversations about renewable and nonrenewable resources, valuable natural resources, and the globalization and exploitation of soil and mineral resources.

2. Other than providing support and a growing medium for plants, what role does soil play in gas exchange, water filtration, and temperature moderation? What kinds of minerals, organic matter, and microorganisms can students gather together to create a well-balanced, healthy potting soil?

3. What indicators can the class look for to determine if their potting soil will be a good planting medium? Color, smell, texture, and the amount of time it takes water to pass through are some examples.

> "All natural resources must be used carefully and respectfully."[5]

Activity (30 minutes)

1. Individually, or in pairs, the students will be gathering available materials from the garden to create a soil-based planting medium that will be the foundation for their future planting activity. They can store ingredients in a bucket or a designated pile for future use. The students should work together to thoroughly mix the ingredients that they gathered and use shovels, trowels, and their hands to accomplish the task.

2. These groups should be the same partnerships that worked on the previous research project. In the spring, these students will be building planting containers together and will transplant their heirloom starts into this soil.

3. Materials can include recycled potting soil from past containers, rotten sawdust, rotted leaf mulch, sand, and screened garden soil. The class should also add screened compost from the previous gardening activity to their mixtures. Some gardeners would express concern about compost being the bulk of the planting medium (due to potentially harmful bacteria and microbes), but it is up to each gardener to determine, based on their site and material availability.

4. As they mix the ingredients, the groups should pay attention to the consistency of particle size and make sure there is a diversity of small to medium-sized materials. When the groups feel confident in their potting soil mixes, the class can tour other student mixtures to feel the consistency of different soils, compare it to their own mixture, and make any last changes they would like to their potting soil.

5. Finally, the groups can return any tools they used and clean up, before enjoying any seasonal taste tests from the garden.

Assessment (5 minutes)

1. How were students able to work as a team to accomplish the activity? What issues did they problem-solve as they mixed the materials?

2. What is the water-draining properties of the potting soil, and what materials assist in regulating water flow?

Preparation

1. Screen/sifting tray
2. Buckets: five-gallon or bigger
3. Trowels
4. Potting soil materials

Resources

- Fred Magdoff and Harold van Es. *Building Soils for Better Crops.*

8 Leadership: Planting Heirloom Seeds

Time Frame: 45 minutes for older students, 30 minutes for a younger class

Overview: The older students will lead a planting activity with a younger class and teach them about the value of heirloom seeds.

Objective: The older class will learn through teaching and pass on the cultural importance of saving seed and the genetic lines of plants.

Introduction (10 minutes)

1. Gather the older class together, before the younger class joins them, and have the students reflect on their message and goal for the activity. What is the most important information that the older students can teach their partners about heirloom seeds and the seed variety they will be planting? What values do they want to pass on? How can the class accomplish these goals and be Garden Leaders during the activity?

2. Review with the class what tools and materials they will need and what steps to follow, such as gathering soil, planting, labeling, and watering. If there is additional time, the older students can prepare their workspace before the younger class joins them.

3. When the younger students arrive in the garden, they can be partnered with an older student and review their expected behaviors.

Activity (30 minutes)

1. In pairs, the older students will lead their younger partner in a planting activity. The groups will gather their supplies, fill planting containers with their homemade potting soil, label them with the seed's variety and planting date, plant the seeds, and then gently water them. When they are finished, the groups can transport their pots to a planting tray to be stored in a safe, sunny location, such as a windowsill, greenhouse, or under growing lights.

2. As they work, the seventh graders will teach their lesson about heirloom seeds, while also captivating and sustaining the attention of their partner. The students can explain how they created the soil and the value of using natural resources in the garden.

3. When all the seeds have been planted or the activity time is close to over, the groups can clean up their work area, wash their hands, and enjoy a taste test of hardy winter foods together. The older students can guide the younger ones on how to harvest these foods and discuss flavor and texture with them.

Assessment (5 minutes)

1. Gathering the older students back together, have them reflect upon their leadership experiences. How well did they accomplish the goals of the day? What successes and challenges did they face in sharing their lesson? How did they overcome any difficulties?

2. How will the class care for these seeds over the next few weeks? How long will it take the seed varieties to germinate?

Preparation

1. Potting soil
2. Planting containers
3. Watering can
4. Labels and markers
5. Planting trays

9 Garden Skills: Mulching Green Manure

Time Frame: 45 minutes

Overview: The class will dig in the cover crops they planted in the fall and participate in cycling energy and nutrients through the garden soil and ecosystem.

Objective: The students will prepare the garden beds for spring planting and continue engaging with soil-building and cultivating the garden's valuable resources.

Introduction (10 minutes)

1. Review with the class what kinds of cover crop seeds they planted in the fall. What were the benefits of covering the soil with cold hardy plants over the winter? What cover crops grew best or most vigorously? Which plants were most successful at weed suppression and controlling erosion? What environmental conditions may have affected plant growth?

2. Cover cropping and mulching green manure are good soil-building practices because they increase soil structure, fertility, and organic matter while also loosening soil, reducing weeds, and potentially attracting beneficial insects. Encourage the class to discuss how

While I aspire to tend to an entirely no-till garden, the reality of many school and community gardens is that the soil has been initially sterilized and imported to fill raised beds, or the gardens' soils have been converted from lawns and need time to build soil vitality. In these cases, occasional soil disturbance, such as hand digging green manure into the soil, can help facilitate this soil-building process. I do not recommend using mechanical tillers for this work. Not only is human-powered activity a way to get children active and involved with soil systems, but it eliminates a conventional fuel-based practice that does little to increase the long-term vitality of the soil ecosystem.

Digging in cover crops is only effective in areas that are designated for annual plantings. In spaces with perennial plants and valuable root systems, cover crops will need to be hand-pulled or cut down near the soil surface. Double digging is a process that can occasionally occur on a large scale with a spade, or a small scale with a trowel (see Resources). Educators may want to alternate between different methods every year and build towards entirely no-till practices.

these techniques also relate to carbon sequestration and storage.

3. Have the class compare and contrast different methods for incorporating green manure into the soil. What are the pros and cons of surface-level mulching with no soil disturbance, tilling, and double digging? What are the different methods that students can use to incorporate the plants into or on top of the soil? Which method will work the best for this garden site or for soil health?

3. Demonstrate the activity to the students and review tool and personal safety expectations before beginning the activity.

Activity (30 minutes)

1. Individually or in pairs, the students will carefully mulch or double dig cover crop plants into the soil. If the garden space is small, the students can dig the plants in with small shovels or trowels. If the plants are larger, then the students may be able to pull them out by hand first, before mulching them into the soil.

2. The students should practice teamwork, respect the soil ecosystem to the best of their ability, and respect their own labor and that of others. The class should also work to avoid compacting the soil during their work. The class can use clippers, their hands, and shovels to accomplish this task, according to the needs to the community.

3. When the green manure has been processed into the soil, the students can return their tools and clean up before enjoying a seasonal taste test from the garden.

Assessment (5 minutes)

1. What members of the soil and garden ecosystem did the students encounter during the activity? What were student experiences in mulching and digging in the crops? Was the activity easier or more difficult than they expected?

2. Would the students recommend planting this cover crop again? Why or why not?

Activity Extensions

In-class conversations and studies could include research on the impact commercial tilling practices have on soil ecosystems and alternative ways to build soil health from small gardens to large biodynamic farms.

Resources

- Brian Barth. "Double Digging: How to Build a Better Veggie Bed." (online—see Appendix B).
- Fred Magdoff and Harold van Es. "Reducing Tillage" in *Building Soils for Better Crops.*
- Barbara Pleasant. "Use Cover Crops to Improve Soil—Organic Gardening." (online—see Appendix B).

10 Garden Skills: Separating Perennial Plants

Time Frame: 45 minutes

Overview: The class will learn how to separate perennial plants to increase food, pollination, and medicinal production.

Objective: Students will practice a valuable skill and propagate important perennial resources.

Introduction (10 minutes)

1. Brainstorm with the class what kinds of perennial plants they know grow in the garden: What are the uses and value of these plants? What are perennial and annual plants? Can perennial plants be considered a renewable resource, why or why not? What is the purpose and value of separating perennial roots and replanting new plants within the garden?

2. Review with the class what types of roots, such as fibrous, tap, or tuber, they might encounter in the garden. Demonstrate the activity and how to gently separate root crowns or tubers from each other in order to propagate more plants.

3. How can the students behave in ways that respect themselves, each other, and the plants, as well as demonstrate tool safety?

Activity (30 minutes)

1. In pairs or small groups, the students will gently dig up the root balls of selected perennial plants (such as rhubarb, thyme, chives, yarrow, tickseed, aster, sage, sorrel, artichokes, asparagus, onions, and sunchokes). They will work carefully and diligently to dig around the plant, remove the roots, and separate the plant (depending on its structure).

2. The students should transplant part of the perennial plant back into the original site and can either replant the other sections in new parts of the garden or move them into planting containers for future use.

3. The groups can continue this practice with multiple plants for the activity period, making sure to interact with various types of root structures (fibrous, tap, tuber).

4. Before the activity time is over, the students can return their tools, clean up, and enjoy a seasonal taste test from the garden.

Assessment (5 minutes)

1. What were student experiences in separating the roots? Did they find that some types of root systems are easier to separate than others?

2. How many plants were they able to spread and replant over the course of the activity?

Preparation

1. Shovels
2. Gloves
3. Examples of perennial root systems

Resources

- Eric Toensmeier. *Perennial Vegetables: From Artichoke to Zuiki Taro, a Gardener's Guide to Over 100 Delicious, Easy-to-grow Edibles.*

Spring: Using Small and Slow Solutions

1–2 Garden Skills: Building Planting Containers

Time Frame: 1½ hours (2 × 45 minutes)
Overview: The students will build small-scale raised beds or planting containers from local materials.
Objective: Students will value creating small and slow solutions while also building their engineering and physical skills.

I recommend having the students construct planting containers for this activity, though they can also build or rebuild garden beds if needed. There are many innovative, creative, and unique designs or materials that can be used to construct a planting container. These can add beauty and curiosity in the school garden and community.

Introduction (10 minutes)

1. Depending on the space and material availability, as well as community need, the students will be constructing small planting beds or containers to cultivate their heirloom plants.

2. Brainstorm with the class what the dimensions of their planting containers need to be for their variety of heirloom plant. What are the average depths of the plant's roots and other space requirements? What materials are easily accessible for constructing planting containers?

3. Discuss with the class what it means to starts small and dream big. How does building planting containers and fostering heirloom plants contribute to the value of *Small and Slow Solutions*? See Holmgren's *Permaculture: Principles & Pathways.*

Class Period #1

1. In pairs, the students will design a planting container to grow their heirloom plant from available local resources. These containers should be simple to build and easy to maintain. The groups can use available online and book resources to design, construct, or plan on how to acquire their ideal container (see Resources).

2. In their research, the students can also look into what companion plants might support the growth of their heirloom variety, without outcompeting them. How might the students make space for companion plants?

Class Period #2

1. The student groups will construct a planting container from gathered materials or available resources. The containers should be durable, well-crafted, and high-quality. They can include simple or innovative designs, but should fulfill the goal of providing a good space to cultivate plants for future heirloom seed saving activities. The potential container materials could include old beehive boxes, recycled fabrics, scrap wood, tightly woven branches, troughs, wash bins, and baskets.

2. Some of these containers, especially if they are reclaimed containers, may need additional drainage holes drilled into them or linings, such as burlap, to keep soil contained. When they have finished, the students can fill their planting containers with the remaining potting soil they created in the winter.

3. Because of their compact size, these planting containers are ideal for placement along fence lines, walls, or raised beds. They can be used to create a new portable gardening area with very little land use requirements, other than light and water access.

Assessment (5 minutes)

1. In what ways did the students practice craftsmanship with their containers? What challenges did they face and overcome?

2. What did they learn about their plant's growth requirements? Did any students need to reevaluate the sustainability of their containers as they built them?

Preparation

1. Drill
2. Various building materials
3. Online and book resources
4. Staples and staple gun
5. Hammer
6. Nails
7. Shovels

Resources

- Editors of Fine Gardening. *Container Gardening: 250 Design Ideas & Step-by-Step Techniques.*

3 Leadership: Transplanting Heirlooms

Time Frame: 45 minutes for older students, 30 minutes for a younger class

Overview: The older students and their partners will transplant heirloom starts into the containers or beds that they constructed.

Objective: Older students will pass on the garden skill of transplanting and bring together the work they have done over the past few months.

Introduction (10 minutes)

1. While the older students wait for their partners to arrive, have them brainstorm what the key message and goals of the day are: What skills and gardening techniques can they teach younger students? What should they know about heirlooms, especially the ones each older student has profiled? What is the value of having other community members interact and understand the value of heirlooms?

2. Have the older students prepare their work area by moving their planting containers to the space they will be at for the growing season. They can gather any tools that the groups need for transplanting.

3. When the younger students join them in the garden, they can be partnered with an older student and should review the activity and behavior expectations before beginning.

Activity (30 minutes)

1. With their partners, the students will transplant heirloom plant starts into the garden containers they built. The older students will teach their young partners how deeply to dig a hole, how to carefully remove a plant from its container, and how to plant with root orientation and care in mind. They can also pass on the important history and contribution their plant brings into the garden, its spacing requirement in relation to other plants, and its life cycle.

2. If possible, the groups can also plant companion starts or seeds with their heirlooms, as a way to promote a more supportive botanical environment and deter potential pests. For example, a student may plant a few garlic bulbs along the edge of their container to promote the health of their heirloom eggplant or pepper plant.

3. After the starts have been transplanted, the students can deeply water them and then install informational signs in the beds or containers, affixed to them, or nearby. The younger student can help with this process and be introduced to the unique traits and value of the plant as well as how to read and interact with the sign.

4. Finally, the groups can clean up, collect and return any tools or planting containers they have, and enjoy a taste test from the garden together.

Assessment (5 minutes)

1. Before the younger class leaves, conduct a brief survey on what they know about heirloom plants. What is an heirloom? What varieties of heirlooms did they plant today? What makes these plants valuable and special?

2. Encourage the older students to reflect on their developing leadership skills. How are their experiences in leading young students changing during the year? What skills and techniques have they developed to help them? How do they feel they accomplished the goal of the day and communicated their message?

Preparation

1. Trowels
2. Heirloom plant starts

3. Companion plant seeds or starts
4. Watering can
5. Informational signs

Resources

• Sarah Israel. "An In-depth Companion Planting Guide." (online—see Appendix B).
• Louise Riotte. *Carrots Love Tomatoes: Secrets of Companion Planting for Successful Gardening.*

4 Garden Skills: Garden-based Fertilizers

Time Frame: 45 minutes
Overview: The class will harvest plants from the garden and create natural fertilizers for their heirloom plants.
Objective: Students will practice creating organic and slow solutions to their plant's nutritional needs.

Introduction (10 minutes)

1. In previous years of gardening, some students may have created homemade organic fertilizers for the garden's plant starts. Brainstorm with the class what kinds of fertilizers they have made and what materials they used: Do any students remember the process of making these liquid fertilizers? What kinds of fertilizers are used in many conventional gardens, and where do commercial fertilizers come from? What are the differences between liquid fertilizers and solid organic products?

2. When growing in containers, plants have a finite amount of nutrients to utilize in their growing space. Fast-acting organic liquid fertilizers can give these plants an extra boost during certain times of year, for example, after sprouting, transplanting, or at the end of the growing season to give the final crop additional food. Organic and homemade liquid fertilizers also tend to have a better natural balance of NPK, depending on the materials used.

3. Brainstorm with the class what role liquid fertilizers have in an organic garden. What is the long-term effect of water passing through soil and leaching nutrients? What materials grow in the garden and can be used as sources of potassium, phosphorous, and nitrogen? Record these organic materials and their individual benefits on a poster (see Resources). The students can use this chart as a reference when making their fertilizers.

Activity (30 minutes)

1. In pairs, the students will create their own natural fertilizers by harvesting certain garden plants and compost items. Grass clippings, coffee grounds, banana peels, comfrey leaves, stinging nettle, and lamb's-quarters are all potential materials, depending on local resources.

2. The students will rip these organic materials into small pieces and gather the pieces in a jar or small bucket. Once their container is packed with organic items, they should fill the jar with water, cover it with a coffee filter, and steep the organic matter as a tea for up to three days, preferably in the sun. Additionally, they can store this tea in a sunny protected space. The groups should label their jars with their names before the activity is done. These teas will be highly concentrated and can be diluted with water before the students begin adding them as fertilizer to their plants.

3. Three days after the students have made their mixtures, have the students water their starts in the morning to help open up the plant roots to receiving the fertilizers and keep them from being burned. Later in the day, the students should remove and compost the organic materials and dilute their fertilizers with water. Then, every two weeks, they will pour small amounts of their liquid fertilizers onto their plant starts, being sure to get as little as possible on the young plant leaves.

Assessment (5 minutes)

1. How does building soil health through natural and organic materials help to *Value Small and Slow Solutions*?

2. How can homemade liquid fertilizers be a valuable resource for the local community? In what other ways can they be used?

Preparation

1. Grass clippings
2. Banana peels
3. Comfrey leaves
4. Coffee grounds
5. Compost
6. Watering can
7. Jars with lids
8. Tape and markers
9. Coffee filters
10. Rubber bands

Resources

- Patrick Lima. "Natural Alternatives" in *The Natural Food Garden: Growing Vegetables and Fruits Chemical-Free.*
- Barbara Pleasant. "Free, Homemade Liquid Fertilizers." (online—see Appendix B).
- Barbara Pleasant. "Homemade Fertilizer Tea Recipes." (online—see Appendix B).
- Jonathon Engels. "4 All-Natural Liquid Fertilizers and How to Make Them." (online—see Appendix B).
- "Here's the scoop on chemical and organic fertilizers." Oregon State University Extension Service. (online—see Appendix B).
- Heide Hermary. "Organic Fertilizers" in *Working with Nature: Shifting Paradigms: The Science and Practice of Organic Horticulture.*

5-8 Self-directed Projects

Time Frame: 45 minutes each class period
Overview: The class will engage in four self-directed projects that benefit the garden and the community.
Objective: For students to interact with the principle of building slow and small solutions by accomplishing work for the garden and the wider community.

Introduction (10 minutes)

1. Before beginning student activities, have them discuss examples of what *Valuing Small and Slow Solutions* look like in the garden. In what ways have the students already been engaging with this principle through seeds and soils? How can students identify spaces and practices that need help and develop realistic and sustainable solutions to them? How can the work students do also fulfill the principle of *Using and Valuing Renewable Resources and Services*?

2. What projects would students like to continue working on? What new projects would they like to explore? How can they continue to be Garden Leaders, and what behaviors will students exhibit to respect each other, the garden space, and the freedom of self-directed work?

Activity (30 minutes)

1. Before each garden activity time, the students should be able to communicate their plan for the day and the goals they want to accomplish. Some students may want to continue working in a specific gardening bed as Student Gardeners, others may want to engage with engineering and design work, and still others may want to design new methods to distribute surplus resources throughout the community. Students may also want to consider better ways of using local resources, rather than importing or buying them.

2. For students who need additional direction, seasonal garden work that they can do includes sowing seeds for pollinator habitat, building additional containers for increased small-scale food production, separating perennials, revitalizing a struggling part of the garden, harvesting compost, or cultivating natural fertilizers.

Assessment (5 minutes)

1. In what ways did the students accomplish work that supports the principles of *Valuing Slow and Small Solutions* and *Using and Valuing Renewable Resources and Services*?

2. How do these projects enhance the garden and the larger community? What lessons did students learn through their work, and what would they do differently next time?

9 Garden Skills: Designing Trellises

Time Frame: 45 minutes

Overview: The class will research and design trellises that will offer support for their heirloom plants or other garden vegetables.

Objective: Students will set their plants up for summertime success, study trellising designs, and create innovative variations that use local resources.

Introduction (10 minutes)

1. With the class, brainstorm what kinds of trellises the students may have built in the past: What structural shapes did they design, what materials did they use, and what plants were they able to support? What is the purpose of creating trellises? How can these structures assist with the growth and harvest potential of the heirloom plants?

2. What are local or reusable materials that the students have access to in the garden or the community? What materials are realistic to use, safe to build with, and sturdy enough the support the weight of a mature plant?

3. Provide a few examples of well-used and loved trellis shapes, such as A-frame, tower, pyramid, and ladder. Have the students observe and discuss how these shapes were constructed.

Activity (30 minutes)

1. Using available online and book resources, students will research a trellis shape that will support their heirloom plants. They should consider the potential height and width of their plants and the shapes of their containers.

2. Students whose plants don't need additional support can identify an additional garden vegetable that will need support and design a structure for it (such as peas, beans, and tomatoes).

3. The trellises should be realistic and constructed using materials that are easy the gather and from local resources. The students should also consider safety concerns around built structures and how to craft a trellis that the community can interact with in a positive, safe way.

4. When each student has created their design, it can be peer-reviewed by a classmate or an adult. Then, each student will make of list of the types and amount of materials that they will need. If there is additional time, they can begin gathering these materials (branches, poles, wire, string).

Assessment (5 minutes)

1. What designs did the students determine to be best for the future growth of their plant? What designs would not have been a good fit?

2. What materials do the students need to build these trellises? How can they acquire them locally?

Resources

- "Garden Trellis Ideas." Seed Savers Exchange. (online—see Appendix B).
- Editors of Cool Springs Press. *Trellises, Planters & Raised Beds: 50 Easy, Unique, and Useful Projects You Can Make with Common Tools and Materials.*

10 Leadership: Building Trellises

Time Frame: 45 minutes for older students, 30 minutes for a younger class

Overview: In their final class, the older students will guide their younger partners in building a trellis.

Objective: Older students will teach a new generation about garden design and engineering methods.

Introduction (10 minutes)

1. Review with the older class what materials they will need to build simple and sturdy trellises for their plants and the garden. How can the students be Garden Leaders during the activity, encourage their partner to assist them in every step, and ensure that their partner has a safe experience? How can they practice craftsmanship during this time?

2. The students should have a clear idea of the materials and resources that they can use, where these are, and the spaces where they will be working. If there is additional time, they can gather the materials they need and set up their workstations.

3. When the younger students join them in the garden, they should review their expected behaviors and discuss how to be safe and engaged participants in the activity. Then, they can be partnered with an older student and begin.

Activity (25–30 minutes)

1. With their partners, the students will be building small-scale trellises to support their heirloom plants. The older students will guide their younger partner through the steps necessary to assemble and secure the structure. As they work, the older students will explain why it is important to support these plants and why they chose this design.

2. Some groups will work very quickly by assembling simple designs with on-hand materials, while others may work on one design for the whole activity period. Groups that work quickly can continue building more trellises for the garden, but should still focus on craftsmanship and safety over production.

3. Before the activity time is over, the students can clean up their work areas and return any tools or extra materials. Then, the older and young partners can enjoy a taste test from the spring garden. The older students should continue being Garden Leaders by guiding the students on how to harvest certain foods and how to talk about taste.

Assessment (5–10 minutes)

1. When the younger students have left, gather the seventh graders back together to reflect on their leadership experience and their learning in the garden over the year. What lessons have they learned about being Garden

Leaders? What skills and techniques have they developed to assist them in guiding and teaching young students? What did they accomplish this year with young students that they were proud of?

2. What were the most memorable activities that the students did this year? What would they like to know more about and what do they look forward to in gardening next year?

Preparation

1. Twine
2. Assorted sizes of branches
3. Chicken wire
4. Staples/staple gun
5. Nails
6. Hammer
7. Wood pallets

Resources

- "Garden Trellis Ideas." Seed Savers Exchange. (online—see Appendix B).
- Editors of Cool Springs Press. *Trellises, Planters & Raised Beds.*

EIGHTH GRADE—GRADE EIGHT

LEADERSHIP AND STEWARDSHIP

In their final year of studying *The School Garden Curriculum*, the students will celebrate their accomplishments, embrace being Garden Leaders, and explore new language and design principles that will inspire and guide them beyond the garden. The opportunities for activity extensions and community engagement are as vast as the imagination of the class and its teachers. In every season, students will have the opportunity to take control of their own learning and continue taking the steps to be ecologically minded changemakers in a world that needs them.

During the fall, the class will explore the principle of *Sharing the Surplus* through community engagement, preparing culturally im-

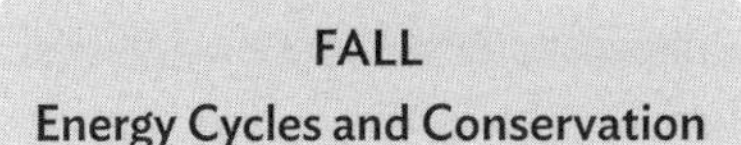

FALL
Energy Cycles and Conservation

1. Garden Skills: Harvesting
2. Garden Skills: Share the Surplus
3. Leadership: Taste Testing and Nutrition
4. Garden Skills: Food Preservation
5. Garden Skills: Produce No Waste
6. Leadership: Mulching or "Putting the Garden to Bed"

7–10. Self-directed Projects

WINTER
Designing Systems

1. Garden Skills: Small-scale, Intensive Systems
2. Garden Skills: Food Forests and Layering
3. Garden Skills: Designing from Patterns to Details
4. Garden Skills: Plant Propagation

5–7. Self-directed Projects

8–9. Leadership: Plan and Teach a Lesson

10. Leadership: Planting Pea Seeds

SPRING
Supporting Biodiversity

1. Garden Skills: Supporting Healthy Habitats
2. Leadership: Installing Habitat Boxes

3–4. Garden Skills: Use and Value Diversity

5–8. Self-directed Projects

9. Leadership: Spring Foraging and Foods
10. Garden Skills: Legacy

portant and familial foods, teaching younger students about taste and nutrition, and learning about food preservation. They will apply the design principle of *Produce No Waste* in the garden and design self-directed projects that support and extend from these two values. In the winter, the students will begin a research and design project where they will map an area of the garden, a bed, or a container and learn how to plan *Small-scale, Intensive Systems,* research potential food forests by mimicking local ecosystems, and learn how to *Design from Patterns to Details.* The class will practice different methods of propagating plants for spring and develop a lesson plan to teach younger students.

In springtime, their lessons will be applied in the garden and focused on supporting its ecosystem function and biodiversity. From building habitat boxes, enhancing the garden's food production, and passing on the values of the garden to a new generation, the eighth grade class will be instrumental in promoting the health and well-being of the garden and community. When the students leave the garden at the end of the school year, they will have left behind a legacy as experienced gardeners and community leaders.

© TinPong / Adobe Stock

Next Generation Science Standards

As the students research local ecosystems, study patterns, and mimic designs in the garden, they will be enhancing their mapmaking abilities and drawing to scale. They will improve their math skills by drawing subjects proportionally, exploring ratios, and developing higher-level research skills. The students will also have the opportunity to compare and contrast the experience of designing on paper and the reality of applying these designs, and to develop arguments supporting or criticizing certain design features. Further in-depth, classroom studies and readings can follow themes or units on how environmental factors impact the stability of ecosystems and the growth of organisms, the effects of resource availability on populations within an ecosystem, the patterns of interactions among organisms across many ecosystems, the cycle of energy and matter through the living and nonliving elements of an ecosystem, and different designs and solutions for supporting biodiversity and ecosystem stability.

Permaculture Principles and Perspectives

- Share the Surplus
- Produce No Waste
- Intensive, Small-scale Systems
- Designing from Patterns to Details
- Use and Value Diversity

Fall: Energy Cycles and Conservation

1 Garden Skills: Harvesting

Time Frame: 45 minutes
Overview: Students will help gather foods for taste testing and future activities, while exploring the garden changes.
Objectives: To carefully harvest selected foods and review behavior expectations in the garden.

Introduction (10 minutes)

1. Welcome the students back to the garden and encourage them to share some of the changes they have already observed inside the space. Are there plants that students remember planting last spring? Which plants thrived over the summer and which did not survive?

2. Review behavior expectations for students in the garden. Now that the students are older, they are expected to be Garden Leaders. What does this mean, and what behaviors do leaders exhibit? What responsibilities do leaders have?

3. Introduce students to the selected foods they will harvest from the garden. Review with them what these foods look and feel like when they are ripe. Some plants need to be twisted, cut, or snapped off, usually with two hands. Plants to pick could include: tomatillos, tomatoes, squash, beans, flowers, herbs, pumpkins, peppers, melons, and corn.

Activities (30 minutes)

1. In pairs or individually, students will harvest selected fruits, seeds, and flowers and gather them in harvest baskets. When students finish harvesting one variety of plant, they can move on to another variety. The gathered foods can be stored in a cool location or brought for cleaning and storage in the school or community's kitchen.

2. If students volunteer, they can wash and prepare some of the foods for later taste testing by making sure to follow expected food handling protocols.

3. At the end of the harvest time, the students can enjoy a taste test of the available garden foods before gathering back together to share their findings.

Assessment (5 minutes)

1. What types of foods did students find that were most prolific? Are there foods that are not ready to harvest? How successful was the plant growth of their heirloom varieties from the previous year?

2. What changes and new discoveries did students make during the activity? What are they most excited about for this year's gardening lessons?

2 Garden Skills: Share the Surplus

Time Frame: 45 minutes
Overview: Students will use the food they gathered in the previous lesson to prepare favorite recipes to share.
Objective: To cook nutritious garden meals and share the surplus with the community.

Introduction (5–10 minutes)

1. Review the types of foods that students harvested in the past activity. What recipes do students like to make with these ingredients?

What are the stories and traditions that accompany them?

Before this activity, ask students to bring in their favorite recipes or family recipes to share with the class or community. The recipes should require minimal time to prepare, assemble, or cook—30 minutes maximum—and focus on using ingredients from the garden. This activity can accompany a fall celebration, festival, or community event. The recipes and food can be shared among the class or to a wider group, according to the community's needs.

2. During the activity, the students will be working together to prepare recipes and cook garden foods. What does safety, communication, and craftsmanship look like during this activity?

Activities and Assessment (35 minutes)

1. In pairs or individually, students will gather the harvested fruits, seeds, and flowers from the garden. Students with similar recipes can work on gathering the necessary ingredients, cooking implements, and tools before following their recipe and creating a dish. They should practice safe, careful cooking and work as a team.

2. After they have assembled their recipes, they can clean up their workstation and compost any plant pieces that remain. When all the dishes have been prepared, the class can share their cooking with the class, either to taste test or tell a story about. What memories do students have with their recipes? What adaptations or changes did groups make according to the seasonal produce available?

3. In *Permaculture Perspectives*, David Holmgren writes, "Redistribute Surplus requires us to share surplus resources to help the earth and people beyond our immediate circle of power and responsibility."[3] Discuss with the class how the garden produce is being utilized by the school and community. In what ways is it effectively being distributed and used? What strategies can students design to better distribute garden surplus?

Activity Extension

This is a great opportunity to become involved with the surrounding community by volunteering the class with a local food bank, businesses that distribute surplus foods, an organization that is involved with addressing hunger and food security, or a gleaning event on a local farm. I recommend engaging in such service activity as often as possible, and more than once. Rather than perform a single day of service, commit the class to a season of service or once-a-month throughout the year to instill a culture of community engagement and greater opportunities for connections and conversations.

Preparation

1. Cooking utensils
2. Bowls
3. Spatulas
4. Oven or stove
5. Dishware and utensils
6. Additional ingredients not from the garden

3 Leadership: Taste Testing and Nutrition

Time Frame: 45 minutes with older students, 30 minutes with a younger class

Overview: The older students will lead a younger class on a taste test tour of the garden and teach a small lesson on nutrition.

Objective: Younger students will begin learning how to talk about flavors and identify plants, while the older students will develop their leadership skills and pass on their knowledge about garden food.

Introduction (10 minutes)

1. Begin with a short introduction with the eighth grade students about group behavior expectations when working with younger partners. How can older students help the younger ones stay focused for the activity? What can they do to make the activity engaging and educational for younger students?

2. How can older students help accomplish the goals of the day? The goal for young students is to identify edible plants in the garden and learn how to carefully harvest them. The goal for the older students is to begin passing on the culture of the garden through teaching them how to gently gather garden foods and to taste them. They can also take this opportunity to teach the younger students about the nutritional benefits of the seasonal garden produce.

3. Introduce the eighth graders to the edible plants they will be harvesting from in the garden. Where did they find these plants during their previous activity? What are the best picking practices with each plant? What stories and memories can the older students share with the younger ones about these plants and how to treat the living things in the garden?

4. When the younger class joins the group, they should be introduced to their older partner and review behavior expectations in the garden and their activity for the day.

Activity (30 minutes)

1. In pairs, the older and younger students will gather taste tests from the garden. The eighth grade student will take the younger student on a tour of the garden and help them identify 3–5 edible plant parts that are available seasonally.

2. The groups can harvest a select amount of taste tests from these plants and wash and prepare them according to school and community procedures. When they have collected their taste tests, the groups can gather back together, sitting with their partners, to share their experiences.

3. As they share their foods, the older students can teach the younger ones about food combinations and how to talk about the flavor of different foods. What words would students use to describe flavor? What behaviors can a student exhibit if they don't care for a type of food or flavor? Where is the best place to put food in the garden if it won't be finished?

4. Facilitate a brief survey of the younger students' experiences in the garden. What foods did students identify during the activity? What foods were new to them? What are their partners' names? What foods would they like to try again? What are nutrients and what nutritional benefits do some garden foods have?

Assessment (5 minutes)

1. When the younger students have thanked their partners and left, gather the older students back together to discuss the successes and challenges of the activity. Where they able to focus on their partners, guide them throughout the garden, and accomplish the goals for the day?

2. What techniques did they learn or use to help them accomplish their goals? What would they do differently next time they worked with younger students?

Preparation

Prior to this activity, the older students should research an easily picked, edible garden food and its nutritional benefits. They will need to think of the best way to communicate these facts to their partner during the activity, in a way that is fun and engaging for younger students.

Resources

- "Oregon Harvest for Schools." Oregon Department of Education. (online—see Appendix B).
- The Collective School Garden Network. "Garden-Enhanced Nutrition Teaching Resources." (online—see Appendix B).
- KidsGardening. "Create & Sustain a Program—Nutrition Education in the Garden."(online—see Appendix B).

4 Garden Skills: Food Preservation

Time Frame: 45 minutes for activity
45 minutes for processing food

Overview: The class will learn how to pickle foods from the garden and discuss the value of food preservation techniques.

Objective: To learn a valuable skill that will provide sustenance months later.

Introduction (10–15 minutes)

1. Brainstorm with the class what they know about the processes of preserving and canning food: Do any students have experience canning or pickling food? What kinds of foods are commonly stored in a can or jar? What are different types of preservation?

2. Introduce the students to the various food preservation methods, such as fermentation, canning, pickling, freezing, and dehydrating. Discuss the process of each and their benefits.

> The method for a class food preservation activity will depend on the resources available to each community. I chose to illustrate a pickling process, because it has clear steps and relies on simpler tools that students can use. I provide basic instructions, because there are plentiful pickling recipes available online and in book resources.

How would preserving food positively impact a person or family for the year? What are some barriers to the process? Demonstrate the food preservation activity for the day and the materials needed for the activity.

3. How will the class work and communicate as a team during this activity? How will they practice personal and food safety? What does this behavior look like?

Activity (30 minutes)

1. In pairs, the students will clean surplus garden vegetables and prepare them for pickling (cucumbers, beans, carrots, or celery will all work well). The class may have to harvest these vegetables first or connect to a local farm or market. They can use the available tools, such as cutting boards and knives to trim the vegetable stems or undesirable pieces, wash them in a strainer under water, and set them in a clean area for packing into jars. At this time, they can also prepare any seasoning or spices that can be added, such as pickling spices, salt, garlic, and dill leaves.

2. When the vegetables have been prepped, the partners will sterilize one or two canning jars in a cleaning station. Then, they can assemble their jar of pickles by gently stacking vegetables into each jar, with the spices. If I have them on hand, my preferred ingredients are a teaspoon of salt added to the bottom of the jar, as well as picking spices, a clove of garlic, a teaspoon of cayenne, a dill blossom or leaves, and a grape leaf. The vegetables shouldn't be packed too tightly, and there should be 1½ inches of headroom left in the jar.

3. The small groups can then pour the vinegar pickling mix into the jars. I like to make a water and vinegar mixture with a ratio of 1:1. The pickling liquid can be boiled at the beginning of the activity as to be ready when the students need it. Carefully, they will use a funnel to pour the liquid in their jars, leaving 1½ inches of headroom available.

4. The groups will insert a slender knife into the jar to remove any trapped air bubbles. Then, they will wipe the rim of the lid and place the canning lid and ring on the jar. The canning lids can sit in warm to hot water until this stage, so that they seal is flexible. The jars should be tightened firmly.

5. The class can clean up their stations and the materials they used. During this time, the jars can be placed in a water bath pot and covered with an inch of water. When brought to a boil, the jars will be processed according to the elevation requirements of your region (see Resources).

6. When the jars have been processed, the students can carefully remove them from the water with a jar lifter and set them out to seal. The class can check on these jars periodically to make sure they have sealed and, when cooled, remove the metal rings from them. Finally, they can label their jars with their names and the date.

Assessment (5 minutes)

1. What did students experience during the pickling process? Was it easier or more complicated than they expected?

2. What are the benefits and setbacks of home preservation? Would students like to do more preservation, and what types of foods would they like to try?

Preparation

1. Glass canning jars (one or two for each group)
2. Canning rings and lids
3. Ladle
4. Funnel
5. Water bath canner
6. Jar lifter
7. Vinegar and water
8. Pickling spices and salt
9. Vegetables for pickling
10. Food safety equipment
11. Sink and stove
12. Cleaning supplies

Activity Extensions

This activity can be extended into science and history studies by looking at the life and work of Louis Pasteur and the microbiology, biology, and chemistry around food preservation and pasteurization. This topic can also be related to the commercial food system and culturally raw and fermented foods.

Resources

- Sandor Ellis Katz. *The Art of Fermentation: An In-Depth Exploration of Essential Concepts and Processes from Around the World.*
- National Center for Home Food Preservation. "Preparing and Canning Fermented and Pickled Foods." (online—see Appendix B).
- Ellie Topp and Margaret Howard. *The Complete Book of Small-Batch Preserving: Over 300 Recipes to Use Year-Round*, 2nd rev. ed.
- Sherry Brooks Vinton. *Put 'em Up!: A Comprehensive Home Preserving Guide for the Creative Cook, from Drying and Freezing to Canning and Pickling.*

5 Garden Skills: Produce No Waste

Time Frame: 45 minutes

Overview: The students will gather *green* and *brown* mulching materials and learn about the ecological design principle *Produce No Waste.*

Objective: To prepare for the next leadership activity, participate in seasonal garden chores, and engage with a new design principle in a hands-on way.

Introduction (10 minutes)

1. Brainstorm with the class what they know about how so-called *waste* is cycled through the garden: What are seasonal surplus materials? How does the community handle these surpluses, what uses do they have, and what potential uses could there be for them? What does waste look like in wild spaces and other ecosystems? How is waste processed in these systems?

2. Introduce the students to the ecological or permaculture design principle, referenced in *Permaculture Pathways* by David Holmgren, *Produce No Waste*. What value does this principle have when looking at the human, industrial, or agricultural waste stream? What creative solutions can students generate to address specific local waste concerns?

3. Compared to larger social and environmental wastes, organic garden waste is a simple concern. But processing this by-product of the growing season is a valuable activity in capturing the plant's remaining nutrients and enhancing the garden's energy cycle. This activity builds soil health, plant nutrition, and supports the garden ecosystem. It is a small but valuable step when considering this ecological principle.

4. Demonstrate to the class the activity of the day and the tools they will be using. How can students practice safety with the tools and behave as careful, considerate gardeners? What do students remember about different *brown* and *green* materials when they studied composting in the fifth grade? What do these materials look like, and how can students identify the plants that need to be broken down for mulch?

Activity (30 minutes)

1. As a class, the students will be gathering *brown* and *green* materials from the garden and cutting them into small pieces to create mulch for a later activity. They will first gather the organic material in the garden, such as old tomato plants, squash leaves, or dried tree leaves.

2. Then, they will sort through their pile and look for any evidence of diseases or pests at work. Diseased material should be separated and not mulched back into the garden soil. Finally, the students will chop the remaining organic matter into small pieces and add this mulch to a pile to be used later.

3. The students can work all together or rotate through jobs of harvesting, sorting, and cutting. When the activity time is finished, the students can clean up and put their tools away.

Assessment (5 minutes)

1. Did the students find any evidence of disease or pest activity?

2. If the students were to consider how many *green* materials and *brown* materials they harvested, what would they say the ratio of the two materials would be?

NGSS and Activity Extensions

There are many studies that can evolve from the principle of *Produce No Waste*, including studies on the environmental and social impact of a consumer culture or value system, mapping a student's personal waste stream, and research into renewable resources and energy. The students could also delve into models that "describe the cycling of matter and flow of energy among living and nonliving parts of an ecosystem" (MS-LS2-3).

Preparation

1. Clippers and handsaws
2. Wheelbarrows
3. Gloves

6 Leadership: Mulching or "Putting the Garden to Bed"

Time Frame: 45 minutes with older students, 30 minutes with a younger class

Overview: The eighth graders will guide a younger class on how to mulch organic materials into the garden soil using the mulch they created during the previous lesson.

Objective: For older student to continue developing their leadership skills and to learn about mulching by teaching about it.

Introduction (10 minutes)

1. Begin with a five-minute introduction and review with the middle school students about the expectations of working with the younger students, what they remembered and learned about working with them last activity, and what the goal is of today's activity.

2. During this activity, the middle school students will work alongside their younger partners to lay the premade mulch onto designated garden beds or containers. The older students' goals are to demonstrate to their partners how to spread the mulch around living plants and spread it out over the soil, while also teaching them the importance of this seasonal task and the composition of the mulch. Their final goal is to limit waste and mess by doing slow, careful gardening and teaching this attitude and behavior to younger students.

3. When the younger class joins the group, the students will also receive a short introduction about the activity of the day, the materials they will use, and behavior expectations. Then, they will break into new partner groups before they get started.

Activity (30 minutes)

1. The partner groups can gather the mulch into buckets and go into the garden to mulch one bed at a time. They should carefully handle the materials and gently work around established plants, focusing on "tucking in" the plants. The groups can continue this activity until the mulch is used or the activity time is done.

> Some gardeners would advise caution when mixing organic matter from different plant varieties together, such as nightshades and brassicas (tomatoes and kale). The reasoning is to limit any potentially non-beneficial interactions in the composition of these families. By working with small spaces and regularly rotating crops, I have never experienced or observed any negative effects. Like in all gardening matters, each decision depends on each gardener's experiences, climate, and condition.

2. After the mulch has been laid, the partner groups can put their tools away and enjoy taste testing in the garden, with the younger students now showing their partners what they liked to eat and the older students guiding their partners through the garden.

Assessment (5 minutes)

1. At the end of the activity, gather the middle school students together to reflect, share their stories, and explain the teaching techniques that worked well for them over the course of this activity.

2. How did students keep in constant communication with their partner? How did the older students manage or make use of transition times? Do they feel like they effectively communicated the day's message or goal to their partners?

Preparation

1. Premade mulch
2. Five-gallon buckets

7-10 Self-directed Projects

Time Frame: 45 minutes for each class period
Overview: The class will develop self-directed projects that will enhance the community's experiences in the garden.
Objectives: For students to take control of their own learning, develop their gardening skills, and contribute to the wellness of the garden community.

Introduction (10 minutes)

1. Many of the lessons over this season have been supporting the role and responsibility of eighth graders to be leaders in their community, to give back, and to share the surpluses of the garden. With this in mind, the students will design a project, or a series of projects, that will continue to support the values of *Share the Surplus* and *Produce No Waste* (as described in Holmgren's *Permaculture Pathways*).

2. What aspects of the past lessons have inspired students to want to know more about them or to do more within their community? What are realistic projects that students can do in the garden to support these initiatives?

© Lisa F. Young / Adobe Stock

Activity (30 minutes)

1. For the next four lessons, the class will design and accomplish individual or small-group projects. The work may be centered in the garden (e.g., Student Gardeners) or extend beyond the garden through work in the community. Some students may want to plan a community event, host a workshop or lesson, or start a new initiative at school. For example, a student may be interested in developing more outlets for school garden food to be distributed into the larger community.

2. Additional seasonal self-directed projects could include invasive plant removal, pruning trees, seeding root crops, mulching with compost, planting cold-loving plants, and planting bulbs.

Assessment (5 minutes)

1. How have student projects benefited the school and community? How did the chosen projects support the ecological and permaculture principles of *Share the Surplus* and *Produce No Waste*?

2. How would students like to see their work mature and grow?

Winter: Designing Systems

1 Garden Skills: Small-scale, Intensive Systems

Time Frame: 45 minutes

Overview: The students will begin a mapping project of a garden bed or container through the lens of ecological and permaculture principles and practices.

Objective: The students learn about and interact with some of the foundational language and perspectives of ecologically minded gardening and design.

Introduction (10 minutes)

1. Introduce the class to the concept of permaculture and the permaculture principle for ecological design, *Small-scale, Intensive Systems*, as found in Toby Hemenway's book *Gaia's Garden* (see Resources). What do small-scale intensive systems look like in the garden? What examples exist in nature already? What concerns or ideas do students have about mimicking these systems in the garden?

2. What are the benefits of annual plants over perennial? What effect do these plants typically have on available nutrient quantities in their soil? What perennial plants do students see in ecological spaces, and what effect do they have on their ecosystems?

3. Review with the class how to measure and create a model of a space on grid paper. What tools and skills do students have to assist them in creating realistic models? What should they consider about scale of the site?

Activity (30 minutes)

1. Individually or in pairs, the students will use the provided tools to map a realistic diagram of their assigned garden bed or planting container. A topographic view is preferable for future use of the map. The goal of this first part of the activity is to create an accurate model of the site.

2. After the site is mapped, the students will identify the perennial plants that live in the site and recall the annual plants that were grown in it last season. There may be evidence of these plants still in the soil to assist them, or they can ask other students and staff members. Reference books on plant identification can also help to identify perennial plants. When they have identified these plants, the students will add them to their map, to scale, and create a key/legend to indicate the species.

3. The students should pay attention to spaces with open soil and the relationship of plants to each other in the space. At the end of the activity, the students can gather back together to share their findings. The class should keep these maps for the follow-up lesson. They will build off these maps and, in the spring, will apply their designs as best they can and as resources are available in order to increase the food production, layering, and biodiversity of the garden.

Assessment (5 minutes)

1. What perennial plants exist in the garden beds? What annual plants did the students find evidence of? If students were to estimate a ratio of annual to perennial plants in the site, what would they guess? What can be added to increase food production in the garden?

2. How can the class mimic successful ecosystem models in the garden beds? Are there places where small intensive systems are not a good design plan?

Preparation

1. Grid paper
2. Clipboards
3. Pencils
4. Measuring tapes
5. Plant identification books and resources

Activity Extensions

This lesson is a small piece of the depth of studies that students can explore about *Small-scale, Intensive Systems.* The principle ties into studies on agricultural and food systems and offers opportunities for students to compare and contrast small-scale (family farms) and large-scale (industrial) systems. For example, students could explore energy outputs within different systems and their effect on natural resources and natural systems. Further extensions might involve the effects of an industrial food system on factory and farm laborers, human health and nutrition, the use of commercial herbicides and pesticides, and the resource and fuel use associated with shipping food.

Resources

- H. C. Flores. *Food Not Lawns.*
- Toby Hemenway. *Gaia's Garden: A Guide to Home-Scale Permaculture*, 2nd ed.
- Permacultura México. "Permaculture Design Certificate." Video produced by Bajio Community Foundation. (online—see Appendix B).
- Deep Green Permaculture. "7. Small-Scale Intensive Systems." (online—see Appendix B).
- Wayne Weiseman et al. *Integrated Forest Gardening: The Complete Guide to Polycultures and Plant Guilds in Permaculture Systems.*

2 Garden Skills: Food Forests and Layering

Time Frame: 45 minutes

Overview: The class will learn about methods of layering plants to create a food forest and design one in their garden bed/container.

Objective: To study practical applications of ecological principles and explore how intensive garden planting can increase production with minimal energy output and larger energy conservation.

Introduction (10 minutes)

1. Brainstorm with the class what they imagine when they hear the term *food forest*: What plants do they imagine inhabiting this space? How are these plants growing and interacting with each other? What does the ecosystem of this imagined space look like; how are creatures and insects interacting with each other?

2. The layers of a food forest include the tall-tree layer, low-tree layer, shrub layer, herb layer, ground cover layer, vine layer, and root layer (see Resources). Where can these layers be found in ecosystems outside of the garden? What examples of plants can students think of that fill these niches?

3. Introduce the class to examples of food forests with layered plants. What observations can they make about the order or structure of the space. Are these wild places, or is there a design and system? Where in nature or wild spaces would they see similar growth patterns?

4. What are the benefits of this type of vertical system over a horizontal design? What is a *microclimate*? How do layered plants support each other and the garden ecosystem? What effect does this type of planting have on the soil?

Activities (30 minutes)

1. Individually or with their partners, the students will return with their maps to the garden site they studied. Using what they know about food forests and layering, the students will identify which layers exist in the garden site (tall-tree, low-tree, shrub, herb, ground cover, vine, or root). The students may have to dig into the soil to discover some of the layers and will add these plants and their layer identification to their maps.

2. Then, the students will identify any missing layers to their garden site and make a list of these on the back of their map or on an additional paper. Keeping in mind the climate and site requirements of the space, the students will use available resources and books to research potential perennial garden plants to fill these niches. They can also make observations about other plants in the garden that could fill these spaces. They can write these down and make recommendations for future garden design and planting.

Assessment (5 minutes)

1. In what ways is the school or community now gardening horizontally but could also be gardening vertically?

2. What layers are not realistic considering the growth and site requirements of that layer (large trees, vines). Have the students share their ideas and begin to discuss the feasibility of certain recommendations and plant varieties.

Preparation

Plant identification books and resources

Activity Extensions

Further studies could include the students reading a chapter about Mark Shepherd's food forest and application of the seven layers in *Restoration Agriculture*. Students can also explore the prevalence of monocrops in commercial agriculture, hydroponics and aquaculture, and other innovative solutions to producing food vertically or in soilless conditions.

Resources

- Ron Finley. "A Guerilla Gardener in South Central LA." (online—see Appendix B).
- "Designing a Forest Garden: The Seven-Layer Garden." (online—see Appendix B).
- Jonathan Engels. "Planting in Pots and Other Ways of Playing with Permaculture in the Big City." (online—see Appendix B).
- Toby Hemenway. *Gaia's Garden.*
- "Small-Scale Intensive Systems." Deep Green Permaculture. (online—see Appendix B).
- Mark Shepard. *Restoration Agriculture.*
- Wayne Weiseman et al. *Integrated Forest Gardening.*

3 Garden Skills: Designing from Patterns to Details

Time Frame: 45 minutes

Overview: The class with study local natural, urban, or traditional ecosystems and compare the territory of their maps to these established and functional spaces.

Objective: For students to practice designing from patterns to details while learning how to mimic natural successful spaces.

Introduction (10 minutes)

1. Brainstorm with the class about what know of local ecosystems: What plants, animals, insects, water cycles, and environments are a part of it? What ecosystems are successful, and what ones are struggling for stability?

2. What lessons can the students, as gardeners and designers, take from these successful systems and mimic in the garden? What plants within these ecosystems fulfill the layers of a food forest? Are there wild foods that the students would like to see growing in the garden?

3. What does it mean to *Design from Patterns to Details* (see *Permaculture Pathways*, page 127). What skills and perspectives will help the students learn from local ecosystems?

Activity (30 minutes)

1. In partner groups, the students will perform an in-depth study on one local and traditional ecosystem within their landscape. They can use online resources, books, and local experts to create a list or map of different layers within this ecosystem, the plant growth requirements within it, and edible native plants that might be good additions in the school garden.

2. If possible, a field trip to a natural and wild space would provide hands-on opportunities and observations for the students. They can continue mapping in these spaces, using plant identification books, and identifying different layers at work in the space.

3. Inviting an expert or elder into the learning space is another excellent opportunity to learn about the local plants, their histories, and their uses that may interest students.

4. The students should pay attention to the relationship of plants with one another. For example, what relationship does a large conifer have with the grasses, shrubs, and flowers below it? What other species rely on these plants?

5. The students should seek to answer the questions: How is balance, resilience, and stability reflected in these relationships? What patterns can students identify within this ecosystem, and how can they be mimicked in the garden?

6. The students should keep these maps until the springtime when they can begin to implement their ideas.

Assessment (5 minutes)

1. Have the students reflect on their findings and the patterns they identified in the ecosystems. What local ecosystem did the students investigate, and what lessons did they learn from it?

2. What would these patterns look like in the garden? What plants could mimic or play a similar role than in the local ecosystem? Are there plants that students would like to see added into the garden?

Preparation

1. Knowledgeable community members
2. Online and book resources

Resources

- Commission for Environmental Cooperation. "North American Environmental Atlas." (online—see Appendix B).
- US Environmental Protection Agency. "EnviroAtlas." (online—see Appendix B).
- US Environmental Protection Agency. "Ecoregions of North America." (online—see Appendix B).

4 Garden Skills: Plant Propagation

Time Frame: 45 minutes

Overview: The class will begin propagating edible and pollinator-friendly plants from cuttings and rhizomes.

Objective: To learn new techniques for growing free and beneficial plants for the garden.

Introduction (10 minutes)

1. Discuss with the students about what they may know about *plant propagation.* What does this term mean? Are there any indoor or outdoor plants that they have they planted from cuttings before or seen propagated? How do perennial garden plants, such as rhubarb, raspberries, and strawberries, propagate themselves?

2. What kinds of edible and pollinator-friendly plants are in the garden that students would like to see more of? What plants would they like to add in the garden?

3. Other than saving seeds, there are three common perennial propagation techniques that the students can interact with: clippings, roots/rhizome separation, and clones. Demonstrate where these occur in the garden and the processes for harvesting them. Show how to collect and cut plants clippings, as well as which plants the students will be harvesting during the activity. New perennial plants gathered in these ways can be distributed into new areas of the garden and increase food production.

4. Perennial plant examples that the students could potentially gather from include:

- Clippings: rosemary, sage
- Roots/rhizomes: rhubarb, lamb's ears, catnip, mint, chives, perennial flowers
- Rooted clones: strawberries, raspberries

See Resources for information on how to gather these three types.

Activities (30 minutes)

1. In small groups, the students will rotate through three stations and spend about 10 minutes in each space. The stations will consist of the three plant propagation techniques that the class has learned about: clippings, roots/rhizomes, and rooted clones. In each station, the students will practice the techniques of gathering these potentially viable future plants, before moving on to the next technique. See Resources for examples of how to perform each technique.

2. Station 1: Clippings: First, the students should identify the parts of the stem of the provided plants and then use clippers to carefully cut the stems just below a bud. They should follow the directions of the adult volunteer or provided resources to make sure their clippers are clean and to cut at the correct angle. I recommend having the students take clippings from plants that can be propagated in water. This way, the students can gather their clippings into a clear jar of water and watch them root. See Resources for more detailed directions on this activity.

3. Station 2: Roots/rhizomes: The small groups will dig into the soil and remove a perennial plant, gently separating the roots from the soil. They should work together to identify and carefully separate the multiple plants in the bunch and plant them in separate potting containers filled with soil. The groups can gently transplant the roots or rhizomes and then label and water the plants afterward.

4. Station 3: Rooted clones: The small group will identify the cloned offspring of plants such as strawberries and raspberries. They will identify the new growth and gently dig up the plants before transplanting them into planting containers. They will label and water these plants afterward.

5. The harvested plants can be stored in a safe space until spring planting. The clippings in water can be displayed in a window sill and the water freshened as needed. They should all be monitored over the next few weeks until they are ready to be transplanted into the garden.

Assessment (5 minutes)

1. What technique did students prefer during this activity? What techniques would they like to know more about?

2. What additional uses can students generate for this surplus of plants? What are the benefits of harvesting more plants from the garden?

Preparation

1. Clippers
2. Plant examples
3. Adult volunteers, if possible
4. Small potting containers
5. Soil
6. Watering can
7. Plant marker or tape for labeling

Adaptations

If the students aren't able to harvest plant clippings in their local environment, then work to gather a few piles of branches or vines before the class time. A parent, guardian, staff member, or community member may have access to pruned clippings that the students can work with.

Resources

- Tara Luna. "Vegetative Propagation" in *Nursery Manual for Native Plants: A Guide for Tribal Nurseries.* (online—see Appendix B).
- Lewis Hill. *Secrets of Plant Propagation: Starting Your Own Flowers, Vegetables, Fruits, Berries, Shrubs, Trees, and Houseplants.*
- Alan Toogood, ed. *American Horticultural Society Plant Propagation: The Fully Illustrated Plant-by-Plant Manual of Practical Techniques.*

5-7 Self-directed Projects

Time Frame: 45 minutes for each class period

Overview: The students will identify and design a solution for a part of the garden that needs improvement, while also increasing the health, safety, or community experience in the space.

Objective: To apply their garden skills and past lessons in the garden in order to create a healthier ecosystem and to apply trial and error to their learning.

Introduction

1. During the next three sessions of garden class, the students will be applying their past lessons to address a need in the garden space. They should be able to articulate clearly how they will address this need, what materials they will use, and a goal for each work day.

2. Brainstorm with the class about what self-directed projects they would like to pursue. Have any of the past lessons inspired the students to address a need in the garden or community? Are there changes they would like to see occur in the garden?

3. Review behavior and social expectations for self-directed work. How can students respect their own time and work and that of others over the next few weeks? What do these behaviors look like?

4. What projects would students like to do that support the three lessons and permaculture values or practices they have just learned (such as *Planning Small Intensive Systems*, food forests and layering, and *Designing from Patterns to Details*).

Activity

1. Before each lesson, provide a context for the class of what they will be working on. Take this time to check in with each student and small group before they begin working. This provides the opportunity to identify students who need additional tasks, help, or inspiration and those that need supplies or additional support.

2. The students should be encouraged to work in the garden beds or containers they mapped as often as possible. What solutions do they have to address issues in these spaces? Some students may want to extend their learning from the previous mapping project and apply solutions in these spaces (e.g., by separating perennial root systems or preparing the beds for early spring planting).

3. Examples of simple self-directed work include: removing perennial "weeds" from garden beds, pruning shrubs and fruit trees (semi self-directed), laying bark mulch on the pathways, laying down compost, assisting with structural improvements, and planting fruit trees.

Assessment

1. What changes are occurring in the garden through student work? What success and challenges have they faced in accomplishing their goals?

2. What additional changes and solutions would the class like to see happen?

8|9 Leadership: Plan and Teach a Lesson

Overview

The students will construct and execute a lesson plan about their self-directed projects for younger students.

Objective: For older students to articulate the ecological and permaculture principles and practices that inspired their projects while passing on the cultural values of the garden to younger students.

Activity 8

Time Frame: 15 minutes

Introduction (10 minutes)

1. Brainstorm with the class about themes from previous this year: What does it mean to use small intensive systems, create food forests and plant layering, and use and value diversity? How have their self-directed projects enhanced these principles and practices or created new opportunities for them?

2. What lessons have the students learned as they accomplished their projects? What challenges and successes did they experience? Did they engage with an activity that they really enjoyed or were excited about?

3. Introduce the class to their upcoming challenge: creating a lesson plan about their self-directed project and teaching an activity about it to a younger student.

4. What are techniques to gather and keep a younger student's attention? What transition times may occur that will distract the younger student? From their past experiences, what advice do students have for each other about how to be a leader and a teacher?

Activity (30 minutes)

1. In partners or individually, the students will plan a 30-minute lesson to teach a younger student. The lesson plan will include an introduction, key vocabulary, activity, and reflection. The goal of the lesson is to teach a younger student about the older student's self-directed work and how to do a task or activity associated with it. The lesson and activity should be something the older student is excited about so that the younger student is inspired and excited too.

2. The lesson should also use the language of one of the previous ecological and permaculture subjects: (1) small intensive systems;

(2) food forests and layering; and (3) use and value diversity.

3. The students will write out their lesson plan and break down the expected time frame of each section. Afterward, students can work with another student/small group to review each other's lesson plans for any missing pieces or questions. Or, they can present their lesson plan to the class for a group discussion on the feasibility and strength of the lesson.

Assessment (5 minutes)

1. What activities are students excited to lead? What materials will they need to accomplish their activities?

2. What concerns or questions do they have about being a leader and a teacher?

Activity 9

Time Frame: 45 minutes with older students, 30 minutes with younger students

Overview: The older students will teach their lesson plan to a younger student in the garden.

Objective: Older students will develop their leadership skills and learn through teaching.

Introduction (5–10 minutes)

1. Before the younger students join the older class in the garden, review expectations for the activity. The older students are expected to accompany their partner at all times, assist with transitions, provide clear and supportive directions, and show enthusiasm for the activity. What final questions or concerns do older students have about their lesson plan?

2. Provide a few minutes for the students to review their lesson plans, gather their materials and tools, and stage the area in the garden where they will be working.

Activity (30 minutes)

1. When the younger students arrive in the garden, then they can be paired with an older student for their garden lesson. Then, the older student will guide their partner through their lesson plan and activity.

2. If the older student needs any assistance, then they can ask an adult or staff member who is available. The groups should practice safety, communication, and careful gardening.

3. Before the end of the activity, the classes can be informed about the time and begin ending their activity and cleaning up. When they are finished, the older students can lead their partner through a short reflection about the activity.

Assessment (5–10 minutes)

1. When the younger class has thanked their partners and left, the older students can finish cleaning up their station and putting away their tools.

2. Gathering back together as a class, have the eighth grade students reflect on their experiences teaching and guiding the activity. What unexpected challenges and successes occurred during the lesson? How did the younger student interact with the activity?

3. What would the older students do differently if they wrote another lesson plan? How successful did they feel about accomplishing the goal of the lesson and passing on one of the three ecological and permaculture principles and practices?

10 Leadership: Planting Pea Seeds

Time Frame: 45 minutes with older students, 30 minutes with younger class
Overview: The students will teach a younger group about how to plant cold-hardy seeds.
Objective: Older students will take the lessons they learned in the previous leadership activity and apply it to their teaching.

Introduction (10 minutes)

1. Begin the activity by briefly introducing the eighth grade students to expectations about working with younger partners and the goals of their joint activity. In planting peas seeds, the students will focus on reiterating and reinforcing the lesson of how deeply to plant seeds and how to identify good places to plant.

2. Brainstorm the lessons and skills they learned in the last activity, Design and Teach a Lesson: What would they do differently when working with younger students? What challenges and successes did they face? How can the older students keep their partners active and engaged during the whole activity time?

3. Review with the students where good planting spaces are in the garden and give them time to identify the planting spaces they will use during this activity.

4. When the younger class joins the group, partner each student with a middle schooler and have the young students review the behavior expectations for the activity.

Activity (30 minutes)

1. In their mixed-age pairs, the students will gather their pea seeds and go out into the garden to carefully plant them. The students should spread and find their own quiet spaces to focus on planting with their partners, if possible.

2. The groups can plant multiple handfuls of seeds and water them with watering cans as they go. Additionally, the groups can mark their planting areas with plant markers and label the pea variety in each spot.

3. About five minutes before the end of the activity, the groups can return any extra seeds, and the middle school students can take their partners on a taste test tour of the garden, showing them where to find edible early spring foods.

Assessment (5 minutes)

1. Gathering the older students back together, have them discuss the success and challenges of this activity compared to the previous one. Were they able to focus on their partners, guide them throughout the garden, and accomplish the goal of the day?

2. How did they demonstrate how deeply to plant a seed and how far apart to space them? What challenges did students overcome? What are valuable leadership lessons and skills that they learned and will carry away from this activity?

Preparation

1. Pea seeds
2. Plant markers
3. Permanent pens
4. Watering can

Spring: Supporting Biodiversity

1 Garden Skills: Supporting Healthy Habitats

Time Frame: 45 minutes
Overview: The students will build habitat boxes for a variety of local species.
Objective: To increase habitat opportunities and ecosystem relationships in the garden.

Introduction (10 minutes)

1. What types of habitat have students built, planted, or encouraged as they have learned in the garden? What species did these systems support, and why are those species important for the garden ecosystem? How do ecological designs, such as layering plants, benefit these species too?

2. For the activity, the students will be building bird or bat boxes to be installed in or around the garden. Depending on the garden site, the students may be able to install these boxes in or near the garden beds they've been working on, or beside their classroom. The goal of this activity is to create well-constructed, usable habitat boxes and increase habitat opportunities for local species.

Swallows

I see the swallows flying over my garden

of green. I see they like it here, I see.

I watch them fly over head. I watch them

settle down to rest in their twiglike bed.

I watch them dive and float back up,

I watch them fly away, farther and farther,

when every minute, the sky gets darker.

I see them return back to bed, good

night swallows, sweet dreams, go ahead.

— Addie

After years of studying gardening with this curriculum, the students may have built mason bee and leaf cutter bee homes, as well as amphibian houses, and planted many wildflower seeds to promote pollinator health.

3. Demonstrate the building activity and what tools and materials are available for the students. What behaviors will the class use to be safe during this activity? How will they work with their partner to communicate and accomplish the tasks?

Activity (30 minutes)

1. In pairs, preferably the partnership they have been in for the mapping project, the students will assemble the provided materials and build a bird or bat box. The bird box should be for a local bird species, depending on the region.

2. The students will practice safety when handling the tools and wear any necessary eye protection. The complexity of student work depends on regulations in each school or community. Some institutions may allow students to measure and saw wood, others may require the use of precut pieces.

3. The groups should follow the directions for assembling the pieces and work as a team to accomplish the goal. If possible, each group should make two habitat boxes.

4. When the groups have finished assembling

their boxes, they can store them in a safe area for the next lesson and clean up their station. Then, they can use available resources to research the nesting preferences for their bird or bat species and determine a few potential spots in install the boxes in the garden or surrounding community.

Assessment (5 minutes)

1. How well did the groups work as a team for this activity? What did they do successfully, and what challenges did they experience?

2. Why is it important to support these species?

3. What other activities or changes can students make in the garden to support them?

Preparation

1. Bird and bat box materials
2. Nails
3. Saw
4. Hammers
5. Resources and books on local bird and bat species

NGSS and Activity Extensions

Further in-class study can include research into local bird and bat species, their habitat range and requirements, and any concerns or issues facing them regionally. Students can explore the life of a local bird or bat and discuss what responsibility the class has to addressing any concerns. Migration patterns are another great extension, as is research and gathering evidence to argue how changes to ecosystems affect populations (MS-LS2-4).

Resources

• "All About Birdhouses." NestWatch, Cornell Lab of Ornithology, 2018. (online—see Appendix B).
• "Educator's Guide to Nest Boxes." BirdSleuth K-12, Cornell Lab of Ornithology. (online—see Appendix B).
• Russell Link. *Landscaping for Wildlife in the Pacific Northwest.*

2 Leadership: Installing Habitat Boxes

Time Frame: 45 minutes + 15 minute follow-up activity
Overview: The eighth grade students will guide a younger class in decorating and installing habitat boxes in the garden or around the community.
Objective: Older students can pass on their knowledge of local bird and bat species and communicate why supporting habitat is important to a new generation of land stewards.

Introduction (10 minutes)

1. Before the younger students arrive, review behavior expectations for the older students. How will they act as leaders during this activity? How can they help the younger students be engaged and stay on task? What message is most important for the younger students to walk away with after the activity? How will the older students communicate this?

2. When the younger students join the group, then they can be partnered with an older student and review their own expected behaviors as a class before embarking on the activity.

Activity (30 minutes)

1. Each partner group will gather one habitat box and the necessary supplies to decorate it. They can be encouraged to draw or paint images that represent the species that may inhabit the box, or provide words or poems to add meaning to the piece.

2. The groups should work together and focus on communication and craftsmanship. This is an opportunity for the older students to lead discussions on the bird or bat species and the value of supporting ecosystem members.

3. When the groups have finished, then they can store their box in a safe place to dry and clean up their supplies and station. Afterward, the partner groups can go into the garden to consider and discuss where good places to install the habitat box might be. The older students should already have a good idea of potential sites and guide the younger student in conversations about the nesting preferences of different species.

4. When the groups have found the space they want to install the box, the older student can mark this space or install a nail to later hang the box.

Assessment (5 minutes)

1. When the younger students have finished the activity, gather the older students together to share their experiences. How are their experiences being leaders now different from their experiences at the beginning of the year?

2. Do the students feel they effectively communicated the message of the activity? How will the younger students interact with the habitat boxes in later years?

Preparation

1. Paint
2. Markers
3. Paintbrushes
4. Cups and plates

Follow-up Activity (10–15 minutes)

When the paint has dried and been sealed, the older students can return to install the boxes in the garden or a designated community site. They can also sign their name and their partner's name on the box before they hang it.

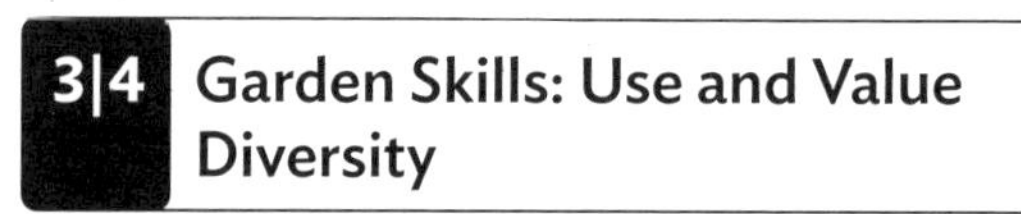

3|4 Garden Skills: Use and Value Diversity

Time Frame: 45 minutes for each class period

Overview: The students will review their maps from the winter research project, identify different plants that can be added to their food forests, and begin adding perennial plants to the garden.

Objective: Students will engage with the principle of *Using and Valuing Diversity* by looking at how different plants within an ecosystem can support each other through beneficial relationships.

Activity 3

Introduction (10 minutes)

1. Introduce the class to the permaculture design principle *Use and Value Diversity* as described in David Holmgren's *Permaculture Principles* (see Resources). What images does this term evoke when students think about it in the garden? What could it mean outside of the garden—in their classrooms, homes, and communities?

2. What does diversity look like within a natural space, and how do diverse members of an ecosystem interact? What role does diversity have in the current concerns and issues raised by global climate change? What role does competition have in ecosystems and in conversations about diversity?

3. Discuss with the class how plant and species diversity impacts the garden, even on a small scale. Students should understand that *diversity* is not necessarily about the number of elements in a garden system, but the number of

> "Diversity provides alternative pathways for essential ecosystem functions in the face of changing conditions."[4]

beneficial relationships and connections that can exist between those elements.

4. Introduce the students to their activity and the expectations around it. The class will be using available resources to research and design a potential garden ecosystem with the beds they have been mapping. The goal will be to focus on promoting stability through diversity while also keeping in mind the layers of a food forest. What does responsible and realistic designing look like during this activity?

Activity (30 minutes)

1. Individually or with their partners, the students will research different plant species that support biodiversity, stability, and resilience to add to their garden bed or container. They should keep in mind the layers of a food forest and what elements are missing from their site, as well as the lessons they learned when investigating local ecosystems during previous activities. Students can use available resources to develop thorough, realistic, and well-researched ideas. They can also use the plant lists they developed in previous lessons to do even more research on which plants would contribute most to the ecosystem.

2. The students can illustrate these findings and designs on their maps, being sure to indicate the plants that exist already in the bed and the hypothetical plants that could thrive there. Additionally, the students can list these suggested new plants with sentences justifying or arguing for their addition.

3. Encourage the class to research plants that exist in other parts of the garden or local natural spaces, to increase the possibility of gathering and adding these plants to the bed or container. The class should know the local climate zone and plant requirements to help them focus their research on realistic varieties.

4. This is a chance for students to experiment with plant groupings and space, research polycultures and companion plants, as well as challenging popular paradigms about garden order and patterns.

Assessment (5 minutes)

1. What exciting possibilities did students discover during their research? What plants do they recommend adding to the garden? What are the possibilities and concerns about acquiring these plants?

2. What layers of the forest garden are missing in student's garden spaces, and what plants did they discover to fill these niches? What perennial plants would students recommend acquiring and why?

Preparation

Plant identification books and online resources

Activity 4

Overview: The students will transplant the plants they cultivated during the plant propagation activity in the winter.

Objective: For students to increase the food production of the garden beds they mapped and apply their research on plant spacing, relationships, and ecosystem-mimesis into the garden.

Introduction (10 minutes)

1. Have the students recall previous lessons on plant propagation. What plants did they propagate from cuttings, rhizomes/roots, and rooted clones? What plants survived to springtime and are thriving? What roles do these plants play in the food forest layers? Which garden beds or containers would benefit from these plants?

2. Introduce the planting activity to the students and the tools that are available. What does careful gardening look like during this activity, and how can students respect the work they have done in designing these spaces?

Activity (30 minutes)

1. In their mapping groups, the students will plant varieties in their garden bed or container to build the food forest design they created. They should follow their maps and research to plant the species they propagated in reasonable quantity.

2. In addition to the propagated plants, supplementary plants and seeds can be provided to support this layering and to increase food production. The groups can deeply water their transplants afterward.

3. Students with extra activity time can create plant markers or simple signs detailing the purpose of the layered plants and their relationships to each other.

4. When the activity time is over, the students can clean up their areas, return the tools, and gather a taste test from the garden.

Assessment (5 minutes)

1. How well did students do in following their designs, maps, and research? What changes did they have to make as they accomplished the activity?

2. What additional plants could the garden beds use? How will these plants grow and interact with each other?

NGSS and Activity Extensions

The extensions for this principle go far beyond the garden into social and political studies and conversations. Within ecologically minded gardens, diversity is considered a cornerstone of stability and resiliency for plant, animal, and insect species. This lesson can be extended into research on other examples of ecosystem resiliency, global climate change, and industrial and commercial monocultures, as well as social connections and political perspectives on human diversity. Students could also develop and compare different design solutions for "maintaining biodiversity and ecosystem services" (MS-LS2-5).

Resources

- "Companion Planting." Deep Green Permaculture (online—see Appendix B).
- "Diversity." Deep Green Permaculture (online—see Appendix B).
- Toby Hemenway. *Gaia's Garden.*
- Paul Alfrey. "How Much Food Can You Grow in a Polyculture?" (online—see Appendix B).
- "Designing Polycultures for a Garden Setting." Learn, Garden & Reflect with Cornell Garden-Based Learning. Cornell University (online—see Appendix B).
- Wayne Weiseman, et al. *Integrated Forest Gardening.*

5-8 Self-Directed Projects

Time Frame: 45 minutes for each class period
Overview: The class will develop a self-directed project, or multiple ones, that support and extend the spring lessons they have experienced.
Objective: Students will actively support increased biodiversity and the principle of *Using and Valuing Diversity* in the garden.

Introduction (10 minutes)

1. During the next four lessons, the students will design and implement a self-directed project that will support the community and exemplify the values of *Using and Valuing Diversity.* These projects can be focused in the garden and extend from previous lessons, or promote these values socially in the larger community.

2. Students can be encouraged to work individually or in partners and should be prepared to begin each work day with a detailed plan of action.

Activity (30 minutes)

1. The class will work to support the garden values and actively pursue their interests during each class period. They should support one another and respect their own work by staying focused, setting small and manageable goals, and communicating as a team.

2. Seasonal garden work could include: planting perennial and biennial plants, increasing food production for struggling garden beds or containers, trellising vine plants, maintenance of garden infrastructure, creating habitat boxes for local species.

Assessment (5 minutes)

1. In what ways did students support or enhance the lessons they have learned in the garden?

2. How will their work benefit the garden and community after the students have graduated?

9 Leadership: Spring Foraging and Foods

Time Frame: 45 minutes for older students
30 minutes for the younger class

Overview: The eighth graders will lead a younger class through plant identification and taste testing new foods.

Objective: For older students to pass on garden cultural values and knowledge by teaching a younger generation.

Introduction (10 minutes)

1. Before the younger class joins the group, review leadership and behavior expectations with the students. What leaderships skills and lessons has the class learned through working with younger students? What responsibilities do they have as leaders during this activity?

2. The class will be leading the younger students though foraging, plant identification, and taste test experiences in the garden. What seasonal plants are ready to be harvested? What stories and lessons can the older students teach about them? What values do the students want to pass on about taste testing and trying new foods? How will they pass on these messages?

3. When the younger class arrives, they can review the group behavior expectations around care for self, others, and the land and be partnered with an older student.

Activities (25 minutes)

1. With their partners, the students will embark on a taste test tour in the garden. The older students will help the younger ones identify certain edible spring foods. The groups can gather all the plants first and then wash and plate them for an in-depth taste testing experience. This activity will be similar to the fall nutrition activity, but should be focused on spring seasonal foods.

2. The older students will lead their partners in a discussion about garden food flavors and pass on the values of the garden and the lessons they have learned. They can also quiz their partners on the names of plants and how to harvest them.

Assessment (10 minutes)

1. When the younger students have left, gather the older class together and reflect on their experiences. Did they succeed at passing on the lessons and experiences they have had in the garden? What challenges did they face, and how did they overcome them?

2. How have the students grown and developed as leaders since the beginning of the school year?

10 Garden Skills: Legacy

Time Frame: 45 minutes
Overview: The class will create a legacy project to go into the garden or community and celebrate their work.
Objective: To acknowledge student learning and work over their years gardening and supporting their community and to consider possible next steps in their learning.

Introduction (10 minutes)

1. This is the last class the students will have in the garden space. It is a time to celebrate their learning and accomplishments and to reflect upon their experiences and legacy. Future classes may not see all the work this class has accomplished, but their hands have touched all parts of it.
2. What is the most memorable part of gardening for the students? How do students want their legacy to be remembered in the garden or in their community?

> This legacy project can be planned before this activity and, if possible, determined through class consensus. This activity is just a simple way for students to leave behind a visual imprint of themselves, but there are many other possibilities that can be tailored to fit into the needs and resources available to the community. For example, students could install a new design feature in the garden or plant a new wave of perennial plants.

Activity (25 minutes)

1. For this activity, the students will be marking their handprints within the garden space using paint and markers. The handprints should be placed in a visible area so that they can be covered with sealant and preserved. I have had students put their handprints on garden fence posts or underneath the area where they built habitat boxes—or even on their habitat boxes. Garden doorways or designated sign boards are other options.
2. Taking turns, the students will gently lay their hands in paint and imprint them onto a space in the garden. Afterward, they can wash their hands and enjoy a taste test from the garden while the paint dries. When it is dried, they can write their names with permanent marker at the center of the hand/hands. Then they, or an adult, can cover the hands with an outdoor sealant.
3. When the students have finished leaving their marks, they can help clean up the paint station and gather back together to enjoy their final taste test and to reflect upon their gardening experiences.

Assessment (10 minutes)

1. What were some of the challenges and successes they experienced in the garden over the years? What do they think future middle schoolers should know about gardens and gardening class? What kinds of activities should future middle schoolers do?
2. What are the next steps students will take in studying gardening, beyond this garden? What do they want to know more about?

Preparation

1. Fast-drying paint
2. Permanent markers
3. Sealant

Endnotes

Introduction

1. Bill Mollison. *Introduction to Permaculture*, 2nd ed. Tagari, 1994, p. 1.
2. David Holmgren. *Permaculture: Principles & Pathways Beyond Sustainability*. Permanent Publications, 2011, p. 16.
3. Oregon State University Extension Service. *Oregon Environmental Literacy Program*. [online]. [cited Spetember 13, 2018]. oelp.oregonstate.edu.
4. Rachel Carson. *The Sense of Wonder: A Celebration of Nature for Parents and Children*. Harper Perennial, 2017, p. 44.
5. Holmgren. *Permaculture*, p. 22.
6. Ibid., p. 217.
7. Carson. *The Sense of Wonder*, p. 57.
8. Holmgren. *Permaculture*, p. 18.
9. Ibid., p. 14.
10. Ibid., p. 223.
11. Heide Hermary. *Working with Nature: Shifting Paradigms: The Science and Practice of Organic Horticulture*. Gaia College, 2007, p. 35.
12. Ibid, p. 202.
13. Holmgren. *Permaculture*, p. 61.
14. Ibid., p. 8.

Third Grade—Grade Three

1. Heide Hermary. *Working with Nature: Shifting Paradigms: The Science and Practice of Organic Horticulture*. Gaia College, 2007, p. 45.
2. Patrick Lima. *The Natural Food Garden: Growing Vegetables and Fruits Chemical-free*. Prima, 1992, p. 7.
3. Hermary. *Working with Nature*, p. 45.
4. Ibid., p. 35.
5. Lima. *The Natural Food Garden*, p. 10.
6. Ibid., p. 9.
7. David Holmgren. *Permaculture: Principles & Pathways Beyond Sustainability*. Holmgren Design Services, 2002, p. 15.
8. Hermary. *Working with Nature*, p. 106.

Fourth Grade—Grade Four

1. Hermary. *Working with Nature*, p. 11.
2. Lima. *The Natural Food Garden*, p. 14.

Fifth Grade—Grade Five

1. Holmgren. *Permaculture: Principles & Pathways*, p. 111.
2. Ibid., p. 122.
3. Lima. *The Natural Food Garden*, p. 10.

Sixth Grade—Grade Six

1. Hermary. *Working with Nature*, p. 47.
2. Holmgren. *Permaculture: Principles & Pathways*, p. 138.
3. Ibid., p. 127.
4. Ibid., p. 107.

Seventh-Eighth Grade Preface

1. Holmgren. *Permaculture: Principles & Pathways*.
2. Ibid., p. 83.

Seventh Grade—Grade Seven

1. Holmgren. *Permaculture: Principles & Pathways*, p. 29.
2. Ibid., p. 42.

3. Ibid., p. 27.
4. Ibid., p. 35.
5. Ibid., p. 94.

Eighth Grade — Grade Eight

1. Holmgren. *Permaculture: Principles & Pathways.*
2. Ibid., p. 83.
3. Ibid., p. 9.
3. Ibid., p. 205.

School Garden Curriculum Worksheets
are available for free download at:
https://tinyurl.com/SGC-Worksheets

Appendix A:
Next Generation Science Standards

I chose to include NGSS into this curriculum in an effort to make it more accessible to educators who use these standards. Personally, they have provided a structure to many of my lesson plans, and I hope they will also make gardening education more realistic to other educators, especially those who are teaching in a standards-driven system. My hope is that educators who have to fulfill rigorous state or regional standards won't feel as if gardening is an add-on to their already tight schedules, but rather offers supplemental, supportive, and fulfilling science experiences for their students.

Kindergarten

K-PS3-1 Energy: Make observations to determine the effect of sunlight on the Earth's surface.

K-LS1-1: From Molecules to Organisms: Structures and Processes: Use observations to describe patterns of what plants and animals (including humans) need to survive.

K-ESS2-1: Earth's Systems: Use and share observations of local weather conditions to describe patterns over time.

K-ESS2-2: Earth's Systems: Construct an argument supported by evidence for how plants and animals (including humans) can change the environment to meet their needs.

K-ESS3-1: Earth and Human Activity: Use a model to represent the relationship between the needs of different plants or animals (including humans) and the places they live.

K-ESS3-3: Earth and Human Activity: Communicate solutions that will reduce the impact of humans on the land, water, air, and/or other living things in the local environment.

First Grade—Grade One

1-LS1-1: From Molecules to Organisms: Use materials to design a solution to a human problem by mimicking how plants and/or animals use their external parts to help them survive, grow, and meet their needs.

1-LS1-2: From Molecules to Organisms: Read texts and use media to determine patterns in behavior of parents and offspri ng that help offspring survive.

1-LS3-1: Heredity: Make observations to construct an evidence-based account that young plants and animals are like, but not exactly like, their parents.

Second Grade—Grade Two

2-PS1-1: Matter and Its Interactions: Plan and conduct an investigation to describe and classify different kinds of materials by their observable properties.

2-PS1-2: Matter and Its Interactions: Analyze data obtained from testing different materials to determine which materials have the properties that are best suited for an intended purpose.

2-LS2-1: Ecosystems: Plan and conduct an investigation to determine if plants need sunlight and water to grow.

2-LS2-2: Ecosystems: Develop a simple model that mimics the function of an animal in dispersing seeds or pollinating plants.

2-LS4-1: Make observations of plants and animals to compare the diversity of life in different habitats.

Third Grade—Grade Three

3-LS3-1: Heredity: Analyze and interpret data to provide evidence that plants and animals have traits inherited from parents and that variation of these traits exists in a group of similar organisms.

3-LS3-2: Heredity: Use evidence to support the explanation that traits can be influenced by the environment.

3-ESS3-1: Earth and Human Activity: Make a claim about the merit of a design solution that reduces the impacts of a weather-related hazard.

3-LS2-1: Ecosystems: Construct an argument that some animals form groups that help members survive.

3-LS1-1: From Molecules to Organisms: Develop models to describe that organisms have unique and diverse life cycles but all have in common birth, growth, reproduction, and death.

Fourth Grade—Grade Four

4-LS1-1: From Molecules to Organisms: Construct an argument that plants and animals have internal and external structures that function to support survival, growth, behavior, and reproduction.

4-LS1-2: From Molecules to Organisms: Use a model to describe that animals receive different types of information through their senses, process the information in their brain, and respond to the information in different ways.

Fifth Grade—Grade Five

5-PS1-1: Matter and Its Interactions: Develop a model to describe that matter is made of particles too small to be seen.

5-PS1-2: Matter and Its Interactions: Measure and graph quantities to provide evidence that regardless of the type of change that occurs when heating, cooling, or mixing substances, the total weight of matter is conserved.

5-PS3-1: Energy: Use models to describe that energy in animals' food (used for body repair, growth, motion, and to maintain body warmth) was once energy from the sun.

5-LS1-1: From Molecules to Organisms: Support an argument that plants get the materials they need for growth chiefly from air and water.

5-LS2-1: Ecosystems: Develop a model to describe the movement of matter among plants, animals, decomposers, and the environment.

Middle School: Grade Six to Grade Eight

MS-LS1-5: From Molecules to Organisms: Construct a scientific explanation based on evidence for how environmental and genetic factors influence the growth of organism.

MS-ESS2-1: Earth's Systems: Develop a model to describe the cycling of Earth's materials and the flow of energy that drives this process.

MS-ESS2-2: Earth's Systems: Construct an explanation based on evidence for how geoscience processes have changed Earth's surface at varying time and spatial scales.

MS-ESS2-4: Earth Systems: Develop a model to describe the cycling of water through Earth's systems driven by energy from the sun and the force of gravity.

MS-ETS1-1: Define the criteria and constraints of a design problem with sufficient precision to ensure a successful solution, taking into account relevant scientific principles and potential impacts on people and the natural environment that may limit possible solutions.

MS-ETS1-2: Evaluate competing design solutions using a systematic process to determine how well they meet the criteria and constraints of the problem.

MS-ETS1-3: Analyze data from tests to determine similarities and differences among several design solutions to identify the best characteristics of each that can be combined into a new solution to better meet the criteria for success.

MS-ETS1-4: Develop a model to generate data for iterative testing and modification of a proposed object, tool, or process such that an optimal design can be achieved.

Seven/Eight Grade — Grade Seven/Eight

MS-LS1-6: From Molecules to Organisms, Structures and Processes: Construct a scientific explanation based on evidence for the role of photosynthesis in the cycling of matter and flow of energy into and out of organisms.

MS-LS2-3: Ecosystems, Interactions, Energy, and Dynamics: Develop a model to describe the cycling of matter and flow of energy among living and nonliving parts of an ecosystem.

MS-LS2-4: Ecosystems, Interactions, Energy, and Dynamics: Construct an argument supported by empirical evidence that changes to physical or biological components of an ecosystem affect populations.

MS-LS2-5: Ecosystems, Interactions, Energy, and Dynamics: Evaluate competing design solutions for maintaining biodiversity and ecosystem services.

Appendix B: Resources

Agroecology and Permaculture Perspectives

"9. Diversity." Deep Green Permaculture. [online]. [cited October 14, 2018]. deepgreenpermaculture.com/permaculture/permaculture-design-principles/9-diversity.

"Carbon Sequestration in Soils." Ecological Society of America, Summer 2000. [online]. [cited October 17, 2018]. esa.org/esa/wp-content/uploads/2012/12/carbonsequestrationinsoils.pdf.

Commission for Environmental Cooperation. "North American Environmental Atlas." [online]. [cited October 11, 2018]. cec.org/tools-and-resources/north-american-environmental-atlas/map-files.

"Companion Planting." Deep Green Permaculture. [online]. [cited October 14, 2018]. deepgreenpermaculture.com/companion-planting.

Deep Green Permaculture. "7. Small Scale Intensive Systems." [online]. [cited October 11, 2018]. deepgreenpermaculture.com/permaculture/permaculture-design-principles/7-small-scale-intensive-systems.

"Designing a Forest Garden: The Seven-Layer Garden." Chelsea Green Publishing, May 21, 2014. [online]. [cited October 11, 2018]. chelseagreen.com/blogs/designing-a-forest-garden-the-seven-story-garden.

Engels, Jonathan. "Planting in Pots and Other Ways of Playing with Permaculture in the Big City." Permaculture Research Institute, April 10, 2015. [online]. [cited October 11, 2018]. permaculturenews.org/2015/04/10/planting-in-pots-and-other-ways-of-playing-with-permaculture-in-the-big-city.

Finley, Ron. "A Guerilla Gardener in South Central LA." TEDtalksDirector, March 6, 2013. [online]. [cited October 11, 2018]. youtube.com/watch?v=EzZzZ_qpZ4w&t=16s.

Gliessman, Stephen R. *Agroecology: The Ecology of Sustainable Food Systems*, Vol 1, 3rd ed. CRC Press, 2014.

Permacultura México. "Permaculture Design Certificate." Video produced by Bajio Community Foundation, directed by Martin Gonzalez Licano, 2017. [online]. [cited October 11, 2018]. youtu.be/o3JzVEyb38w?t=4m30s.

Sage Publications. "Compost Can Turn Agricultural Soils into a Carbon Sink, Thus Protecting Against Climate Change." *ScienceDaily*, February 27, 2008. [online]. [cited October 17, 2018]. sciencedaily.com/releases/2008/02/080225072624.htm.

US Environmental Protection Agency. "Ecoregions of North America." US EPA, November 12, 2016. [online]. [cited October 11, 2018]. epa.gov/eco-research/ecoregions-north-america.

US Environmental Protection Agency. "EnviroAtlas." US EPA, September 21, 2018. [online]. [cited October 11, 2018]. epa.gov/enviroatlas.

Weisman, Wayne, et al. *Integrated Forest Gardening: The Complete Guide to Polycultures and Plant Guilds in Permaculture Systems*. Chelsea Green, 2014.

Beyond Food Production

"All About Birdhouses." NestWatch, Cornell Lab of Ornithology, 2018. [online]. [cited October 14, 2018]. nestwatch.org/learn/all-about-birdhouses.

Bruton-Seal, Julie and Matthew Seal. *Backyard Medicine: Harvest and Make Your Own Herbal Remedies*. Skyhorse, 2009.

Caduto, Michael J. and Joseph Bruchac. *Native American Gardening: Stories, Projects, and Recipes for Families.* Fulcrum, 1996.
Crain, Rhiannon. "Habitat Connectivity in the Yard." Cornell Laboratory of Ornithology, 2015. [online]. [cited October 6, 2018]. content.yardmap.org/learn/habitat-connectivity-2.
Dumroese, R. Kasten, Tara Luna, and Thomas D. Landis, eds. *Nursery Manual for Native Plants: A Guide for Tribal Nurseries—Vol. 1*: Nursery Management.USDA Forest Service Agriculture Handbook 730, 2009, pp. 153–175.
"Educator's Guide to Nest Boxes." BirdSleuth K-12, Cornell Lab of Ornithology. [online]. [cited OPctober 14, 2018]. birdsleuth.org/an-educators-guide-to-nestboxes.
Gladstar, Rosemary. *Medicinal Herbs: A Beginner's Guide.* Storey, 2012.
Vesely, David and Gabe Tucker. "A Landowner's Guide for Restoring and Managing Oregon White Oak Habitats." USDI Bureau of Land Management, Salem District, Oregon Department of Forestry.

Books for Young Children

Aston, Dianna Hutts. *A Seed Is Sleepy.* Chronicle, 2014.
Carle, Eric. *The Tiny Seed.* Little Simon, 2018.
Carle, Eric. *The Very Busy Spider.* Grosset & Dunlap, 1985.
Carle, Eric. *The Very Hungry Caterpillar.* Longman, 2015.
Cherry, Lynne. *How Groundhog's Garden Grew.* Scholastic, 2003.
Elpel, Thomas J. *Botany in a Day: The Patterns Method of Plant Identification: An Herbal Field Guide To Plant Families of North America,* 6th ed. Hops Press, 2013.
Fleischman, Paul. *Seedfolks.* HarperTrophy, 2004.
Krauss, Ruth. *The Carrot Seed.* Harper Collins, 2004.
Graham, Margaret Bloy. *Be Nice to Spiders.* Collins, 1967.
Mallett, David. *Inch by Inch: The Garden Song.* Harper Colins, 1997.
Patent, Dorothy Hinshaw. *Plants on the Trails With Lewis & Clark.* Clarion, 2001.
Portman, Michelle Eva. *Compost, By Gosh!: An Adventure with Vermicomposting.* Flower Press, 2003.
Rabe, Tish. *My, Oh My—A Butterfly!: All About Butterflies.* Random House, 2007.
Richards, Jean. *A Fruit Is a Suitcase for Seeds.* Paw Prints, 2008.
Robbins, Ken. *Seeds.* Pearson/Scott Foresman, 2007.
Siddals, Mary McKenna. *Compost Stew: An A to Z Recipe for the Earth.* Dragonfly, 2014.
Swope, Sam. *Gotta Go! Gotta Go!* Scholastic, 2013.
Tierra, Lesley, et al. *A Kid's Herb Book.* Robert Reed, 2010.

Botanical Illustration

Botanical Artists of Canada. "History of Botanical Illustration." [online]. [cited October 17, 2018]. botanicalartistsofcanada.org/about/history-of-botanical-illustration.aspx.
Willis, Katie and Katie Scott. *Botanicum: Welcome to the Museum.* Big Picture Press, 2016.

Building a Garden

Alfrey, Paul. "How Much Food Can You Grow in a Polyculture?" *Permaculture Magazine*, December 27, 2014. [online]. [cited August 22, 2018]. permaculture.co.uk/articles/how-much-food-can-you-grow-polyculture.
Barth, Brian. "Double Digging: How to Build a Better Veggie Bed." *Modern Farmer*, March 7, 2016. [online]. [cited October 17, 2018]. modernfarmer.com/2016/03/double-digging.
"Designing Polycultures for a Garden Setting." Learn, Garden & Reflect with Cornell Garden-Based Learning, Cornell University. [online]. [cited September 26, 2018]. gardening.cals.cornell.edu/garden-guidance/design/#Designing%20Polycultures%20for%20a%20Garden%20Setting.
Editors of Cool Springs Press. *Trellises, Planters & Raised Beds: 50 Easy, Unique, and Useful Projects You Can Make with Common Tools and Materials.* Cool Springs Press, 2013.
Editors of Fine Gardening. *Container Gardening: 250 Design Ideas & Step-by-Step Techniques.* Taunton, 2009.
"Garden Trellis Ideas." Seed Savers Exchange.

[online]. [cited October 18, 2018]. seedsavers.org/garden-trellis-ideas.

Israel, Sarah. "An In-Depth Companion Planting Guide." *Mother Earth News*, May/June 1981. [online]. [cited August 16, 2018]. motherearthnews.com/organic-gardening/companion-planting-guide-zmaz81mjzraw.

Kloos, Scott. *Pacific Northwest Medicinal Plants: Identify, Harvest, and Use 120 Wild Herbs for Health and Wellness*. Timber, 2017.

Langelloto-Rhodaback, Gail A. *Growing Your Own*. Oregon State University Extension Publication #EM9027, April. 2011. [online]. [cited October 3, 2018]. catalog.extension.oregonstate.edu/em9027.

Lima, Patrick. *The Natural Food Garden: Growing Vegetables and Fruits Chemical-free*. Prima, 1992.

Link, Russell. *Landscaping for Wildlife in the Pacific Northwest*. University of Washington, 1999.

Pojar, Jim and Andy MacKinnon. *Plants of the Pacific Northwest Coast: Washington, Oregon, British Columbia & Alaska*. Partners Publishing, 2016.

Riotte, Louise. *Carrots Love Tomatoes: Secrets of Companion Planting for Successful Gardening*. Storey, 1998.

Seattle Tilth Alliance. "What Is a Maritime Climate?" in *Maritime Garden Guide, 2012*.

Composting

"All About Organic Mulch." Bonnie Plants Organics. [online]. [cited August 21, 2018]. bonnieorganics.com/all-about-organic-mulch.

"Approximate Time It Takes for Garbage to Decompose in the Environment." New Hampshire Department of Environmental Services. [online]. [cited August 23, 2018]. des.nh.gov/organization/divisions/water/wmb/coastal/trash/documents/marine_debris.pdf.

Bell, Neil at al. "Improving Garden Soils with Organic Matter." Oregon State Extension Service, May 2003. [online]. [cited August 29, 2018]. catalog.extension.oregonstate.edu/ec1561.

Chaskey, Scott. "Behold This Compost!" in *This Common Ground: Seasons on an Organic Farm*. Viking, 2005.

Composting for the Homeowner. "The Science of Composting." University of Illinois Extension, 2018. [online]. [cited October 4, 2018]. web.extension.illinois.edu/homecompost/science.cfm.

Dawling, Pam. "Screen Compost Now to Make Your Own Seed Compost for Spring." *Mother Earth News*, July 5, 2016 [online]. [cited October 17, 2018]. motherearthnews.com/organic-gardening/screen-compost-zbcz1607.

"Do the Rot Thing: A Teacher's Guide to Compost Activities." Central Vermont Solid Waste Management District, January 2007. [online]. [cited August 29, 2018]. cvswmd.org/uploads/6/1/2/6/6126179/do_the_rot_thing_cvswmd1.pdf.

Healthy Youth Institute. "Compost Bin Identification." Oregon State University, Linus Pauling Institute. [online]. [cited August 28, 2018]. lpi.oregonstate.edu/sites/lpi.oregonstate.edu/files/pdf/hyp/lessons-manuals/K12/K5/grade_two_compost_bin_exploration.pdf.

Hermary, Heide. "Working With Soil: Compost" in *Working with Nature: Shifting Paradigms: The Science and Practice of Organic Horticulture*. Gaia College, 2007. gaiacollege.ca/working-with-nature-shifting-paradigms.html.

How to Compost homepage. Howtocompost.org, 2013. [online]. [cited August 28, 2018]. howtocompost.org.

Lima, Patrick. "Sifting Compost" in *The Natural Food Garden: Growing Vegetables and Fruits Chemical-free*. Prima, 1992.

Lowenfels, Jeff, and Wayne Lewis. *Teaming with Microbes: The Organic Gardener's Guide to the Soil Food Web*, rev. ed. Timber Press, 2010.

Martin, Deborah L., ed. "The Benefits of Compost" in *The Rodale Book of Composting*, new rev. ed. Rodale, 1992.

Philbrick, John and Helen Philbrick. "The Uses of Ripe Compost" in *Gardening for Health and Nutrition; An Introduction to the Method of Bio-dynamic Gardening*. Steiner Publications, 1971.

"The Science of Composting." University of Illinois Extension—Composting for the Homeowner. [online]. [cited August 1, 2018]. web.extension.illinois.edu/homecompost/science.cfm.

Siddals, Mary McKenna. *Compost Stew: An A to Z Recipe for the Earth*. Dragonfly, 2014.

Squire, David. *The Compost Specialist: The Essential Guide to Creating and Using Garden Compost, and Using Potting and Seed Composts*. New Holland, 2009.

Texas A&M Agrilife Extension. "Chapter 1, The Decomposition Process | Earth-Kind® Landscaping." Earth-Kind®. [online]. [cited August 29, 2018]. aggie-horticulture.tamu.edu/earthkind/landscape/dont-bag-it/chapter-1-the-decomposition-process.

"Trash Timeline: Exploring the Biodegradability of Trash." Alice Ferguson Foundation. [online]. [cited August 23, 2018]. fergusonfoundation.org/teacher_resources/trash_timeline_lesson_plan_10-06.pdf.

Trautmann, Nancy and Elaina Olynciw. "Compost Microorganisms." Cornell Waste Mangement Institute, 1996. [online]. [cited August 30, 2018]. compost.css.cornell.edu/microorg.html.

Ulsh, Christine Zieglar and Paul Hepperly. "Good Compost Made Better." Rodale Institute, April 13 2006. [online]. [cited August 16, 2018]. newfarm.rodaleinstitute.org/depts/NFfield_trials/2006/0413/compost.shtml.

US Environmental Protection Agency. "Composting at Home." [online]. [cited August 28, 2018]. epa.gov/recycle/composting-home.

Wheeler, Philip A. and Ronald B. Ward. *The Non-Toxic Farming Handbook*. Acres USA, 1998.

"Worm Composting." Oregon Metro, Yard and Garden. [online]. [cited August 21, 2018]. oregonmetro.gov/tools-living/yard-and-garden/composting/worm-composting.

Food Preservation and Sovereignty

"Food Sovereignty." Navdanya. [online]. [cited October 17, 2018]. navdanya.org/earth-democracy/food-sovereignty.

Slow Food Foundation for Biodiversity. "Ark of Taste." [online]. [cited October 17, 2018]. fondazioneslowfood.com/en/what-we-do/the-ark-of-taste.

Katz, Sandor Ellix. *The Art of Fermentation: An In-Depth Exploration of Essential Concepts and Processes from Around the World*. Chelsea Green, 2012.

National Center for Home Food Preservation. "Preparing and Canning Fermented and Pickled Foods." NCHFP Publications, August 2018. [online]. [cited October 10, 2018]. nchfp.uga.edu/how/can_06/prep_foods.html.

Topp, Ellie and Margaret Howard. *The Complete Book of Small-Batch Perserving: Over 300 Recipes to Use Year-Round*, 2nd rev. ed. Firefly, 2007.

Vinton, Sherri Brooks. *Put 'em Up!: A Comprehensive Home Preserving Guide for the Creative Cook, from Drying and Freezing to Canning and Pickling*. Storey, 2010.

Insects, Bugs, and Decomposers

American Museum of Natural History. "How Do Spiders Hunt?" [online]. [cited August 8, 2018]. amnh.org/explore/news-blogs/on-exhibit-posts/how-do-spiders-hunt.

Arnett, Ross H. and Richard L. Jacques. *Simon & Schuster's Guide to Insects*. Simon & Schuster, 1994.

BBC Earth. "Spider With Three Super Powers." from "The Hunt," July 2, 2017. [online]. [cited August 8, 2018]. youtube.com/watch?v=UDtlvZGmHYk.

The Bug Chicks. "Order Mantodea, Mantid." [online]. [cited August 8, 2018]. vimeo.com/39801755.

Cranshaw, Whitney and David Shetlar. *Garden Insects of North America: The Ultimate Guide to Backyard Bugs*, 2nd ed. Princeton, 2017.

Eaton, Eric R. and Kenn Kaufman. *Kaufman Field Guide to Insects of North America*. Houghton Mifflin Harcourt, 2007.

Evans, Arthur V. *National Wildlife Federation Field Guide to Insects and Spiders & Related Species of North America*. Sterling, 2007.

"Farming for Pest Management." Xerces Society of Invertebrate Conservation. [online]. [cited October 4, 2018]. xerces.org/wp-content/uploads/2008/09/farming_for_pest_management_brochure_compressed.pdf.

Go Science Girls. "Mushroom Spore Prints." Go Science Girls website, June 28, 2017. [online]. [cited August 7, 2018]. gosciencegirls.com/mushroom-spore-prints.

Imes, Rick. *The Practical Entomologist*. Fireside, 1992.

Kidzone. "The Body of a Spider." Spider Facts.

[online]. [cited August 8, 2018]. kidzone.ws/lw/spiders/facts02.htm.

Loomis, J. and H. Stone. "Lady Beetle (Hippodamia convergens)." Oregon State University Extension Service, Doc. #EC 1604, April 2007. [online]. [cited September 23, 2018]. catalog.extension.oregonstate.edu/sites/catalog/files/project/pdf/ec1604.pdf.

Mason, Robert. "Snakes Slither Through the Garden." Oregon State University Extension Service. [online]. [cited August 1, 2018]. extension.oregonstate.edu/node/80916.

National Geographic. "Praying Mantis." [online]. [cited August 8, 2018]. nationalgeographic.com/animals/invertebrates/p/praying-mantis.

Spider ID. "Argiope Aurantia (Black and Yellow Garden Spider)." [online]. [cited August 8, 2018]. spiderid.com/spider/araneidae/argiope/aurantia.

Spider ID. "Spiders in Oregon." [online]. [cited August 8, 2018]. spiderid.com/locations/united-states/oregon.

Vlach, Joshua. "Slugs and Snails in Oregon: A Guide to Common Land Molluscs and Their Relatives." Oregon Department of Agriculture, 2016. [online]. [cited August 1, 2018]. oregon.gov/ODA/shared/Documents/Publications/IPPM/ODAGuideMolluscs2016ForWeb.pdf.

Lesson Plans/Curricula/NGSS

Eat.Think.Grow: Lessons for the School Garden. [online]. [cited August 4, 2018]. eatthinkgrow.org.

Guerraro, Amoreena. "Increasing Inclusion in the School Garden: A Resource Packet for Garden Educators." School Garden Project of Lane County. [online]. [cited October 2, 2018]. schoolgardenproject.org/download/increasing-inclusion.

Hicks, Steven. "Seed Sensation: Exploring and Sorting Seeds." Scholastic. [online]. [cited July 31, 2018]. scholastic.com/teachers/lesson-plans/teaching-content/seed-sensation-exploring-and-sorting-seeds.

Laws, John Muir et al. Opening the World Through Nature Journaling: Integrating Art, Sciences, and Language Arts, 2nd ed. California Native Plant Society. [online]. [cited August 7, 2018]. store.cnps.org/products/opening-the-world-through-nature-journaling-cnps-curriculum.

Life Lab. [online]. [cited October 2, 2018]. lifelab.org.

Lovejoy, Sharon. *Sunflower Houses.* Workman, 2001.

Lovejoy, Sharon. *Roots, Shoots, Buckets & Boots: Gardening Together with Children.* Workman, 1999.

Oregon Agriculture in the Classroom Foundation. [online]. [cited October 2, 2018]. oregonaitc.org.

Philips, Sarah and Jules Montes. "Grow Your Own Assessments." Sauvie Island Center. Oregon Farm to School & School Garden Summit. [online]. [cited October 24, 2018]. docs.wixstatic.com/ugd/274b86_89d4409005894ed887a6146bae74fbb9.pdf.

Project Learning Tree. *PreK-8 Environmental Education Activity Guide,* 11th ed. American Forest Foundation, 2005.

School Garden Project of Lane County. [online]. [cited October 2, 2018]. schoolgardenproject.org/resources/curriculum.

Team Nutrition. Dig-In! Standards-Based Nutrition Education from the Ground Up. USDA Food and Nutrition Service, April 2013. [online]. [cited October 2, 2018]. fns.usda.gov/tn/dig-standards-based-nutrition-education-ground.

University of Nebraska Lincoln. "Images of Plants and Maps." Journals of the Lewis and Clark Expedition. [online]. cited September 23, 2018]. lewisandclarkjournals.unl.edu/images/plants_animals.

Williams, Dilafruz R. et al. "Science in the Learning Gardens (SciLG): A Study of Students' Motivation, Achievement, and Science Identity in Low-income Middle Schools." *International Journal of STEM Education* Vol. 5#8 (2018). [online]. [cited October 2, 2018]. stemeducationjournal.springeropen.com/track/pdf/10.1186/s40594-018-0104-9.

Permaculture Perspectives and Ecological Gardening

Berry, Wendell. *Our Only World: Ten Essays.* Counterpoint, 2016.

Bowers, Chet A. and David J. Flinders. *Responsive Teaching: An Ecological Approach to Classroom Patterns of Language, Culture, and Thought.* Teachers College Press, 1990.

Carson, Rachel. *The Sense of Wonder: A Celebration of Nature for Parents and Children.* Harper Perennial, 2017.

Coleman, Eliot. *Four-Season Harvest: Organic Vegetables from Your Home Garden All Year Long,* 2nd ed. Chelsea Green, 1999.

Coleman, Eliot. *The New Organic Grower: A Master's Manual of Tools and Techniques for the Home and Market Gardener,* rev. and expanded ed. Chelsea Green, 1995.

Facciola, Stephen. *Cornucopia II: A Source Book of Edible Plants,* 2nd. ed. Kampong Publications, 1998.

Flores, H.C. *Food Not Lawns.* Chelsea Green, 2006.

Frey, Darrell. *Bioshelter Market Garden: A Permaculture Farm.* New Society, 2011.

Fukuoka, Mansanobu. *The One Straw Revolution: An Introduction to Natural Farming.* NYRB Classics, 2009.

Gilekson, Linda. *Backyard Bounty: The Complete Guide to Year-Round Organic Gardening in the Pacific Northwest,* revised and expanded 2nd ed. New Society, 2018.

Hemenway, Toby. *Gaia's Garden: A Guide to Homescale Permaculture,* 2nd ed. Chelsea Green, 2009.

Hemenway, Toby. "Zones and Sectors." Denver Permaculture Guild Permaculture Design Course. [online]. [cited October 24, 2018]. denverpdc.files.wordpress.com/2014/10/14-11-zone-and-sector.pdf.

Holmgren, David. *Permaculture: Principles & Pathways Beyond Sustainability.* Holmgren Design Services, 2002.

Jeavons, John and Carol Cox. *The Sustainable Vegetable Garden: A Backyard Guide to Healthy Soil and Higher Yields.* Ten Speed Press, 1999.

Kneen, Brewster. "Progress It Is Not" in *Farmageddon: Food and the Culture of Biotechnology.* New Society, 1999.

Louv, Richard. *The Nature Principle: Reconnecting with Life in a Virtual Age.* Algonquin, 2012.

Meadows, Donella H. and Diana Wright. *Thinking in Systems: A Primer.* Chelsea Green Publishing, 2008.

Mollison, Bill. *Permaculture: A Designer's Manual.* Tagari, 1988.

Mollison, Bill. *Introduction to Permaculture,* 2nd ed. Tagari, 1994.

Nabhan, Gary Paul. *Renewing America's Food Traditions: Saving and Savoring the Continent's Most Endangered Foods.* Chelsea Green, 2008.

Oregon State University Extension Service. Oregon Environmental Literacy Program. [online]. [cited Spetember 13, 2018]. oelp.oregonstate.edu.

Orion, Tao. *Beyond the War on Invasive Species: A Permaculture Approach to Ecosystem Restoration.* Chelsea Green, 2015.

Proctor, Peter and Gillian Cole. "Introduction" in *Grasp the Nettle: Making Biodynamic Farming and Gardening Work.* Steiner Books, 1997.

Shepard, Mark. *Restoration Agriculture.* Acres USA, 2013.

Shiva, Vandana. "Biodiversity: A Third World Perspective" in *Monocultures of the Mind: Perspectives on Biodiversity and Biotechnology.* Zed, 1993.

Sobel, David. *Childhood and Nature: Design Principles for Educators.* Stenhouse, 2008.

Sobel, David. *Beyond Ecophobia: Reclaiming the Heart in Nature Education.* Orion Society, 1999.

Stone, Michael K. and Zenobia Barlow, eds. *Ecological Literacy: Educating Our Children for a Sustainable World.* Sierra Club Books, 2005.

Thomashow, Mitchell. *Ecological Identity: Becoming a Reflective Environmentalist.* MIT, 1996.

Wagner, Tony. *Creating Innovators: The Making of Young People Who Will Change the World.* Scribner, 2012.

Weaver, William Woys. *Heirloom Vegetable Gardening: A Master Gardener's Guide to Planting, Growing, Seed Saving, and Cultural History,* rev ed. Voyageur Press, 2018.

White-Kaulaity, Marlinda. "The Voices of Power and the Power of Voices: Teaching with Native American Literature." *The ALAN Review,* Vol. 34#1 (Fall 2006). [online]. [cited August 16, 2018]. scholar.lib.vt.edu/ejournals/ALAN/v34n1/kaulaity.pdf.

Jon Young, et al. *Coyote's Guide to Connecting with Nature*, 2nd ed. Owlink Media, 2010.

Plant Propagation

Hill, Lewis. *Secrets of Plant Propagation: Starting Your Own Flowers, Vegetables, Fruits, Berries, Shrubs, Trees, and Houseplants*. Storey, 1985.

Luna, Tara. "Vegetative Propagation" in *Nursery Manual for Native Plants: A Guide for Tribal Nurseries*. [online]. [cited October 11, 2018]. fs.fed.us/rm/pubs_other/wo_AgricHandbook730/wo_AgricHandbook727_153_175.pdf.

Toogood, Alan, ed. *American Horticultural Society Plant Propagation: The Fully Illustrated Plant-by-Plant Manual of Practical Techniques*. DK, 1999.

Pollinators

"12 Plants to Entice Pollinators to Your Garden." Oregon State University Extension Service. [online]. [cited July 31, 2018]. extension.oregonstate.edu/node/81056.

Borren, Bob. "Building an Insect Hotel." Permaculture Research Institute, October 8. 2013. [online]. [cited August 13, 2018]. permaculturenews.org/2013/10/08/building-insect-hotel.

"The Bug Chicks: Butterflies & Moths." [online]. [cited August 1, 2018]. https://vimeo.com/86540285.

Conrad, Ross. "Working With the Hive" in *Natural Beekeeping: Organic Approaches to Modern Apiculture*. Chelsea Green, 2013.

Everyday Mysteries. "How Can You Tell the Difference Between a Butterfly and a Moth?" Library of Congress. [online]. [cited August 1, 2018]. loc.gov/rr/scitech/mysteries/butterflymoth.html.

"Farming for Pollinators." The Xerces Society of Invertebrate Conservation. [online]. [cited October 2, 2018]. xerces.org/wp-content/uploads/2008/09/farming_for_pollinators_brochure.pdf.

Fisher, Rose-Lynn. *Bee*. Princeton Architectural Press, 2012.

The Free Library, Science Service 1986. "Bee Navigation: The Eyes Have It." *Science News* Vol. 130#14 (1986), p. 214 [online]. [cited August 10, 2018]. thefreelibrary.com/Bee+navigation%3a+the+eyes+have+it.-a04460303.

"Gardening for Pollinators." USDA Forest Service. [online]. [cited August 1, 2018]. fs.fed.us/wildflowers/pollinators/gardening.shtml.

Girl Next Door Honey. Educational Bee Posters. [online]. [cited August 10, 2018]. girlnextdoorhoney.com/product/educational-posters.

Harrap, Michael J. M. et al. "The Diversity of Floral Temperature Patterns, and Their Use by Pollinators." ELife Sciences Publications (December 19, 2017). [online]. [cited August 10, 2018]. elifesciences.org/articles/31262.

Havenhand, Gloria. *Honey: Nature's Golden Healer*. Firefly Books, 2011.

Honeybee Conservancy website. [online]. [cited August 13, 2018]. thehoneybeeconservancy.org.

Horton, Robin Plaskoff. "How to Design a Bug Hotel to Attract Beneficial Insects & Bees." *Urban Gardens*, February 29, 2016. [online]. [cited August 1, 2018]. urbangardensweb.com/2016/02/27/how-to-design-a-bug-hotel-to-attract-beneficial-insects-and-bees.

"Invertebrate Conservation Guidelines. Pollinator Conservation: Three Easy Steps to Help Bees and Butterflies." Xerces Society, 2011. [online]. [cited August 1, 2018]. xerces.org/wp-content/uploads/2010/11/pollinator-three-steps_fact_sheet2.pdf.

Kincaid, Sarah. Common Bee Pollinators of Oregon Crops. Oregon Department of Agriculture Guide, rev. ed. April 2017. [online]. [cited August 10, 2018]. oregon.gov/ODA/shared/Documents/Publications/IPPM/ODABeeGuide.pdf.

Mader, Eric, et al. *Managing Alternative Pollinators: A Handbook for Beekeepers, Growers, and Conservationists*. USDA Sustainable Agriculture Research and Education (SARE), 2010. [online]. [cited October 4, 2018]. sare.org/Learning-Center/Books/Managing-Alternative-Pollinators.

Miller, Jeffrey and Paul C. Hammond. *Lepidoptera of the Pacific Northwest: Caterpillars and Adults*. Forest Health Technology Enterprise Team, 2003.

Papiorek, S., et al. "Bees, Birds and Yellow Flowers: Pollinator-Dependent Convergent Evolution of UV Patterns." *Plant Biology* Vol. 18#1 (January 2016), pp. 46–55. [online]. [cited August 10,

2018]. onlinelibrary.wiley.com/doi/abs/10.1111/plb.12322.

PBS. "How Bees Can See the Invisible." *It's Okay To Be Smart*, April 22, 2013. [online]. [cited August 10, 2018]. youtube.com/watch?v=NITUDFCOwjY.

Roth, Sally. *Attracting Butterflies & Hummingbirds to Your Backyard*. Rodale, 2002.

Sarniñas, Hillary. "Getting to Know Our Native Northwest Bees." Washington Park Arboretum Bulletin, Summer 2016. [online]. [cited August 10, 2018]. arboretumfoundation.org/about-us/publications/bulletin/bulletin-archive/pacific-northwest-bees.

Schmitt, Klaus. "Photography of the Invisible World." [online]. [cited August 10, 2018]. photographyoftheinvisibleworld.blogspot.com.

Seeley, Thomas D. *Honeybee Democracy*. Princeton, 2010.

Shepherd, Matthew. "Pacific Northwest Plants for Native Bees." Xerces Society Invertebrate Conservation Fact Sheet. [online]. [cited October 4, 2018]. xerces.org/wp-content/uploads/2008/11/pnw_plants_bees_xerces_society_factsheet1.pdf.

Stokes, Donald W. and Lillian Q. *The Hummingbird Book: The Complete Guide to Attracting, Identifying, and Enjoying Hummingbirds*. Little Brown, 1989.

Xerces Society. "Meet the Pollinators" in *Attracting Native Pollinators: Protecting North America's Bees and Butterflies*. Storey, 2011.

Xerces Society. *Xerces Bee Monitoring Tools*. [online]. [cited September 23, 2018]. xerces.org/xerces-bee-monitoring-tools.

Rain Catchment and Rain Gardens

"4. Zones and Sectors—Efficient Energy Planning." *Deep Green Permaculture*. [online]. [cited September 3, 2018]. deepgreenpermaculture.com/permaculture/permaculture-design-principles/4-zones-and-sectors-efficient-energy-planning/.

Babb, Terry. *Harvest Rain: The Movie*. [online]. [cited September 3, 2018]. harvesth2o.com/harvest_rain_dvd.shtml.

Corkran, Charlotte C. and Chris Thoms. *Amphibians of Oregon, Washington and British Columbia: A Field Identification Guide*, rev. and updated ed. Lone Pine, 2006.

"Curriculum" in *Rainwater Harvesting for Drylands and Beyond* by Brad Lancaster. [online]. [cited September 3, 2018]. harvestingrainwater.com/rainwater-harvesting-inforesources/water-harvesting-curriculum/.

Daily, Cado and Cyndi Wilkins. "Passive Water Harvesting: Rainwater Collection." University of Arizona College of Agriculture and Life Sciences-Cooperative Extension, October 2012. [online]. [cited September 3, 2018]. extension.arizona.edu/sites/extension.arizona.edu/files/pubs/az1564.pdf.

Deppe, Carol. "Gardening in an Era of Wild Weather and Climate Change" in *The Resilient Gardener: Food Production and Self-Reliance in Uncertain Times*. Chelsea Green, 2010.

Emanuel, Robert et al. *The Oregon Rain Garden Guide: A Step-by-step Guide to Landscaping for Clean Water and Healthy Streams*. Oregon Sea Grant, Oregon State University, 2010. [online]. [cited September 4, 2018]. seagrant.oregonstate.edu/sgpubs/oregon-rain-garden-guide.

Hermary, Heide. "Working With Water: Wetlands, Bog Gardens and Rain Gardens" in *Working with Nature: Shifting Paradigms: The Science and Practice of Organic Horticulture*. Gaia College, 2007. gaiacollege.ca/working-with-nature-shifting-paradigms.html.

King Conservation District. "Rain Garden Care: A Guide for Residents and Community Organizations." 12,000 Rain Gardens. [online]. [cited September 27, 2018]. 12000raingardens.org/wp-content/uploads/2013/03/RainGardenCareGuideComplete.pdf.

Lancaster, Brad. *Rainwater Harvesting for Drylands and Beyond*. Rainsource, 2008.

Link, Russell. *Landscaping for Wildlife in the Pacific Northwest*. University of Washington, 1999.

Nast, Phil. "Teaching with Maps: Lessons, Activities, Map-Making Resources, and More." US National Education Association. [online]. [cited September 28, 2018]. nea.org/tools/teaching-with-maps.html.

National Wildlife Federation. "Attracting Amphibians." [online]. [cited September 10, 2018]. nwf.org/Garden-for-Wildlife/Wildlife /Attracting-Amphibians.

National Wildlife Federation. "Schoolyard Habitats." [online]. [cited September 10, 2018]. nwf .org/Garden-for-Wildlife/Create/Schoolyards.

"Rain Gardens in Grass Lawn Park." City of Redmond. [online]. [cited September 3, 2018]. redmond.gov/common/pages/UserFile.aspx ?fileId=22925.

Seattle Public Utilities. *RainWise Program.* [online]. [cited September 27, 2018]. seattle.gov/util/Envir onmentConservation/Projects/GreenStorm waterInfrastructure/RainWise/index.htm.

"Stormwater in the Desert: Middle School Activity Book." Storm website. [online]. [cited September 27, 2018]. azstorm.org/home-page /middle-school-activity-book.

Sustainability Ambassadors. *Low Impact Development Manual for Schools.* [online]. [cited September 28, 2018]. sustainabilityambassadors .org/lid-manual-for-schools.

Walbert, David. "Building Map-Reading Skills and Higher-Order Thinking." Learn NC. University of North Carolina School of Education, 2010. [online]. [cited September 28, 2018]. learnnc.org /lp/editions/mapping/6430 [Note that this link takes you to an archive where you must reenter the link].

"Water Harvesting." Permaculture Research Institute website. [online]. [cited September 3, 2018]. permaculturenews.org/category/water/water -harvesting.

Seeds and Harvesting

Ashworth, Suzanne. *Seed to Seed: Seed Saving and Growing Techniques*, 2nd ed. Seed Savers Exchange, 2002.

Baker Creek Heirloom Seeds. "Growing Guide." [online]. [cited October 17, 2018]. rareseeds.com /resources/growing-guide.

Bonnie Plants website. "Growing Potatoes." [online]. [cited August 6, 2018]. bonnieplants.com /growing/growing-potatoes.

Bubel, Nancy. *The New Seed-Starters Handbook.* Rodale, 1988.

Burr, Chuck. "How to Save Tomato Seeds." Permaculture Research Institute, July 8, 2014. [online]. [cited September 3, 2018] permaculturenews.org /2014/07/08/save-tomato-seeds.

Chavan, J. K. et al. "Nutritional Improvement of Cereals by Sprouting." *Critical Reviews in Food Science and Nutrition* Vol. 28#5 (1989). [online]. [cited July 31, 2018]. tandfonline.com/doi/abs/10 .1080/10408398909527508.

"Community Seed Donations." Seed Savers Exchange. [online]. [cited October 17, 2018]. seed savers.org/seed-donation-program?_ga=2.2655 96684.2062025174.1539457519-601172661.1536 853211.

Community Seed Network. "Resources & Information for Seed Saving, Sharing & Networking." [online]. [cited October 17, 2018]. community seednetwork.org/home.

Conner, Cindy. *Seed Libraries And Other Means of Keeping Seeds in the Hands of the People.* New Society, 2015.

"How to Save Seeds." Seed Savers Exchange. [online]. [cited August 23, 2018]. seedsavers.org /how-to-save-seeds.

Kaldor, Connie. "The Seed in the Ground." *A Duck in New York City*, October, 2003.

Nature. "The Seedy Side of Plants." PBS February, 2008. [online]. [cited August 5, 2018]. pbs.org /wnet/nature/the-seedy-side-of-plants-introduc tion/1268.

"Origami Maple Seed." Jo Nakashima—Origami Tutorials, Youtube, September 29, 2010. [online]. [cited September 20, 2018]. youtube.com/watch ?v=VyPsIQlZH_Q.

"Origami Maple Seed Helicopter." PaperFun Youtube, November 12, 2015. [online]. [cited September 20, 2018]. youtube.com/watch?v =nwHZDzBqInY.

"Parts of a Seed Worksheet." Education.com. [online]. [cited August 5, 2018]. education.com /worksheet/article/parts-of-a-seed-1.

Peterson, Christy. "Seeds on the Move—Seed Dispersal for Kids." Kids Discover, September 26, 2013. [online]. [cited August 5, 2018]. kidsdiscover.com/parentresources/seed -dispersal.

Poles, Tina M. "A Handful of Seeds: Seed Study and Seed Saving for Educators." Occidental Arts and Ecology Center. [online]. [cited October 15, 2018]. oaec.org/wp-content/uploads/2014/10/A-Handful-of-Seeds.pdf.

Schreiber, Andrew. "Making Seedballs: An Ancient Method of No-till Agriculture." Permaculture Research Institute, June 18, 2014. [online]. [cited August 6, 2018]. permaculturenews.org/2014/06/18/making-seedballs-ancient-method-till-agriculture.

Seed Savers Exchange. "Seed Saving Chart." [online]. [cited October 17, 2018]. seedsavers.org/seed-saving-chart.

Seeds of Diversity. "The Ecological Seed Finder." [online]. [cited October 17, 2018]. seeds.ca/seedfinder.

Stewart, Deborah. "Creating a Mini-Greenhouse in Preschool." Teach Preschool, April 17, 2013. [online]. [cited July 31, 2018]. teachpreschool.org/2013/04/17/creating-a-mini-greenhouse-in-preschool.

Toensmeier, Eric. *Perennial Vegetables: From Artichoke to Zuiki Taro, a Gardener's Guide to Over 100 Delicious, Easy-to-grow Edibles.* Chelsea Green, 2007.

Waheenee and Gilbert L. Wilson. *Buffalo Bird Woman's Garden: Agriculture of the Hidatsa Indians.* Minnesota Historical Society Press, 1987.

Watson, Benjamin A. *Taylor's Guide to Heirloom Vegetables: A Complete Guide to the Best Historic and Ethnic Varieties.* Houghton Mifflin Harcourt, 1996.

Soil Vitality

Albert, Steve. "Organic Fertilizers and Soil Amendments." Harvest to Table. [online]. [cited August 16, 2018]. harvesttotable.com/test_ferti.

Chaskey, Scott. "Soils Are Beautiful" in *This Common Ground: Seasons on an Organic Farm.* Viking, 2005.

Coleman, Eliot. "Soil Fertility" in *The New Organic Grower: A Master's Manual of Tools and Techniques for the Home and Market Gardener,* rev. and expanded ed. Chelsea Green, 1995.

Dig Deeper. "Soil Experiments and Hands-On Projects." The Soil Science Society of America. [online]. [cited September 24, 2018]. soils4kids.org/experiments.

Engels, Jonathon. "4 All-Natural Liquid Fertilizers and How to Make Them." One Green Planet, September 1, 2017. [online]. [cited August 15, 2018]. onegreenplanet.org/lifestyle/all-natural-liquid-fertilizers.

"Estimating Soil Moisture By Feel." Appendix 1, Unit 1.2 in Center for Agroecology & Sustainable Food Systems, U of California, Santa Cruz, Teaching Organic Farming and Gardening Skills. [online]. [cited October 24, 2018]. casfs.ucsc.edu/about/publications/Teaching-Organic-Farming/part-1.html.

"Family Activity: Web of Life." Project Learning Tree. [online]. [cited August 14, 2018]. plt.org/family-activity/web-of-life.

"Field Guide to Soil Food Webs." Soil Science Society of America. [online]. [cited August 14, 2018]. soils.org/files/sssa/iys/soil-food-web-field-guide.pdf.

Hemenway, Toby. "The Ultimate Bomb-Proof Sheet Mulch" in *Gaia's Garden: A Guide to Home-scale Permaculture,* 2nd ed. Chelsea Green, 2009.

"Here's the Scoop on Chemical and Organic Fertilizers." Oregon State University Extension Service. [online]. [cited August 16, 2018]. extension.oregonstate.edu/news/heres-scoop-chemical-organic-fertilizers.

Hermary, Heide. "Relationships Within Ecosystems" and "Organic Fertilizers" in *Working with Nature: Shifting Paradigms: The Science and Practice of Organic Horticulture.* Gaia College, 2007. gaiacollege.ca/working-with-nature-shifting-paradigms.html.

Hermary, Heide. "Soil in an Ecosystem" in *Working with Nature: Shifting Paradigms: The Science and Practice of Organic Horticulture.* Gaia College, 2007. gaiacollege.ca/working-with-nature-shifting-paradigms.html.

Hoorman, James and Rafiq Islam. "Understanding Soil Microbes and Nutrient Recycling." Ohio State University Extension, Agriculture and Natural Resources SAG-16 (September 7, 2010). [online]. [cited August 1, 2018]. ohioline.osu.edu/factsheet/SAG-16.

"Hugelkultur Bed Construction." Deep Green Permaculture website. [online]. [cited August 16, 2018]. deepgreenpermaculture.com/diy-instructions/hugelkultur-bed-construction.

"Hugelkultur: The Ultimate Raised Garden Beds." Richsoil website. [online]. [cited August 16, 2018]. richsoil.com/hugelkultur.

Lima, Patrick. "Natural Alternatives" in *The Natural Food Garden: Growing Vegetables and Fruits Chemical-free*. Prima, 1992.

Lima, Patrick. "Natural Fertilizer" in *The Natural Food Garden: Growing Vegetables and Fruits Chemical-free*. Prima, 1992.

Lovelock, James. *Gaia: A New Look at Life on Earth*. Oxford, 1987.

Luster Leaf. Rapitest (#1605) Soil Testing kit. [online]. [cited August 15, 2018]. lusterleaf.com/nav/soil_test.html.

Magdoff, Fred and Harold van Es. *Building Soils for Better Crops*, 3rd ed. USDA Sustainable Agriculture Research and Education (SARE), 2010. [online]. [cited October 4, 2018]. sare.org/Learning-Center/Books/Building-Soils-for-Better-Crops-3rd-Edition.

"The Many Benefits of Hugelkultur." *Permaculture Magazine*, October 17, 2013. [online]. [cited August 16, 2018]. permaculture.co.uk/articles/many-benefits-hugelkultur.

"Nature's Recyclers" in Project Learning Tree. *PreK-8 Environmental Education Activity Guide*, 11th ed. American Forest Foundation, 2005. [online]. [cited August 14, 2018]. plt.org/blog/activity/prek-8-activity-24-natures-recyclers.

"Perkin' Through the Pores: Slip Slidin' Away." Agriculture in the Classroom. Utah State University Cooperative Extension. [online]. [cited August 15, 2018]. soils4kids.org/files/s4k/perkin.pdf.

Pleasant, Barbara. "Free, Homemade Liquid Fertilizers: Organic Gardening." *Mother Earth News*, February/March 2011. [online]. [cited August 15, 2018]. motherearthnews.com/organic-gardening/gardening-techniques/liquid-fertilizers-zm0z11zhun.

Pleasant, Barbara. "Homemade Fertilizer Tea Recipes: Organic Gardening." *Mother Earth News*. [online]. [cited August 15, 2018]. motherearthnews.com/organic-gardening/gardening-techniques/homemade-fertilizer-tea-recipes-zm0z11zkon.

Pleasant, Barbara. "Use Cover Crops to Improve Soil." *Mother Earth News*, October-November 2009. [online]. [cited October 15, 2018]. motherearthnews.com/organic-gardening/cover-crops-improve-soil-zmaz09onzraw.

"Producers, Consumers, Decomposers." PBS Learning Media. [online]. [cited August 14, 2018]. pbslearningmedia.org/resource/tdc02.sci.life.oate.lp_energyweb/producers-consumers-decomposers.

"Safe Bug Trap." Easy Outdoor Activities for Kids, How Stuff Works. [online]. [cited August 14, 2018]. lifestyle.howstuffworks.com/crafts/home-crafts/easy-outdoor-activities-for-kids5.htm.

Salatin, Joel. "Growing Soil" in *The Sheer Ecstasy of Being a Lunatic Farmer*. Polyface, 2010.

Schwenke, Karl. *Successful Small-scale Farming: An Organic Approach*, 2nd ed. Storey, 1991.

"Soil Percolation Test" in Project Learning Tree. *PreK-8 Environmental Education Activity Guide*, 11th ed., American Forest Foundation, 2005.

Soil Science Society of America. "Paint with Soil!" [online]. [cited September 24, 2018]. soils.org/files/sssa/iys/paint-with-soil.pdf.

University of Illinois Extension. "The Color of Soil." [online]. [cited September 24, 2018]. extension.illinois.edu/soil/less_pln/color/color.htm.

USDA Natural Resources Conservation Service. "Painting with Soil." [online]. [cited September 24, 2018]. nrcs.usda.gov/wps/portal/nrcs/detail/soils/edu/?cid=nrcs142p2_054304.

USDA Natural Resources Conservation Service. "Painting with Soil: Jan Lang's Images of the Lewis and Clark Expedition." [online]. [cited September 24, 2018]. nrcs.usda.gov//wps/portal/nrcs/detail/soils/edu/?cid=nrcs142p2_054282.

USDA Natural Resources Conservation Service. "Soil Biology." [online].[cited September 24, 2018]. nrcs.usda.gov/wps/portal/nrcs/main/soils/health/biology.

UN Food and Agriculture Organization. "World Soil Day: Soil Painting Competition." [online].

[cited September 24, 2018]. fao.org/world-soil-day/activities/soil-painting-competition/en.

Taste Testing, Recipes, and Nutrition

The Collective School Garden Network. "Garden-Enhanced Nutrition Teaching Resources." Western Growers Foundation, 2005. [online]. [cited October 10, 2018]. csgn.org/garden-enhanced-nutrition-teaching-resources.

Food and Nutrition Service. "School Meals: School Food and Produce Safety." US Department of Agriculture. [online]. [cited Ovtober25, 2018]. fns.usda.gov/school-meals/school-food-and-produce-safety.

KidsGardening. "Create & Sustain a Program: Nutrition Education in the Garden." [online]. [cited October 10, 2018]. kidsgardening.org/create-sustain-a-program-nutrition-education-in-the-garden.

Love and Lemons. "Pea Tendril & Pistachio Pesto." [online]. [cited September 10, 2018]. loveandlemons.com/pea-tendril-pistachio-pesto.

Martin, Adriana. "Strawberries and Cream a Classic Mexican Treat." *Adriana's Best Recipes*, May 2017. [online]. [cited September 10, 2018]. adrianasbestrecipes.com/2017/05/01/strawberries.

Ohio Smarter Lunchrooms. [online]. [cited October 6, 2018]. ohiosmarterlunchrooms.com.

Oregon Harvest for Schools. Oregon Department of Education. [online]. [cited October 6, 2018]. oregon.gov/ode/students-and-family/childnutrition/F2S/Pages/OregonHarvestforSchools.aspx.

Oregon State University. Food Hero. [online]. [cited October 6, 2018]. foodhero.org.

Saffitz, Claire. "Beet and Ricotta Hummus." Bon Appétit, March 2017. [online]. [cited September 10, 2018]. bonappetit.com/recipe/beet-and-ricotta-hummus.

Serio, Tasha de. "Roasted Radishes with Brown Butter, Lemon, and Radish Tops." *Epicurious, Bon Appétit*, April 2011. [online]. [cited September 10, 2018]. epicurious.com/recipes/food/views/roasted-radishes-with-brown-butter-lemon-and-radish-tops-364609.

Vermicomposting

Angima, Sam et al. "Composting With Worms." Oregon State University Extension Service, October, 2011. [online]. [cited August 22, 2018]. ir.library.oregonstate.edu/downloads/vq27zn937.

Appelhof, Mary et al. *Worms Eat Our Garbage: Classroom Activities for a Better Environment.* Chelsea Green, 2003.

"Build a Mini Worm Bin." Oregon Agriculture in the Classroom Foundation. [online]. [cited August 21, 2018]. oregonaitc.org/lessonplan/build-a-mini-worm-bin.

Harland, Darci J. "Can Worms Smell? Wormy Experiment." STEM Mom, October 23, 2012. [online]. [cited August 21, 2018]. drstemmom.com/2012/10/can-worms-smell-wormy-experiment.html.

"Here's the Scoop on Chemical and Organic Fertilizers." Oregon State University Extension Service. [online]. [cited August 16, 2018]. extension.oregonstate.edu/news/heres-scoop-chemical-organic-fertilizers.

"Niches Within Earthworms' Habitat." Science Learning Hub, June 12, 2012. [online]. [cited August 22, 2018]. sciencelearn.org.nz/resources/7-niches-within-earthworms-habitat.

Payne, Binet. *The Worm Cafe: Mid-scale Vermicomposting of Lunchroom Wastes.* Flower Press, 1999.

"A Simple Way to Make and Use Worm Tea." Uncle Jim's Worm Farm, February 9, 2016. [online]. [cited August 22, 2018]. unclejimswormfarm.com/a-simple-way-to-make-and-use-worm-tea.

"Soil Decomposers." National Wildlife Federation. [online]. [cited August 21, 2018]. nwf.org/~/media/PDFs/Be Out There/National-Wildlife-Week/2011/Soil-Decomposers-K-8.pdf.

UCCE Master Gardener Program. "About Worm Castings." University of California, Agriculture and Natural Resources. [online]. [cited October 24, 2018]. ucanr.edu/sites/mgfresno/files/262372.pdf.

Index

About The Author

Kaci Rae Christopher is an outdoor and garden educator who focuses on life and job skills training for young adults and children. She was previously the School Garden Coordinator for the Springwater Environmental Sciences School and the Outdoor Educator for ERA. Her passion is fostering a healthy land ethic, personal empowerment, and environmental literacy in children of all ages through outdoor immersion and skill building. She lives in Bend, Oregon.

School Garden Curriculum Worksheets
are available for free download at:
https://tinyurl.com/SGC-Worksheets

ABOUT NEW SOCIETY PUBLISHERS

New Society Publishers is an activist, solutions-oriented publisher focused on publishing books for a world of change. Our books offer tips, tools, and insights from leading experts in sustainable building, homesteading, climate change, environment, conscientious commerce, renewable energy, and more—positive solutions for troubled times.

DON'T EAT THIS BOOK *(but you could)*

We're proud to hold to the highest environmental and social standards of any publisher in North America. This is why some of our books might cost a little more. We think it's worth it!

- We print all our books in North America, never overseas
- All our books are printed on 100% **post-consumer recycled paper**, processed chlorine-free, with low-VOC vegetable-based inks (since 2002)
- Our corporate structure is an innovative employee shareholder agreement, so we're one-third employee-owned (since 2015)
- We're carbon-neutral (since 2006)
- We're certified as a B Corporation (since 2016)

At New Society Publishers, we care deeply about *what* we publish—but also about *how* we do business.

Download our catalog at https://newsociety.com/Our-Catalog or for a printed copy please email info@newsocietypub.com or call 1-800-567-6772 ext 111.

New Society Publishers
ENVIRONMENTAL BENEFITS STATEMENT

For every 5,000 books printed, New Society saves the following resources:[1]

49	Trees
4,418	Pounds of Solid Waste
4,861	Gallons of Water
6,341	Kilowatt Hours of Electricity
8,031	Pounds of Greenhouse Gases
35	Pounds of HAPs, VOCs, and AOX Combined
12	Cubic Yards of Landfill Space

[1] Environmental benefits are calculated based on research done by the Environmental Defense Fund and other members of the Paper Task Force who study the environmental impacts of the paper industry.